T0134549

Printing of Graphene and Related 2D Materials

Leonard W. T. Ng • Guohua Hu
Richard C. T. Howe • Xiaoxi Zhu • Zongyin Yang
Christopher G. Jones • Tawfique Hasan

Printing of Graphene and Related 2D Materials

Technology, Formulation and Applications

Leonard W. T. Ng
Cambridge Graphene Centre
University of Cambridge
Cambridge, UK

Richard C. T. Howe
Cambridge Graphene Centre
University of Cambridge
Cambridge, UK

Zongyin Yang
Cambridge Graphene Centre
University of Cambridge
Cambridge, UK

Tawfique Hasan
Cambridge Graphene Centre
University of Cambridge
Cambridge, UK

Guohua Hu
Cambridge Graphene Centre
University of Cambridge
Cambridge, UK

Xiaoxi Zhu
Cambridge Graphene Centre
University of Cambridge
Cambridge, UK

Christopher G. Jones
Novalia Ltd.
Cambridge, UK

ISBN 978-3-030-06257-6 ISBN 978-3-319-91572-2 (eBook)
https://doi.org/10.1007/978-3-319-91572-2

Printed on acid-free paper

This Springer imprint is published by the registered company Springer International Publishing AG part of Springer Nature.
The registered company address is: Gewerbestrasse 11, 6330 Cham, Switzerland

Preface

The field of two-dimensional (2D) materials research has evolved dramatically since the seminal work of Andre Geim and Konstantin Novoselov in 2004, which first reported the creation of atomically thin sheets of graphene. The intervening years have seen an exponential increase in both the level of government and industrial funding and the volume of research undertaken.

The motivation behind this rapid growth is the range of properties which emerge when these materials are reduced to flakes or sheets consisting of a few atomic layers (typically <10). These properties encompass high mechanical strength and flexibility, exotic (opto)electronic properties, and high sensitivity to adsorbed materials or compounds. Within the family of 2D materials, there are also examples of metallic, semiconducting, and insulating types. These complementary properties make them well-suited to forming a platform for a wide range of applications, either through significantly improved cost/performance advantages compared to current solutions or through a new generation of systems and devices.

A significant shift in recent years has been the movement from one-off device demonstrations towards practical applications of graphene and related 2D materials for real-world and mass-producible devices and systems. This has required the development of new methods of production and application for 2D materials, since the first (and still most widely used) mechanical cleavage (or 'scotch tape' method) is inherently non-scalable. Top-down (i.e. exfoliation) and bottom-up (i.e. synthesis) methods have emerged, greatly enhancing the global production capability for graphene flakes.

The next stage, and a key barrier that remains, is the translation of these materials and their properties into useable forms for device and component manufacture. Amongst the different potential methods for achieving this, functional printing, utilising technologies traditionally established for graphics printing, has emerged as a strong contender.

By formulating inks containing 2D materials to work with such widely used technologies in the print and packaging industry, it is anticipated that rapid, low cost, and large-scale production can be achieved. The first report of printing with graphene inks emerged in 2012, using inkjet printing to fabricate transistors. Much has evolved in the field since then, both within academia and in the industry. However, there remains a broad lack of understanding in the literature, in particular into what constitutes an ink and what formulations and processes are feasible for scale-up for existing print processes.

We therefore feel the time is right for a summary of the technological and economic implications of these developments, as well as a more in-depth overview of how to ensure that the potential is fully realised in the transition from lab scale to mass production of 2D material-based printed devices and components.

The book provides a comprehensive description of the current state of the art regarding the printing of 2D materials, including detailed summaries of methods and processes used, applications already demonstrated, and promising directions for future research and investment. It consists of six chapters and includes insights and figures from our own work, as well as that of other researchers. We open the book with an introduction to the field and the associated economic landscape (Chap. 1), a summary of the key 2D materials, their structures and properties (Chap. 2), and methods of production relevant to ink and composite formulation (Chap. 3). In Chaps. 4 and 5, we move to inks (Chap. 4) and printing technologies (Chap. 5), along with their merits and drawbacks, and detail the key parameters that govern ink formulation and successful printing. Having introduced the relevant theory, the final chapter (Chap. 6) provides a review of ways in which 2D material inks have been utilised, and suggests the likely next steps for future development. We believe that this book will be a highly useful overview for students, academics, and industrial researchers new to this area, as well as more experienced readers looking to further enhance their understanding of the latest developments in the field.

We gratefully acknowledge funding from the Royal Academy of Engineering, EPSRC (EP/G037221/1, EP/L016087/1), the Graphene Flagship, and Nippon Kayaku Corporation. We also would like to thank Ethan B. Secor, Adam Kelly, Mark C. Hersam and Jonathan N. Coleman for use of original figures in this book.

Cambridge, UK — Leonard W. T. Ng
Cambridge, UK — Guohua Hu
Cambridge, UK — Richard C. T. Howe
Cambridge, UK — Xiaoxi Zhu
Cambridge, UK — Zongyin Yang
Cambridge, UK — Christopher G. Jones
Cambridge, UK — Tawfique Hasan
April 2018

Contents

Chapter 1
Introduction

Abstract Graphene and related two-dimensional (2D) materials have attracted considerable research interest across a wide range of application fields due to their unique characteristics. However, significant barriers still remain in the large-scale and low-cost fabrication of devices based on these materials. A widely explored route to accomplishing this is in the formulation of 2D material functional inks for use with existing printing technologies that have been traditionally used for graphics printing. In this chapter, we provide an introduction to the key topics that will be introduced later in the book and give an overview of the current and future economic and technological landscape of 2D materials and their potential applications in the context of printing.

This book has been written to describe some basic formulations of inks for graphene and related two-dimensional (2D) materials and to explore their associated print and application technologies. This recently identified family of materials (i.e. materials possessing a planar crystal structure) exhibit many unique properties and have justifiably attracted a considerable research interest across a wide range of application fields [1–5].

Their unique characteristics include conducting, semiconducting, and insulating electronic properties, exotic photonic features, large specific surface area, and high mechanical strength and flexibility [1]. These 2D materials have therefore emerged as a hugely promising material platform for the development of next generation electronics, in particular for flexible and wearable devices, as well as potential applications within high performance energy storage and high strength composite materials [1]. However, as with all new technologies, significant barriers remain in the large-scale, low-cost fabrication of such devices.

One route that has therefore been widely explored is the formulation of 2D material functional inks for use in conjunction with existing printing technologies normally reserved for graphics printing and packaging applications. These printing technologies are a mature, proven method for rapid and low-cost deposition, allowing thin and flexible form factors across a wide range of substrates. Printing, as an additive manufacturing technique has already been adopted for device

L. W. T. Ng et al., *Printing of Graphene and Related 2D Materials*,
https://doi.org/10.1007/978-3-319-91572-2_1

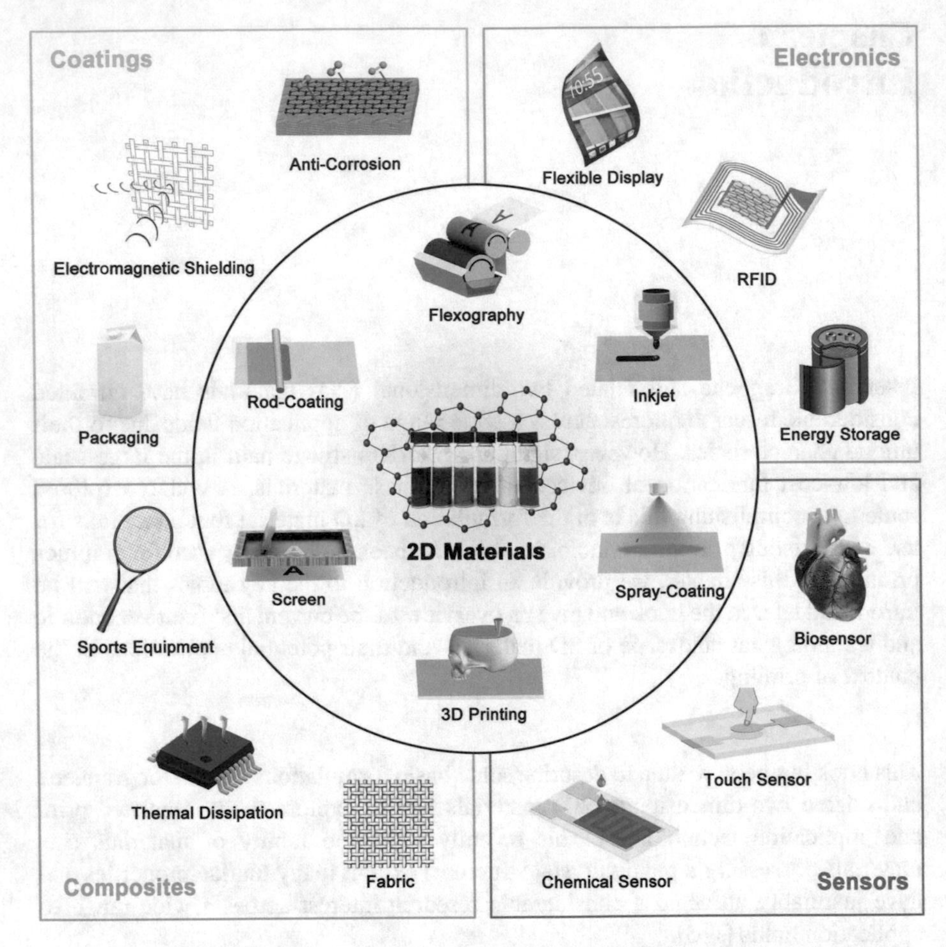

Fig. 1.1 Overview of potential deposition methods of 2D material inks and their applications

fabrication with other functional material systems such as semiconducting polymers to achieve device fabrications in a low-cost, high-throughput environment. There are already demonstrations of 2D material based printed devices, ranging from (opto)electronics, photonics, sensors, and photovoltaics to anti-corrosion coatings, supercapacitors and batteries. These involve a number of printing technologies that include inkjet, screen, flexographic and gravure printing and more recently 3D printing. In this book, we present the key properties and functions of graphene and related 2D materials and discuss the basic strategies for their ink formulations and associated printing technologies with a view towards the scalable development of their applications. In the end, we offer a gestalt of both the demonstrated and potential 2D material based applications relying on printing. Figure 1.1 gives a brief overview of some potential deposition methods of 2D material functional inks and coatings and their relevant applications.

1.1 Functional Printing Technologies

Printing is ubiquitous, and as a manufacturing process is widely exploited in daily life. It delivers consistent, high-volume deposition of inks and pigmented materials over large areas [6–9]. Normally producing transient products, such as labels, advertising materials, packaging, newspapers, and magazines, the printing industry is often overlooked. However, it is estimated to be worth USD $124 billion by 2020 [10]. Printing technologies widely used across the globe include inkjet, screen, flexographic and gravure printing, and offset lithography [6–9]. Aside from using printing for graphics applications, the last 20 years have seen a growing interest in incorporating functional ingredients, such as conducting, semiconducting, and dielectric materials as active pigments to develop products for a wide variety of electronic and optoelectronic applications. These developments allow printing as an additive manufacturing process to fabricate devices over large areas without the need for conventional and expensive vacuum deposition technologies [11].

Functional printing has been gaining momentum as a manner of producing extremely high-volumes of devices at high speed and low cost [9, 11–14]. The prints present thin-form factors, and can also be mechanically flexible and stretchable [9, 11–14] which enables the production of disposable as well as wearable and comformable devices. Demand for such devices has been driven by the increased miniaturisation, technological advancements, and portability needs of electronic products in different sectors such as telecommunications, packaging, automotive and medicine [11, 14] in the twenty-first century.

Examples of devices produced using printing technologies include inkjet printed smart packaging systems [15], gravure printed wireless power transmitters [16], and flexographic printed paper-based chipless RFID tags [17]. The prospective growth of cost efficient functional printing is hallmarked by the evolution of printing technologies from rigid to flexible substrates, printing from cm^2 to m^2 structures, and gradually moving from sheet-to-sheet (S2S) to roll-to-roll (R2R) processes [18]. These advances point to a growth in the functional printing market that is estimated to be worth USD $13.79 billion worldwide by 2020 [19].

Beyond printing of specific patterns or features, the formulated inks can also be used for functional coating, expanding the scope of low-cost, high-speed, and high throughput application development. A coating may be defined as a material (usually a liquid) which is applied onto a surface and appears as a continuous film after drying. The process of application and the resultant thin-film are also regarded as coating [20]. Functional coatings represent a subset of coatings that render additional properties for certain applications [21, 22]. These properties may be diverse, with typical examples being self-cleaning [23], easy clean (anti-graffiti) [24], anti-fouling [25], soft feel [26], antibacterial [27], and smart (e.g. responding to external stimuli). At present, functional coatings are prevalent in home furnishings, cars, laptops, mobile phones and solar panels, and more advanced applications such as medical devices and orthopedic implants, invisible paints, and even in radars and satellites [28]. More recently, there is a strong emphasis on the development of smart coatings for corrosion protection in different technological applications, especially in infrastructure [28].

1.2 Graphene and Related 2D Materials

Layered materials are the bulk counterparts of 2D materials. The majority of them are found widely in nature and include graphite, transition metal dichalcogenides (TMDs), mica and various other silicates [2, 5, 29–33]. Certain other layered materials such as hexagonal boron nitride (*h*-BN) are usually chemically synthesised. 2D materials can be isolated from their layered materials due to weak inter-layer van der Waals forces. Among the 2D materials, graphene, isolated from graphite, is the most widely studied.

Theoretical studies into graphene began 60 years ago [34–36], where it was debated as to whether graphene would be thermodynamically unstable due to thermal fluctuations [30]. In 1999, Ref. [37] reported few-layer graphene flakes cleaved from highly orientated pyrolytic graphite (HOPG) by rubbing on silicon. However, it was not until 2004 that monolayer graphene was irrefutably isolated and studied [38] by repeatedly cleaving a graphitic crystal with adhesive tape. The method used, also known as micromechancal cleavage, is thus colloquially referred to as the 'scotch-tape' method. Graphene's properties, that include monoatomic thickness, optical transparency of 97.7% across the visible and near-infrared wavelengths [39], and very high carrier mobilities [40], promise a new generation of novel devices, especially for flexible and transparent electronics [1, 2].

However, the properties of graphene cannot satisfy the requirements for all applications. The other members of the natural and synthetic 2D materials effectively fill this gap. Indeed, the variety of the layered materials is reflected in their wide spectrum of electronic, optoelectronic, photonic, magnetic and structural properties [2, 3, 5, 41, 42]. This also opens up the possibilities for new exotic properties not found in nature, with many 2D materials exhibiting complementary and at times superior properties to graphene [3, 5, 38, 41, 43–45]. Collectively, this makes the 2D material family a new and exciting material platform for a wide range of applications [3–5, 43, 44, 46, 47]. Inspired by this, academia and industry internationally have focused on developing an understanding of their fundamental properties, production methods, and novel device applications. Some of these are already demonstrated and include devices for electronics [2, 3, 30, 48], optoelectronics [2, 3, 43], photonics [2, 49, 50], energy storage [51, 52], water filtration [53], and even for biomedical applications [54–56].

Typical material production methods are either top-down exfoliation or bottom-up growth [5, 41, 57], with each of these methods offering specific advantages and disadvantages. An example of a top-down exfoliation technique is the 'scotch tape method'. While the quality of the exfoliated crystals using this method is very high, this strategy has an extremely low yield, and cannot be scaled up [57], rendering it unsuitable for large-scale device manufacture. An example of bottom-up growth is chemical vapour deposition (CVD), which produces high quality, large area monolayer 2D materials on to growth substrates [57–61]. However, it is a high temperature growth technique requiring complex equipment, as well as multi-stage processes to transfer the grown film on to the target device substrate. Although there has been much development in production methods of graphene, the extreme production conditions, low yield and high cost of this method are still barriers for most applications [57]. Despite better controlled production processes with higher yields [45, 57, 62, 63], the barriers to large-scale device manufacture with the above material production methods remain very high. This is particularly important for applications requiring large device dimensions and favourable price-performance balance.

Demand for developing 2D material applications is therefore driving the research community to seek alternative production methods and deposition techniques. One very important step towards meeting the industrial requirements is the development of solution processing methods to allow large quantities of 2D materials to be produced at relatively lower or even at room temperatures [57, 62, 63]. These resultant 2D material dispersions can be formulated into functional inks that can be applied through printing technologies to deliver low-cost, high-volume device fabrication [7, 8, 56, 64–67].

1.3 Printing of Graphene and Related 2D Materials

Patents for graphene production were filed as early as 2002 by Jang and Huang [68], even preceding the first verified demonstration of stable graphene by Novosolev et al. from Manchester University in 2004. Unlike Novosolev et al., Jang and Huang described an exfoliation process to achieve large quantities of graphene nanoplatelets [68]. In 2006, Vorbeck Materials, a spin off from Princeton University, released the first graphene products and Vor-inks™ became the first commercially available screen-printable graphene ink. Despite the availability of graphene inks, and the groundbreaking earlier work on liquid phase exfoliation (LPE, a solution processing method) of graphite by Hernandez et al. at Trinity College Dublin [69], it was not until 2012, when Torrisi et al. from the University of Cambridge demonstrated inkjet printing of graphene electronics, that functional 2D material inks rose to prominence [70]. Since then, successful demonstrations of 2D material ink printing have been reported across a range of platforms including screen [71], flexography [72, 73], and gravure [74, 75].

To date, commercial uptake has been slow for 2D materials including graphene. Most early demonstrations are composite applications; whereby 2D materials lend mechanical properties to products such as bicycle tyres, tennis rackets, and lubricants. From an electrical conductivity perspective, smart packaging interconnects and electrostatic dissipative films are closest to market, making them likely first printable graphene applications. Early commercial trials of graphene and other 2D materials for electrodes in energy devices have also shown significant promise. Figure 1.2 provides a broad overview of the status of different solution processed 2D material applications, with graphene being the functional additive in the majority of these.

The printing techniques, and therefore specific ink formulation, used for device applications are directed according to the specific end-application requirements and attributes such as minimum feature size, substrate type and fabrication speed. For instance, inkjet printing is suited to devices requiring sub-100 μm resolution, whereas screen printing is suited for high-thickness (>5 μm) prints where a greater application weight of material might be required [76]. In order to be printable by any particular process, the inks must meet the different rheological and transfer requirements of each printing technique [76]. For example, 2D material dispersions produced via LPE can be readily adapted to inkjet printing without significant modification. However, the low viscosities of these dispersions cannot be used with other printing technologies such as flexography and screen printing where the process requires higher viscosity inks [7, 8, 65, 77, 78].

Certain modifications to such 2D material dispersions therefore are typically required to adapt them for individual printing methods. For example, Secor et al. demonstrated tuning of rheological properties of the LPE dispersions by controlling the graphene loading and using polymeric binders for various printing technologies including inkjet, gravure, flexography and screen printing [64, 79, 80]. Printing also allows the use of hybrid ink systems that contain a combination of different 2D materials and other conventional functional materials. These hybrid combinations can be readily used on existing industrial printing presses. Hasan et al. demonstrated a graphene based hybrid conductive ink that can be printed on a flexographic printing press at industrial manufacturing speed (up to 100 m min^{-1}) [81]. This represented a step-change towards low cost, high-volume and potentially disposable printed electronics.

1.4 Economic Landscape of Graphene and Related 2D Materials

Compared to graphene, the market for other 2D materials is not as developed. Quantitative market analysis on these materials is also virtually non-existent. In Fig. 1.3 we therefore provide a snapshot of the graphene industry only. In 2015, the 'Global Market for Graphene' report by Technavio conservatively postulated the market size for all graphene-related industries (including manufacturing and applications) to be worth USD $24.5 million; Fig. 1.3a. By 2020 the market is expected to be worth USD $126 million, with a compound annual growth rate (CAGR) of 66.4%; Fig. 1.3b [82]. We note that the estimations vary widely, depending on the market research methods employed by analysts. These projections

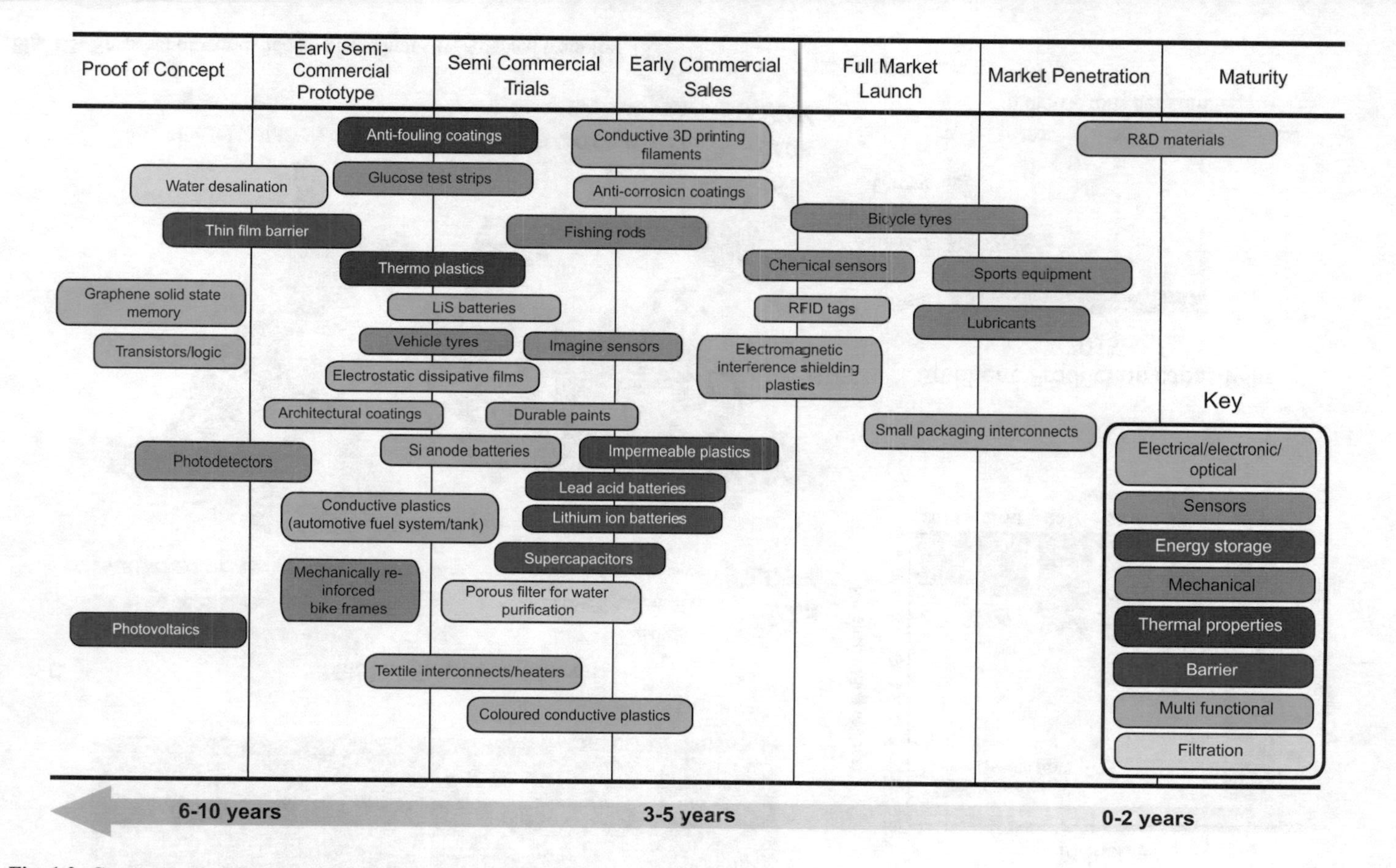

Fig. 1.2 Current stage of development of selected graphene applications. Adapted from Ref. [83]

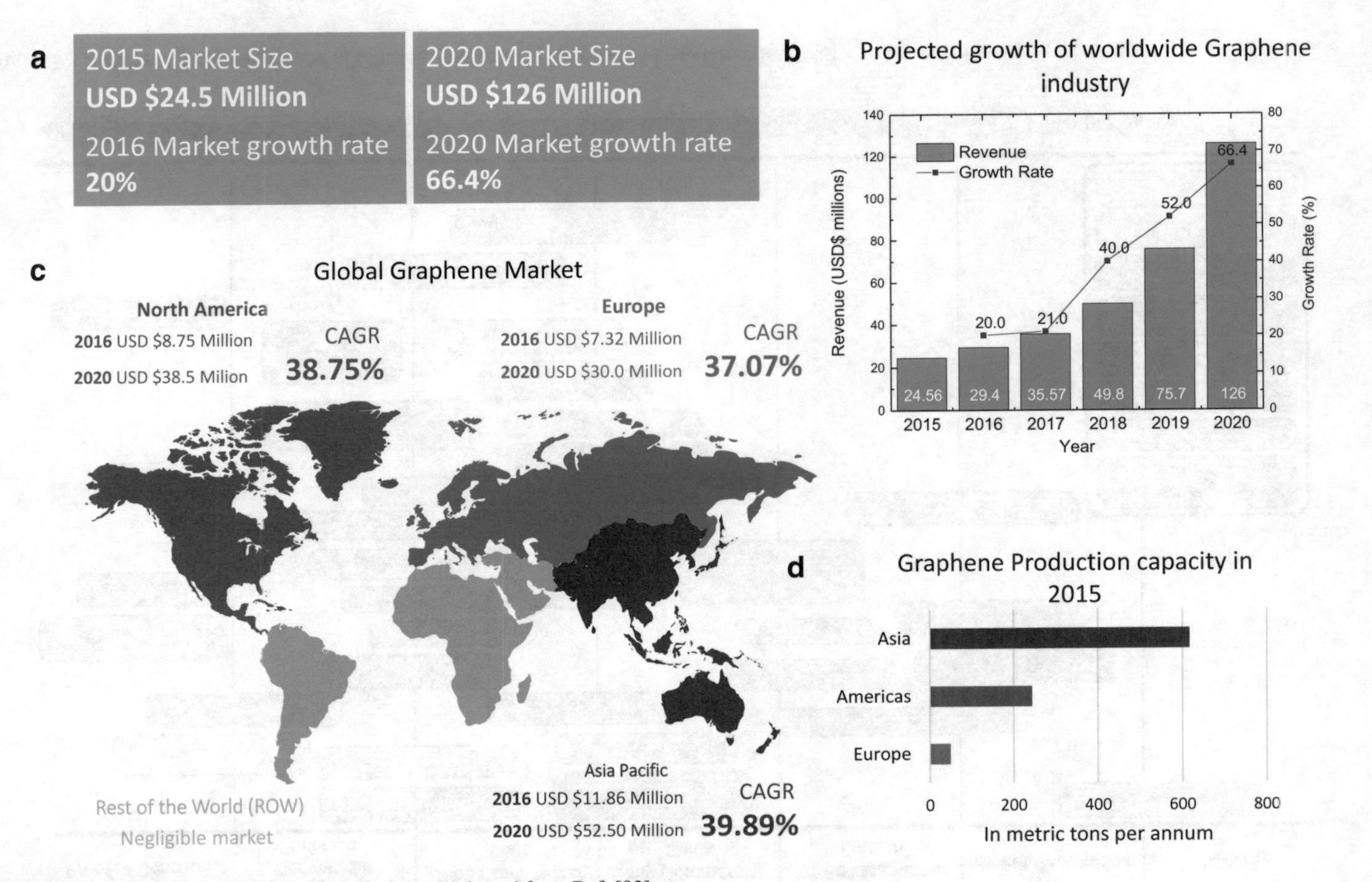

Fig. 1.3 Snapshot of worldwide graphene industry. Adapted from Ref. [83]

point towards a highly bullish graphene market with a great expectation for growth. This is largely due to increasingly competitive research and development efforts from leading universities and large corporations as patents mature into end-user applications. The first products that are starting to appear in the market are composite applications where graphene is incorporated to improve the mechanical performance of the polymers. Already, applications such as graphene tennis rackets from HEAD have started to appear on sale for end-users.

Globally, the Asia Pacific (APAC) region is the largest and the fastest growing graphene market as depicted in Fig. 1.3c. At present, APAC is the world's largest supplier of graphene nanoplatelets (Fig. 1.3d) [82]. This is largely because of the widespread availability of raw graphite in China, aided by some imposed export restrictions, acting as a potential barrier to the growth of industries outside the APAC region [82].

The growth of the graphene industry in Europe at present is driven largely by research and development activities and funding for research groups. To date, the UK government has invested over GBP £100 million in the National Graphene Institute, University of Manchester and GBP £20 million in Cambridge Graphene Centre, University of Cambridge. The European Union is funding Euro €1 billion over a span of 10 years since 2011 in the Graphene Flagship program, to understand and promote graphene and related 2D materials. This program involves approximately 74 organisations from 17 countries and is focused on turning scientific research into commercial applications [82].

By value, the graphene market in North America was USD $8.75 million in 2016 and will reach USD $38.5 million by 2020, growing at a CAGR of 38.75%. There are large deposits of graphite in areas such as Ontario and Quebec in Northern Canada, and resource firms such as Northern Graphite are continuously working to explore graphite in that region. These graphite deposits in Northern Canada could offer a viable, alternative large natural source of graphite to China [82]. Technavio also predicts that the market in North America will be mainly driven by the electrical and electronics industries. The market is led by research and development in graphene-related fields in the universities, spin-off companies from universities and large national laboratories in the United States.

1.4.1 The 2D Material Value Chain

The 2D material supply chain is in its nascency with equipment manufacturers, raw material suppliers, manufacturers and developers looking to exploit early end applications for both materials and processes. Figure 1.4 gives a broad perspective of the value chain of this industry, which is largely dominated by graphite/graphene at present. The scheme shows how different businesses and research organisations receive raw materials, and add value through various processes in intermediate steps towards the development of end-user applications. The supply sector consists

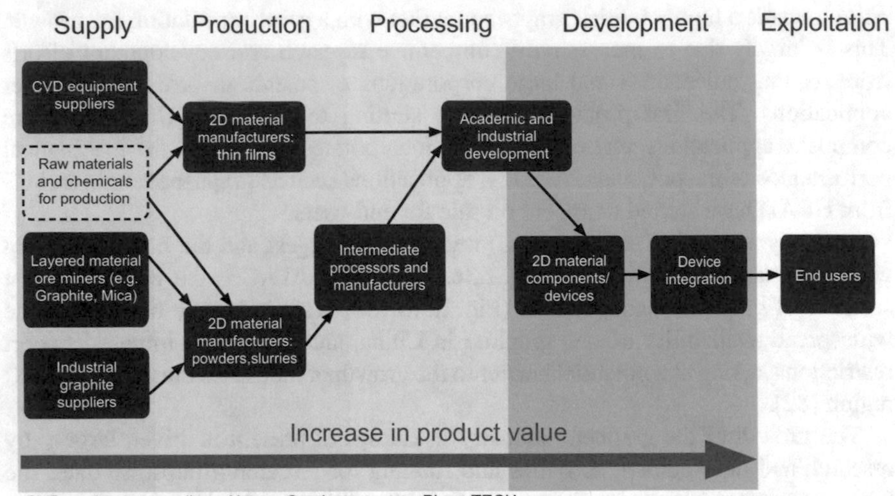

Fig. 1.4 Overview of the 2D material value chain

of layered material ore miners, industrial kish[1] graphite suppliers, as well as raw material and chemical (e.g. for production of boron nitride from boron trioxide or boric acid) suppliers. This industry existed before the interest in 2D materials, but now additionally includes CVD equipment manufacturers. The production sector of the supply chain comprises manufacturers of CVD grown 2D material thin films, typically bound by substrates, and manufacturers of powders or slurries, either chemically from raw materials (e.g. for boron nitride) or physically (e.g. plasma cracking of hydrocarbon for graphene). These slurries and powders can be further processed, for example through their chemical functionalisation, hybridisation with other materials for specific applications (such as metal oxides) or formulation of functional coatings and inks. We have chosen to identify this secondary processing stage in between production and development, although it could legitimately be included within either, as it is at this key stage that materials are refined and additional value added which is distinct from any primary processing. This sector typically supplies materials to academia and industrial development organisations. However, there is some liminality in this space as there is significant attention within academia and industry on further materials processing and development for specific applications. The output from the processing sector feeds the development of 2D

[1]Kish graphite is a byproduct of steel production. It is obtained when carbon crystallises from molten steel during the steel manufacturing process.

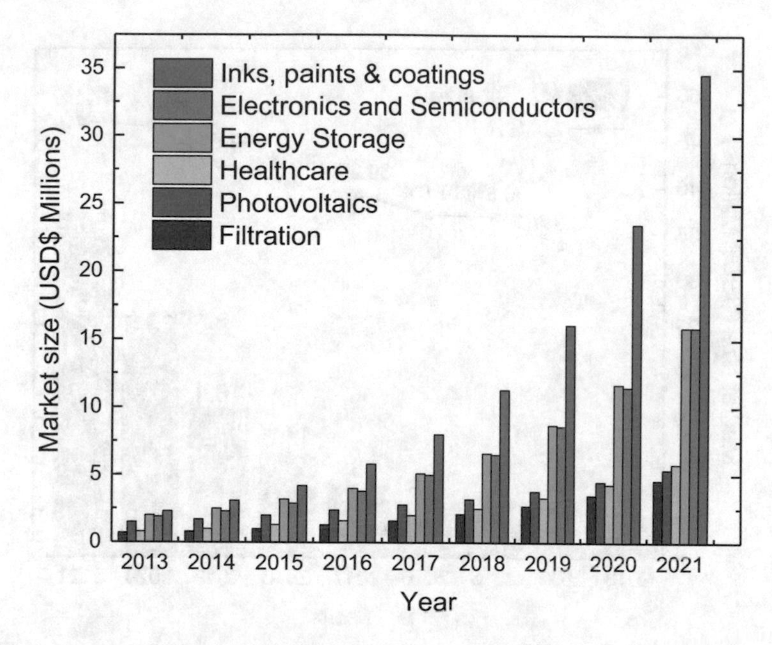

Fig. 1.5 Projected growth of the graphene inks, paints and coatings sector. Adapted from Ref. [84]

material components and devices such as barrier films, composites and battery electrodes. These components are then integrated into complete devices and then full systems directed at end users. At the bottom of Fig. 1.4, we identify example organisations involved in each stage of the value chain. The diagram is intended as an indication of the various organisations at each stage of the 2D materials industry. It is by no means an exhaustive list and skews towards graphite/graphene rather than all 2D materials.

1.4.2 Significance of Graphene Inks, Paints and Coatings

Of the many applications for graphene and its precursors for graphene inks, analysis from the Beige group predicts that paints and coatings are likely to offer the most significant growth; Fig. 1.5. These include applications in electronics and semiconductors, energy storage, healthcare, photovoltaics and filtration. This understandably has led to the projected trend seen in Fig. 1.6. The value of graphene in paint and coating applications has been estimated to be USD $4.2 million in 2015. By 2021, this will likely increase to USD $34.6 million, growing at a CAGR of 47.1%. The growth of the graphene inks and paints industry is largely due to the discovery of new uses for graphene and related 2D materials as functional additives.

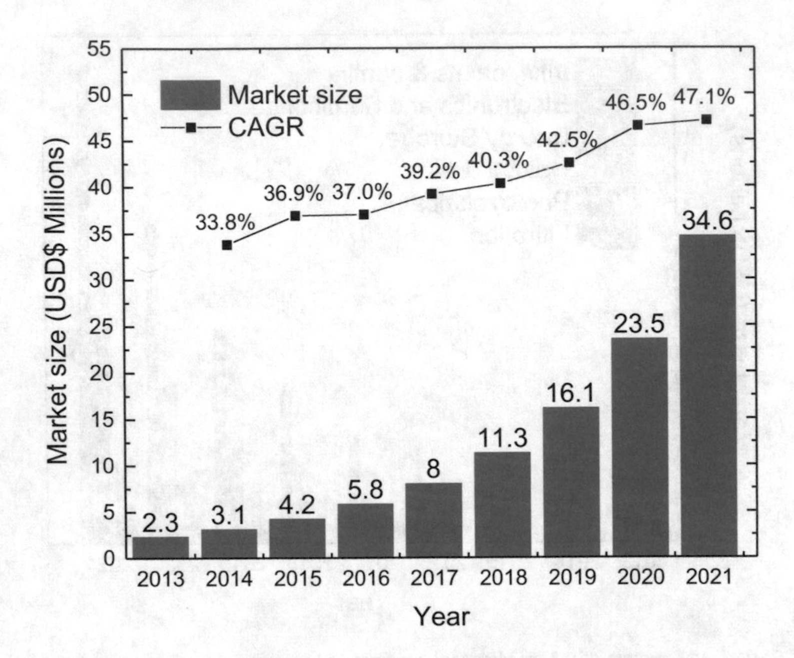

Fig. 1.6 Projected growth in graphene inks, paints and coatings market till 2021. Adapted from Ref. [84]

For instance, graphene oxide can be incorporated into paint formulations to add anti-corrosive and anti-scuff properties, and in membranes for water filtration and desalination [82].

The unique properties of graphene such as high electrical and thermal conductivity and barrier properties are leveraged in the manufacture of a broad range of functional paints and coatings. We note that the low entry barriers and increased intensity of competition in the functional paints and coatings market have also spurred traditional vendors to introduce value-added offerings in their product range incorporating 2D materials. The high rate of growth in the 2D material inks, paints and coatings industry, and the wide variety of applications that they potentially represent, highlights the need for the development of appropriate methods of their solution phase production, functional ink formulation and scalable deposition. In the remaining chapters of this book, we attempt to provide general guidelines to address these steps based on which future 2D material based applications could be developed.

1.5 Conclusion

Functional printing offers a great opportunity for device fabrication due to the ever-increasing demand for low-cost, high throughput fabrication of devices with thin-form factors, mechanical flexibility, stretchability and portability. 2D materials with their unique properties are ideally placed to introduce additional functionalities into this area. Printing as a manufacturing process has been widely exploited with functional materials to develop a growing number of innovative applications. The 2D material family, with properties superior to or even not found in other functional material systems, offers an exciting potential material platform for future applications. Increasing government, industrial and academic interest and financial support is driving new innovations in scale-up production methods for these materials, rapidly forming the platform for the development of practical applications. Important forward steps towards this were the development of solution processing methods and hence functional ink formulations for printing technologies. Lending the superior functionalities of 2D materials to printing therefore holds great potential as a manufacturing method for low-cost, high throughput device fabrication.

References

1. A.C. Ferrari, F. Bonaccorso, V. Fal'ko, K.S. Novoselov, S. Roche, P. Bøggild, S. Borini, F.H.L. Koppens, V. Palermo, N. Pugno, J.A. Garrido, R. Sordan, A. Bianco, L. Ballerini, M. Prato, E. Lidorikis, J. Kivioja, C. Marinelli, T. Ryhänen, A. Morpurgo, J.N. Coleman, V. Nicolosi, L. Colombo, A. Fert, M. Garcia-Hernandez, A. Bachtold, G.F. Schneider, F. Guinea, C. Dekker, M. Barbone, Z. Sun, C. Galiotis, A.N. Grigorenko, G. Konstantatos, A. Kis, M. Katsnelson, L. Vandersypen, A. Loiseau, V. Morandi, D. Neumaier, E. Treossi, V. Pellegrini, M. Polini, A. Tredicucci, G.M. Williams, B.H. Hong, J.-H. Ahn, J.M. Kim, H. Zirath, B.J. van Wees, H. van der Zant, L. Occhipinti, A. Di Matteo, I.A. Kinloch, T. Seyller, E. Quesnel, X. Feng, K. Teo, N. Rupesinghe, P. Hakonen, S.R.T. Neil, Q. Tannock, T. Löfwander, J. Kinaret, Science and technology roadmap for graphene, related two-dimensional crystals, and hybrid systems. Nanoscale **7**(11), 4598–4810 (2015)
2. F. Bonaccorso, Z. Sun, T. Hasan, A.C. Ferrari, Graphene photonics and optoelectronics. Nat. Photonics **4**(9), 611–622 (2010)
3. Q.H. Wang, K. Kalantar-Zadeh, A. Kis, J.N. Coleman, M.S. Strano, Electronics and optoelectronics of two-dimensional transition metal dichalcogenides. Nat. Nanotechnol. **7**(11), 699–712 (2012)
4. A. Castellanos-Gomez, Black phosphorus: narrow gap, wide applications. J. Phys. Chem. Lett. **6**(21), 4280–4291 (2015)
5. M. Xu, T. Liang, M. Shi, H. Chen, Graphene-like two-dimensional materials. Chem. Rev. **113**(5), 3766–3798 (2013)
6. A.A. Tracton (ed.), *Coatings Technology Handbook* (CRC Press, Boca Raton, 2005)
7. R.C.T. Howe, G. Hu, Z. Yang, T. Hasan, Functional inks of graphene, metal dichalcogenides and black phosphorus for photonics and (opto)electronics. Proc. SPIE **9553**, 95530R (2015)
8. F. Bonaccorso, A. Bartolotta, J.N. Coleman, C. Backes, 2D-crystal-based functional inks. Adv. Mater. **28**(29), 6136–6166 (2016)
9. M.A.M. Leenen, V. Arning, H. Thiem, J. Steiger, R. Anselmann, Printable electronics: flexibility for the future. Phys. Status Solidi **206**(4), 588–597 (2009)
10. Technavio, Global printing inks market 2016–2020. Technical report, Infiniti Research Limited (2017)
11. Z. Cui, *Printed Electronics: Materials, Technologies and Applications* (Wiley, Hoboken, 2016)
12. A.C. Arias, J.D. MacKenzie, I. McCulloch, J. Rivnay, A. Salleo, Materials and applications for large area electronics: solution-based approaches. Chem. Rev. **110**(1), 3–24 (2010)

13. G. Grau, J. Cen, H. Kang, R. Kitsomboonloha, W.J. Scheideler, V. Subramanian, Gravure-printed electronics: recent progress in tooling development, understanding of printing physics, and realization of printed devices. Flex. Print. Electron. **1**(2), 023002 (2016)
14. K. Fukuda, T. Someya, Recent progress in the development of printed thin-film transistors and circuits with high-resolution printing technology. Adv. Mater. **29**, 1602736 (2016)
15. J. Miettinen, V. Pekkanen, K. Kaija, P. Mansikkamäki, J. Mäntysalo, M. Mäntysalo, J. Niittynen, J. Pekkanen, T. Saviauk, R. Rönkkä, Inkjet printed System-in-Package design and manufacturing. Microelectron. J. **39**(12), 1740–1750 (2008)
16. H. Park, H. Kang, Y. Lee, Y. Park, J. Noh, G. Cho, Fully roll-to-roll gravure printed rectenna on plastic foils for wireless power transmission at 13.56 MHz. Nanotechnology **23**(34), 344006 (2012)
17. A. Vena, E. Perret, S. Tedjini, G.E.P. Tourtollet, A. Delattre, F. Garet, Y. Boutant, Design of chipless RFID tags printed on paper by flexography. IEEE Trans. Antennas Propag. **61**(12), 5868–5877 (2013)
18. K. Spree, Introduction to organic and printed electronics. Technical report, Holst Centre (2012)
19. Markets and Markets, Functional printing market by materials (substrate, inks), technology (inkjet, screen, flexo, gravure), application (sensors, displays, batteries, RFID, lighting, PV, medical), and geography (North America, Europe, APAC, ROW) - global forecasts and analysis. Technical report, Markets and Markets (2013)
20. Z.W. Wicks Jr., F.N. Jones, S.P. Pappas, D.A. Wicks, *Organic Coatings: Science and Technology*, vol. 5 (Wiley, Hoboken, 1985)
21. N. Rehfeld, V. Stenzel, *Functional Coatings*. European Coating Tech Files (Vincentz Network, Hanover, 2011)
22. U. Riaz, C. Nwaoha, S.M. Ashraf, Recent advances in corrosion protective composite coatings based on conducting polymers and natural resource derived polymers. Prog. Org. Coat. **77**(4), 743–756 (2014)
23. I.P. Parkin, R.G. Palgrave, Self-cleaning coatings. J. Mater. Chem. **15**(17), 1689–1695 (2005)
24. M. Kuhr, S. Bauer, U. Rothhaar, D. Wolff, Coatings on plastics with the PICVD technology. Thin Solid Films **442**(1–2), 107–116 (2003)
25. L.D. Chambers, K.R. Stokes, F.C. Walsh, R.J.K. Wood, Modern approaches to marine antifouling coatings. Surf. Coat. Technol. **201**(6), 3642–3652 (2006)
26. A. Mathiazhagan, R. Joseph, Nanotechnology-a new prospective in organic coating. Int. J. Chem. Eng. Appl. **2**(4), 228–237 (2011)
27. J.C. Tiller, C.J. Liao, K. Lewis, A.M. Klibanov, Designing surfaces that kill bacteria on contact. Proc. Natl. Acad. Sci. U.S.A. **98**(11), 5981–5985 (2001)
28. M.F. Montemor, Functional and smart coatings for corrosion protection: a review of recent advances. Surf. Coat. Technol. **258**, 17–37 (2014)
29. Q.H. Wang, K. Kalantar-Zadeh, A. Kis, J.N. Coleman, M.S. Strano, Electronics and opto-electronics of two-dimensional transition metal dichalcogenides. Nat. Nanotechnol. **7**(11), 699–712 (2012)
30. A.K. Geim, K.S. Novoselov, The rise of graphene. Nat. Mater. **6**(3), 183–191 (2007)
31. F. Annabi-Bergaya, Layered clay minerals. Basic research and innovative composite applications. Microporous Mesoporous Mater. **107**(1), 141–148 (2008)
32. W.G. Kim, S. Nair, Membranes from nanoporous 1D and 2D materials: a review of opportunities, developments, and challenges. Chem. Eng. Sci. **104**, 908–924 (2013)
33. A. Yaya, B. Agyei-Tuffour, D. Dodoo-Arhin, E. Nyankson, E. Annan, D.S. Konadu, E. Sinayobye, E.A. Baryeh, C.P. Ewels, Layered nanomaterials-a review. Glob. J. Eng. Des. Technol. **1**(2), 32–41 (2012)
34. P. Wallace, The band theory of graphite. Phys. Rev. **71**(9), 622–634 (1947)
35. J.W. McClure, Diamagnetism of graphite. Phys. Rev. **104**(3), 666–671 (1956)
36. J. Slonczewski, P. Weiss, Band structure of graphite. Phys. Rev. **109**(2), 272–279 (1958)
37. X. Lu, M. Yu, H. Huang, R.S. Ruoff, Tailoring graphite with the goal of achieving single sheets. Nanotechnology **10**(3), 269–272 (1999)

38. K.S. Novoselov, A.K. Geim, S.V. Morozov, D. Jiang, Y. Zhang, S.V. Dubonos, I.V. Grigorieva, A.A. Firsov, Electric field effect in atomically thin carbon films. Science **306**(5696), 666–669 (2004)
39. R.R. Nair, P. Blake, A.N. Grigorenko, K.S. Novoselov, T.J. Booth, T. Stauber, N.M.R. Peres, A.K. Geim, Fine structure constant defines visual transparency of graphene. Science **320**(5881), 1308 (2008)
40. K.I. Bolotin, K.J. Sikes, Z. Jiang, M. Klima, G. Fudenberg, J. Hone, P. Kim, H.L. Stormer, Ultrahigh electron mobility in suspended graphene. Solid State Commun. **146**(9–10), 351–355 (2008)
41. M. Chhowalla, H.S. Shin, G. Eda, L.-J. Li, K.P. Loh, H. Zhang, The chemistry of two-dimensional layered transition metal dichalcogenide nanosheets. Nat. Chem. **5**(4), 263–275 (2013)
42. T. Juntunen, H. Jussila, M. Ruoho, S. Liu, G. Hu, T. Albrow-Owen, L.W.T. Ng, R.C.T. Howe, T. Hasan, Z. Sun, I. Tittonen, Inkjet printed large-area flexible graphene thermoelectrics. Adv. Funct. Mater. **28**(22), 1800480 (2018)
43. F. Xia, H. Wang, D. Xiao, M. Dubey, A. Ramasubramaniam, Two-dimensional material nanophotonics. Nat. Photonics **8**(12), 899–907 (2014)
44. K.S. Novoselov, D. Jiang, F. Schedin, T.J. Booth, V.V. Khotkevich, S.V. Morozov, A.K. Geim, Two-dimensional atomic crystals. Proc. Natl. Acad. Sci. **102**(30), 10451–10453 (2005)
45. V. Nicolosi, M. Chhowalla, M.G. Kanatzidis, M.S. Strano, J.N. Coleman, Liquid exfoliation of layered materials. Science **340**(6139), 1226419–1226437 (2013)
46. B. Anasori, Y. Xie, M. Beidaghi, J. Lu, B.C. Hosler, L. Hultman, P.R.C. Kent, Y. Gogotsi, M.W. Barsoum, Two-dimensional, ordered, double transition metals carbides (MXenes). ACS Nano **9**(10), 9507–9516 (2015)
47. M. Naguib, V.N. Mochalin, M.W. Barsoum, Y. Gogotsi, MXenes: a new family of two-dimensional materials. Adv. Mater. **26**(7), 992–1005 (2014)
48. A. Nathan, A. Ahnood, M.T. Cole, S. Lee, Y. Suzuki, P. Hiralal, F. Bonaccorso, T. Hasan, L. Garcia-Gancedo, A. Dyadyusha, S. Haque, P. Andrew, S. Hofmann, J. Moultrie, D. Chu, A.J. Flewitt, A.C. Ferrari, M.J. Kelly, J. Robertson, G.A.J. Amaratunga, W.I. Milne, Flexible electronics: the next ubiquitous platform. Proc. IEEE **100**, 1486–1517 (2012)
49. F.J. Garcia de Abajo, Graphene nanophotonics. Science **339**(6122), 917–918 (2013)
50. T. Hasan, F. Torrisi, Z. Sun, D. Popa, V. Nicolosi, G. Privitera, F. Bonaccorso, A.C. Ferrari, Solution-phase exfoliation of graphite for ultrafast photonics. Phys. Status Solidi **247**(11), 2953–2957 (2010)
51. Y. Sun, G. Shi, Graphene/polymer composites for energy applications. J. Polym. Sci. Part B Polym. Phys. **51**(4), 231–253 (2013)
52. H.-J. Choi, S.-M. Jung, J.-M. Seo, D.W. Chang, L. Dai, J.-B. Baek, Graphene for energy conversion and storage in fuel cells and supercapacitors. Nano Energy **1**(4), 534–551 (2012)
53. J. Abraham, K.S. Vasu, C.D. Williams, K. Gopinadhan, Y. Su, C.T. Cherian, J. Dix, E. Prestat, S.J. Haigh, I.V. Grigorieva, P. Carbone, A.K. Geim, R.R. Nair, Tunable sieving of ions using graphene oxide membranes. Nat. Nanotechnol. **12**(6), 546–550 (2017)
54. Y.-Q. Li, T. Yu, T.-Y. Yang, L.-X. Zheng, K. Liao, Bio-inspired nacre-like composite films based on graphene with superior mechanical, electrical, and biocompatible properties. Adv. Mater. **24**(25), 3426–3431 (2012)
55. Y. Lu, B.R. Goldsmith, N.J. Kybert, A.T.C. Johnson, DNA-decorated graphene chemical sensors. Appl. Phys. Lett. **97**(8), 083107 (2010)
56. D. McManus, S. Vranic, F. Withers, V. Sanchez-Romaguera, M. Macucci, H. Yang, R. Sorrentino, K. Parvez, S.-K. Son, G. Iannaccone, K. Kostarelos, G. Fiori, C. Casiraghi, Water-based and biocompatible 2D crystal inks for all-inkjet-printed heterostructures. Nat. Nanotechnol. **12**, 343–350 (2017)
57. F. Bonaccorso, A. Lombardo, T. Hasan, Z. Sun, L. Colombo, A.C. Ferrari, Production and processing of graphene and 2D crystals. Mater. Today **15**(12), 564–589 (2012)

58. S. Bae, H. Kim, Y. Lee, X. Xu, J.-S. Park, Y. Zheng, J. Balakrishnan, T. Lei, H.R. Kim, Y.I. Song, Y.-J. Kim, K.S. Kim, B. Ozyilmaz, J.-H. Ahn, B.H. Hong, S. Iijima, Roll-to-roll production of 30-inch graphene films for transparent electrodes. Nat. Nanotechnol. **5**(8), 574–578 (2010)
59. X. Li, W. Cai, J. An, S. Kim, J. Nah, D. Yang, R. Piner, A. Velamakanni, I. Jung, E. Tutuc, S.K. Banerjee, L. Colombo, R.S. Ruoff, Large-area synthesis of high-quality and uniform graphene films on copper foils. Science **324**(5932), 1312–1314 (2009)
60. K.-K. Liu, W. Zhang, Y.-H. Lee, Y.-C. Lin, M.-T. Chang, C.-Y. Su, C.-S. Chang, H. Li, Y. Shi, H. Zhang, C.-S. Lai, L.-J. Li, Growth of large-area and highly crystalline MoS_2 thin layers on insulating substrates. Nano Lett. **12**(3), 1538–1544 (2012)
61. Y.-H. Lee, L. Yu, H. Wang, W. Fang, X. Ling, Y. Shi, C.-T. Lin, J.-K. Huang, M.-T. Chang, C.-S. Chang, M. Dresselhaus, T. Palacios, L.-J. Li, J. Kong, Synthesis and transfer of single-layer transition metal disulfides on diverse surfaces. Nano Lett. **13**(4), 1852–1857 (2013)
62. J.N. Coleman, Liquid-phase exfoliation of nanotubes and graphene. Adv. Funct. Mater. **19**(23), 3680–3695 (2009)
63. J.N. Coleman, Liquid exfoliation of defect-free graphene. Acc. Chem. Res. **46**(1), 14–22 (2013)
64. E.B. Secor, P.L. Prabhumirashi, K. Puntambekar, M.L. Geier, M.C. Hersam, Inkjet printing of high conductivity, flexible graphene patterns. J. Phys. Chem. Lett. **4**(8), 1347–1351 (2013)
65. F. Torrisi, T. Hasan, W. Wu, Z. Sun, A. Lombardo, T.S. Kulmala, G.-W. Hsieh, S. Jung, F. Bonaccorso, P.J. Paul, D. Chu, A.C Ferrari, Inkjet-printed graphene electronics. ACS Nano **6**(4), 2992–3006 (2012)
66. F. Withers, H. Yang, L. Britnell, A.P. Rooney, E. Lewis, A. Felten, C.R. Woods, V. Sanchez Romaguera, T. Georgiou, A. Eckmann, Y.J. Kim, S.G. Yeates, S.J. Haigh, A.K. Geim, K.S. Novoselov, C. Casiraghi, Heterostructures produced from nanosheet-based inks. Nano Lett. **14**(7), 3987–3992 (2014)
67. D.J. Finn, M. Lotya, G. Cunningham, R.J. Smith, D. McCloskey, J.F. Donegan, J.N. Coleman, Inkjet deposition of liquid-exfoliated graphene and MoS_2 nanosheets for printed device applications. J. Mater. Chem. C **2**(5), 925–932 (2014)
68. B.Z. Jang, W.C. Huang, US 7,071,258 B1 Nano-scaled graphene plates. Technical report 12 (2002)
69. Y. Hernandez, V. Nicolosi, M. Lotya, F.M. Blighe, Z. Sun, S. De, I.T. McGovern, B. Holland, M. Byrne, Y.K. Gun'Ko, J.J. Boland, P. Niraj, G. Duesberg, S. Krishnamurthy, R. Goodhue, J. Hutchison, V. Scardaci, A.C. Ferrari, J.N. Coleman, High-yield production of graphene by liquid-phase exfoliation of graphite. Nat. Nanotechnol. **3**(9), 563–568 (2008)
70. F. Torrisi, T. Hasan, W. Wu, Z. Sun, A. Lombardo, T.S. Kulmala, G.-W. Hsieh, S. Jung, F. Bonaccorso, P.J. Paul, D. Chu, A.C. Ferrari, Inkjet-printed graphene electronics. ACS Nano **6**(4), 2992–3006 (2012)
71. E.B. Secor, T.Z. Gao, A.E. Islam, R. Rao, S.G. Wallace, J. Zhu, K.W. Putz, B. Maruyama, M.C. Hersam, Enhanced conductivity, adhesion, and environmental stability of printed graphene inks with nitrocellulose. Chem. Mater. **29**(5), 2332–2340 (2017)
72. G. Hu, R.C.T. Howe, Z. Yang, L.W.T. Ng, C.G. Jones, K.J. Stone, T. Hasan, WO2017013263A1 Nanoplatelet dispersions, methods for their production and uses thereof (2017)
73. J. Jo, J.-S. Yu, T.-M. Lee, D.-S. Kim, Fabrication of printed organic thin-film transistors using roll printing. Jpn. J. Appl. Phys. **48**(4), 04C181 (2009)
74. M. Jung, J. Kim, J. Noh, N. Lim, C. Lim, G. Lee, J. Kim, H. Kang, K. Jung, A.D. Leonard, J.M. Tour, G. Cho, All-printed and roll-to-roll-printable 13.56-MHz-operated 1-bit RF tag on plastic foils. IEEE Trans. Electron Devices **57**(3), 571–580 (2010)
75. M. Allen, C. Lee, B. Ahn, T. Kololuoma, K. Shin, S. Ko, Microelectronic engineering R2R gravure and inkjet printed RF resonant tag. Microelectron. Eng. **88**(11), 3293–3299 (2011)
76. H. Kipphan (ed.), *Handbook of Print Media* (Springer, Berlin, 2001)
77. G. Hu, J. Kang, L.W.T. Ng, X. Zhu, R.C.T. Howe, M. Hersam, T. Hasan, Functional inks and printing of two-dimensional materials. Chem. Soc. Rev. **47**(9), 3265–3300 (2018)

78. D.J. Finn, M. Lotya, G. Cunningham, R.J. Smith, D. McCloskey, J.F. Donegan, J.N. Coleman, Inkjet deposition of liquid-exfoliated graphene and MoS_2 nanosheets for printed device applications. J. Mater. Chem. C **2**(5), 925–932 (2014)
79. E.B. Secor, M.C. Hersam, Graphene inks for printed electronics, http://www.sigmaaldrich.com/technical-documents/articles/technology-spotlights/graphene-inks-for-printed-electronics.html. Accessed 24 May 2015
80. W.J. Hyun, E.B. Secor, M.C. Hersam, C.D. Frisbie, L.F. Francis, High-resolution patterning of graphene by screen printing with a silicon stencil for highly flexible printed electronics. Adv. Mater. **27**(1), 109–115 (2015)
81. New graphene based inks for high-speed manufacturing of printed electronics, University of Cambridge, http://www.cam.ac.uk/research/news/new-graphene-based-inks-for-high-speed-manufacturing-of-printed-electronics
82. IDTechEx, Graphene, 2D materials, and carbon nanotubes 2017–2027. Technical report (2017)
83. Technavio, Global graphene market 2016–2020. Technical report, Infiniti Research Limited (2017)
84. Beige Market Intelligence, Strategic assessment of worldwide graphene composites market - till 2021. Technical report (2017)

Chapter 2
Structures, Properties and Applications of 2D Materials

Abstract Early scientific investigations into graphene date back to the 1950s. The interest in graphene intensified when Konstantin Novosolev and Andre Geim were awarded the Nobel prize in physics for 'groundbreaking experiments regarding the two-dimensional material graphene'. Since then, other two-dimensional (2D) materials have (re)gained increasing research interest. The most studied 2D materials to date include transition metal dichalcogenides (TMDs), black phosphorus (BP), hexagonal boron nitride (*h*-BN) and transition metal carbides and/or carbonitrides (MXenes). This chapter introduces these key material groups, and provides a review of the current understanding of their structures and properties. In particular, this chapter will focus on the (opto)electronic properties of the individual 2D materials and their potential applications.

2.1 The 2D Material Family

Graphite naturally occurs in metamorphic geology from the reduction of sedimentary carbon materials. The first recorded use of graphite was for ceramic decoration in the *Neolithic* era [1]. By the end of the middle ages the graphite's stability at high temperatures found application as a refractory material for lining casting moulds [2]. As a naturally soft rock, graphite's layered laminar structure was easy to delaminate into separate thin sheets. This particular property led to widespread use as a naturally occurring lubricant. During the investigation on the chemical reactivity of graphite, production of graphite oxide through successive oxidative treatments was first reported by Brodie in 1859 [3]. Subsequently, several strategies were investigated to reduce the graphite oxide to restore the properties of graphite. One such reduction method in alkaline environment was reported in 1962 which produced ultrathin (down to atomic thickness) graphite lamellas [4]. Gradually, scientific research on the laminar structure and monolayers of graphite established parallel works on the theoretical understanding of the structure and properties of graphite in atomically thin form. From an experimental point of view, the major concern was whether such monolayers would be stable enough to exist independently, or whether thermal fluctuations within a monolayer would overcome

L. W. T. Ng et al., *Printing of Graphene and Related 2D Materials*,
https://doi.org/10.1007/978-3-319-91572-2_2

the intralayer covalent bonds, causing the material to disintegrate [5]. Few-layer graphite flakes were first imaged in 1948 and monolayers [6], later to be termed as 'graphene', were eventually reported in 1987 [7]. However, it was in 2004 that graphene was finally demonstrated (with other 2D materials following in 2005) by Novoselov et al., who exfoliated it by repeatedly cleaving graphite (and other layered crystals) with adhesive tape [8, 9]. This method is now widely known as micromechanical cleavage (MC) and is colloquially termed the 'scotch-tape' method [8, 9].

Since 2004, many other materials that share a similar structure with graphite have been studied. Commonly referred to as layered materials, typical examples include transition metal dichalcogenides (TMDs), which occur in nature [10], black phosphorus (BP), which is synthesised from other phosphorus allotropes [11], and hexagonal boron nitride (*h*-BN), which is produced via chemical synthesis [12]. The term 'layered' is used to highlight the crystal structure where one or few atom thick planar layers are bound into a stack with strong in-plane covalent bonds and weak out-of-plane van der Waals (vdW) forces [5, 10, 13, 14]. The difference in the magnitude of these two forces allows delamination or exfoliation of the layered materials into few-layer and even monolayers [9]. Because of their atomic thickness, these are termed as 2D materials [8, 9, 13, 14].

Research into the properties of other 2D materials predate the demonstration of graphene. However, the research interest in graphene stimulated a general renaissance of study in the subject. For instance, although the optical absorption features of exfoliated molybdenum disulfide (MoS_2) [15] were initially reported in 1963 and monolayer MoS_2 was studied as early as in 1986 [16], interest in TMDs was restored only in 2011 when Radisavljevic et al. demonstrated a monolayer MoS_2 based transistor with a I_{ON}/I_{OFF} (i.e. the ratio of currents in the ON- and OFF-state of the transistor) exceeding 10^8 [17]. Similarly, although BP was first successfully synthesised from white phosphorus in 1914 [18], it regained significant attention as a 2D material in 2014 [11].

While 2D materials can typically be exfoliated directly from their bulk form, certain 2D materials have to be artificially produced via processes such as selective etching. A prime example is MXenes such as Ti_3C_2 and Ti_3CN that are produced by selectively etching away the interlayer metal atoms from the corresponding bulk crystals [19, 20].

This diversity amongst the 2D materials illustrates a broad spectrum of distinct and yet complementary optoelectronic properties, enabling a wide range of applications as shown in Fig. 2.1 [10, 13, 19, 21–24]. For the case of graphene, the exceptional combination of atomic thickness, high transparency and high conductivity is promising for the development of flexible transparent conductors, while the high carrier mobility in addition to zero-bandgap (can be engineered to up to <1 eV) can be exploited in high-frequency applications [13, 21, 22, 25, 26]. TMDs (e.g. MoS_2) and BP exhibit large, layer-dependent bandgaps spanning the visible to near-infrared (NIR) range [10, 13, 21]. These properties attract significant attention for electronics, such as thin-film transistors [11, 17, 27], and optoelectronics, for instance photodetectors and light emitters [28–30]. *h*-BN is an insulating 2D material with a bandgap of $\sim$6 eV, which enables ultraviolet (UV)

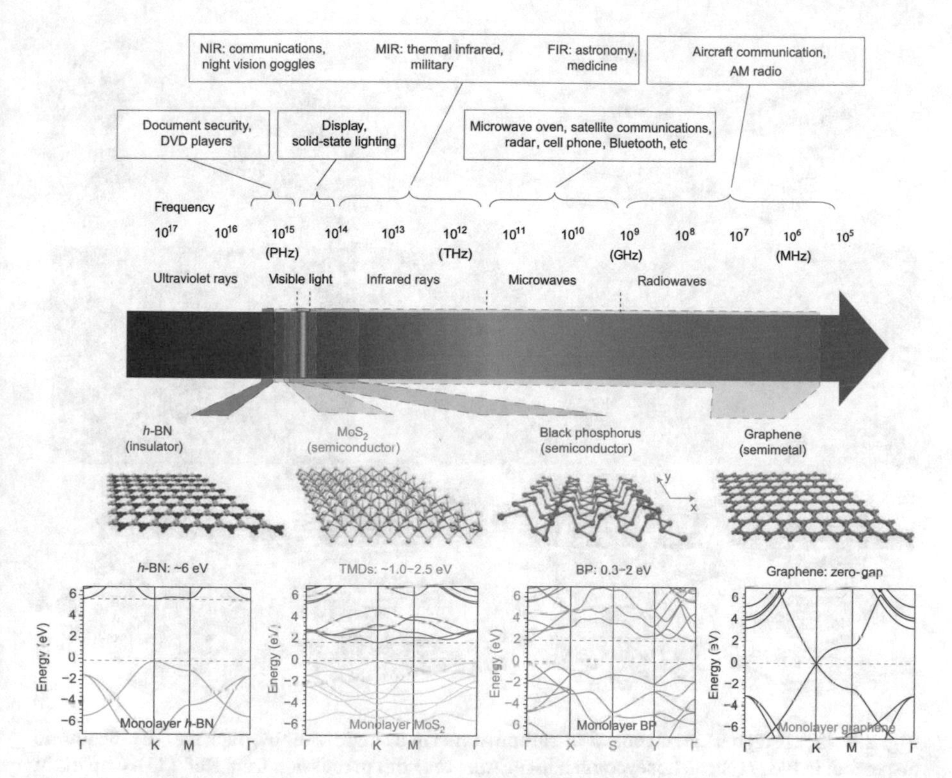

Fig. 2.1 Electromagnetic spectrum covered by selected 2D materials and their potential applications. Adapted with permission from Ref. [21]. Copyright (2014) Nature Publishing Group

light emission [31]. The atomically smooth surface and absence of dangling bonds makes *h*-BN an ideal dielectric substrate for other 2D materials in a wide variety of device applications [32]. Although optoelectronic properties are a convenient way to highlight the diversity the 2D material family can offer, the extent of their applications goes well beyond photonics, electronics and optoelectronics, ranging from composites to sensors through to energy storage and lubricant technologies.

We next discuss in detail the structures, properties and applications of graphene (Sect. 2.2), TMDs (Sect. 2.3), BP (Sect. 2.4), *h*-BN (Sect. 2.5) and MXenes (Sect. 2.6), as well as some other 2D materials (Sect. 2.7).

2.2 Graphene

Graphene is an atomically thin planar carbon sheet. Figure 2.2a schematically shows its honeycomb hexagonal lattice structure. As illustrated, each of the carbon atoms is covalently bonded with three other carbon atoms and as such, three of its four valence electrons are taken to form chemical bonds to the nearest neighbour

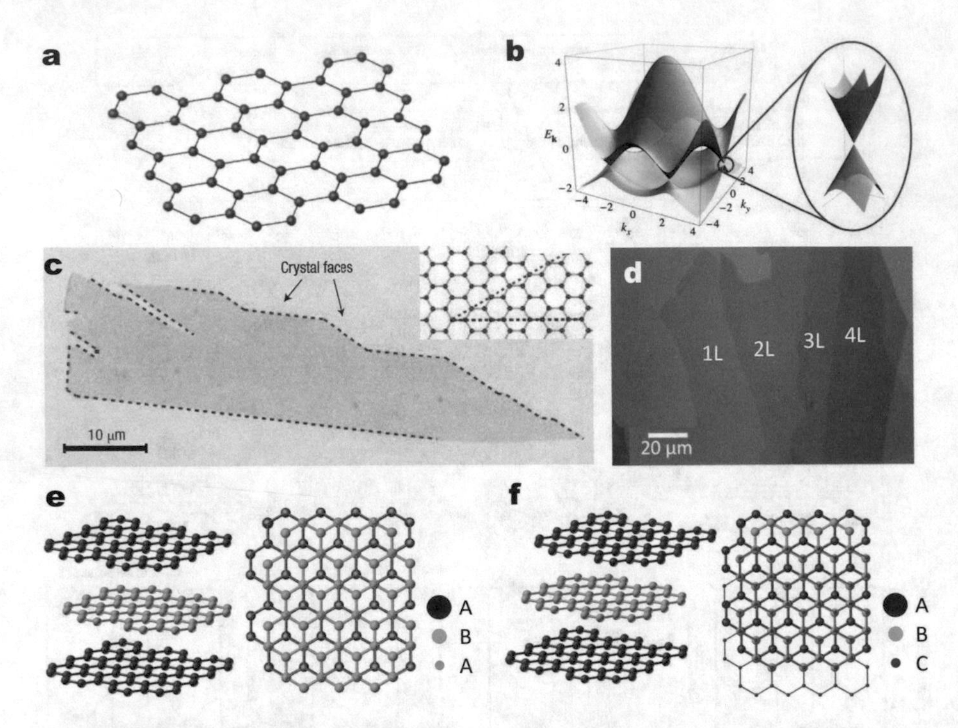

Fig. 2.2 (**a**) Hexagonal arrangement of carbon atoms in a single layer of graphene. (**b**) Electronic dispersion in the graphene honeycomb lattice. Adapted with permission from Ref. [33]. Copyright (2009) American Physical Society. (**c**) SEM micrograph of a large graphene flake, showing the crystal edges. Adapted with permission from Ref. [25]. Copyright (2007) Nature Publishing Group. (**d**) Optical micrograph of a graphene flake. The flake has a thickness varying in steps between 1 and 4 layers. Adapted with permission from Ref. [35]. Copyright (2015) American Chemical Society. (**e**) The most commonly found stacking configurations of multi-layer graphene are typically in ABA stacking configuration. (**f**) ABC stacking is also present on structural defects. The electronic properties of these two stacking variants are qualitatively different. Adapted with permission from Ref. [36]. Copyright (2015) Nature Publishing Group

electrons. However, the fourth electron is free to migrate through the graphene lattice [25, 33]. These freely migrating fourth electrons lead to an exceptionally high carrier concentration (n_c) in graphene [25, 33]. A value exceeding 10^{14} cm^{-2} was reported by Ye et al. in 2011 [34].

Figure 2.2b depicts the electronic dispersion in graphene. The fourth freely migrating electron corresponds to the low energies close to the Fermi energy. As this electron may be subject to a spin-up or spin-down state, the valence band is completely filled, and the conduction band is completely empty. Meanwhile, the electron is subjected to the periodic potential of the graphene honeycomb lattice, giving rise to degeneracies in the valence and conduction bands of graphene. This means that the valence and conduction bands touch, but do not overlap at the Dirac point of the Fermi energy. As a result, graphene behaves as a zero-bandgap semiconductor [25, 33].

This band structure defines the unique optoelectronic properties of graphene [13, 25, 33]. At the degenerate regions, the electrons are dispersed linearly and isotropically as massless Dirac fermions. Graphene therefore exhibits exceptionally high carrier mobility (μ_c). The zero-bandgap nature of graphene enables a good electrical conductivity, with the sheet resistance (R_s) of single and few layer graphene derived according to the relation:

$$R_s = (n_c \mu_c e N)^{-1} \tag{2.1}$$

where N is the number of layers, e is the charge on an electron [13]. Assuming $n_c \approx 0$ for ideal intrinsic graphene, $n_c \mu_c e$ does not go to zero but assumes a constant value of $4e^2/h$ (where h is the Planck constant) [25, 33], giving the R_s of ideal intrinsic monolayer graphene as $6\ \mathrm{k\Omega}\,\square^{-1}$. In practice, however, graphene samples typically exhibit a much higher n_c and hence a much smaller R_s [13]. Typical monolayer chemical vapour deposition (CVD) grown graphene samples may exhibit a sheet resistance as low as $30\ \Omega\,\square^{-1}$ with n_c of 10^{12}–$10^{13}\ \mathrm{cm}^{-2}$ and μ_c of 1,000–20,000 $\mathrm{cm^2\,V^{-1}\,s^{-1}}$ [13]. The sheet resistance can be tuned, for example, using electric field or chemical doping [37, 38].

The linearly dispersed electrons define that the optical transmittance (T) of graphene is independent of the frequency of incident light in the visible to the NIR region [25, 33, 39]. Theoretically, this is related to the fine structure constant ($\alpha = e^2/\hbar c = 1/137$, where c is the speed of light and $\hbar$ is the Dirac constant) [39]. This shows that monolayer graphene absorbs a significant fraction of the incident light ($\pi\alpha = 2.3\%$) [25, 33, 39]. This has been experimentally verified using suspended graphene samples [39]. The high transparency of graphene is largely due to the fact that it is atomically thin. In addition, graphene only reflects <0.1% of the incident light [39]. With such low reflectance between layers, the optical absorption of few-layer graphene may therefore be considered to be proportional to the graphene layer number, i.e. with each additional graphene layer increasing the absorption by 2.3% [39]. Under strong illumination, graphene additionally exhibits nonlinear absorption properties [40], where the absorption saturates, a phenomenon commonly observed in many semiconductors [41].

Figure 2.2c shows the SEM micrograph of a graphene sample. The crystal faces, following the zigzag and armchair direction (inset: Fig. 2.2c) are also indicated by dotted lines. Figure 2.2d shows a graphene sample with different number of layers, as evidenced by their optical contrast. Both these samples are prepared by repeatedly cleaving the face of graphite with an adhesive tape, allowing exfoliation by overcoming the relatively weak interlayer vdW forces [9].

The most common stacking of layers in multilayer graphene follows the so-called Bernal, or ABA configuration; Fig. 2.2e [36]. However, certain graphite crystals can have domains with a less commonly found ABC stacking; Fig. 2.2f. This gives these crystals different electronic properties despite being compositionally the same; [36].

The high electron mobility and carrier density of graphene holds huge promise for the development of transistors [13, 26, 33, 42]. The inset of Fig. 2.3a schematically outlines the structure of a typical top-gated graphene transistor. Such devices

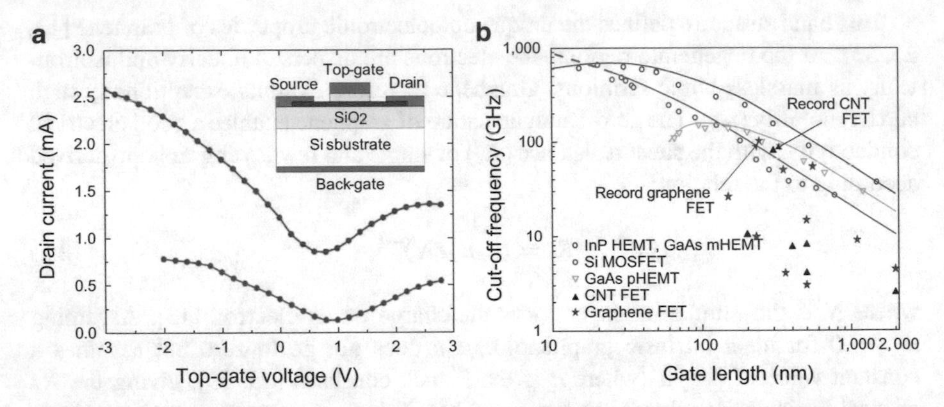

Fig. 2.3 (**a**) Typical graphene transistor transfer characteristic. (**b**) Reported cut-off frequency versus gate length for transistors based on graphene and other materials. Adapted with permission from Ref. [42]. Copyright (2010) Nature Publishing Group

exhibit high carrier mobility, for example 20,000 $cm^2V^{-1}s^{-1}$ reported in Ref. [43]. However, the zero-bandgap of graphene means that the transistors cannot be effectively switched off when conventional device architectures are considered. This seriously limits the I_{ON}/I_{OFF} ratio, typically <100 [26, 42, 44]. For instance, the transfer characteristics in Fig. 2.3a show that both the devices have a I_{ON}/I_{OFF} < 10. Logic electronic applications, however, typically require a sizeable bandgap (>0.4 eV) and excellent switching capabilities ($I_{ON}/I_{OFF} > 10^4$) [42]. Considerable efforts have therefore been devoted to 'opening' the graphene bandgap while avoiding diminishing the electronic properties, for instance, via nanopatterning graphene to create graphene nanoribbons [42]. Unfortunately for practical applications this goal remains elusive. Researchers are therefore investigating alternate device architectures as a way to sidestep this issue. Although the low I_{ON}/I_{OFF} significantly limits the use of graphene transistors in logic electronics, they present exceptionally high cut-off frequencies, as shown in Fig. 2.3b. For example, Ref. [45] reported an intrinsic cut-off frequency of over 100 GHz. The high cut-off frequencies, in combination with the high carrier mobilities, make graphene transistors promising for high frequency applications.

Besides transistors, another promising application opportunity is within the field of transparent conductors, taking the advantages of the high optical transparency and electrical conductivity of graphene. The minimum industry standard for transparent conductors in the majority of applications is $R_s < 100\,\Omega\,\square^{-1}$, $T > 90\%$ [13, 46–48]. CVD graphene with a low R_s and a high transparency has emerged as a promising candidate for the replacement of the current transparent conductor technologies (e.g. indium tin oxide (ITO) and alternatives such as silver nanowires and carbon nanotubes (CNTs)) [13, 46–48]. Indeed, Bae et al. demonstrated a 30 inch CVD graphene transparent conductor with R_s of $125\,\Omega\,\square^{-1}$ at a T of 97.4% [49]. The R_s could be further decreased to $30\,\Omega\,\square^{-1}$ at $T = 90\%$ [49], which would easily surpass current industry standards. As a transparent conductor, graphene offers the additional advantage of flexibility. It is capable of sustaining a bending strain of ~25% without breaking [13, 50]. The current incumbent, ITO, is a brittle material with a crack onset strain of only ~1–1.5%, which restricts the development of next generation flexible devices [13, 47, 48]. Proof-of-concept graphene flexible transparent conductor based applications have already been reported (see Fig. 2.4a), for example in touch screens, photovoltaics and light emitting devices [13, 49, 51]. Transparent graphene films fabricated by other methods (e.g. spray coating, rod coating and self-assembly of solution-processed graphene) instead of CVD have shown promise for the fabrication of transparent electrodes [24, 52–54]. For example, Howe et al. showed a rod-coated graphene-based film with 900 $\Omega\,\square^{-1}$ at $T = 90\%$ [24]. This allows production of flexible transparent graphene conductors with a much lower cost, larger area, lower processing temperature and directly on to the target substrate as opposed to CVD graphene films. Non-transparent printed graphene films can also find uses as electrical conductors [55–59], as a low-cost, high-performance alternative to conventional carbon and metal based printed conductors [59, 60]. Figure 2.4b–d is one such demonstration where graphene was deposited in a high speed commercial flexographic printing press for the first time. Such printed graphene based conductive circuits on flexible substrates can be interfaced with traditional electronics to develop large area applications such as capacitive touchpads [58].

With an atomically thin structure, graphene has a large specific surface area (theoretical value 2,630 $m^2\,g^{-1}$) [61], allowing it to interact strongly with the ambient elements, including moisture, gas, chemicals and biomolecules [62]. The adsorption of such ambient elements can lead to changes in electrical conductivity [33, 63, 64]. The high carrier mobility of graphene enables high sensitivity to any changes, even single molecules [33, 63, 65, 66], making graphene a promising material platform for sensing applications [22, 62, 63, 67]. For example, Schedin et al. demonstrated a transistor architecture based graphene gas sensor that enabled the detection of individual gas molecules of NO_2, NH_3, H_2O and CO [63]. This exceptional sensing capability was attributed to the change in the local carrier concentration of graphene upon the adsorption of gas molecules on the surface of graphene, leading to a step-like change in resistance. However, this approach of exploiting the intrinsic properties of pristine graphene cannot give good selective sensitivities, realising which may require covalent or non-covalent functionalisation of graphene [68–70].

The high specific surface area of graphene is also well-suited for the development of electrodes for energy storage and conversion (e.g. supercapacitors and batteries) and many examples of this effect have been published [71–76]. Using graphene as an electrode material has been reported to enable a decrease in film thickness and an increase in electrode-to-electrolyte contact which, when combined with good electrical conductivity, offers better performance to current technologies. With regard to manufacturing scalability and miniaturisation, El-Kady et al. reported the fabrication of supercapacitors using laser-scribed graphene-based electrodes [72]. The electrode exhibited high electrical conductivity and specific surface area for ultrahigh energy density values (up to 1.36 $mWh\,cm^{-3}$) in different electrolytes while maintaining high power density (~20 $W\,cm^{-3}$) and excellent cycling stability.

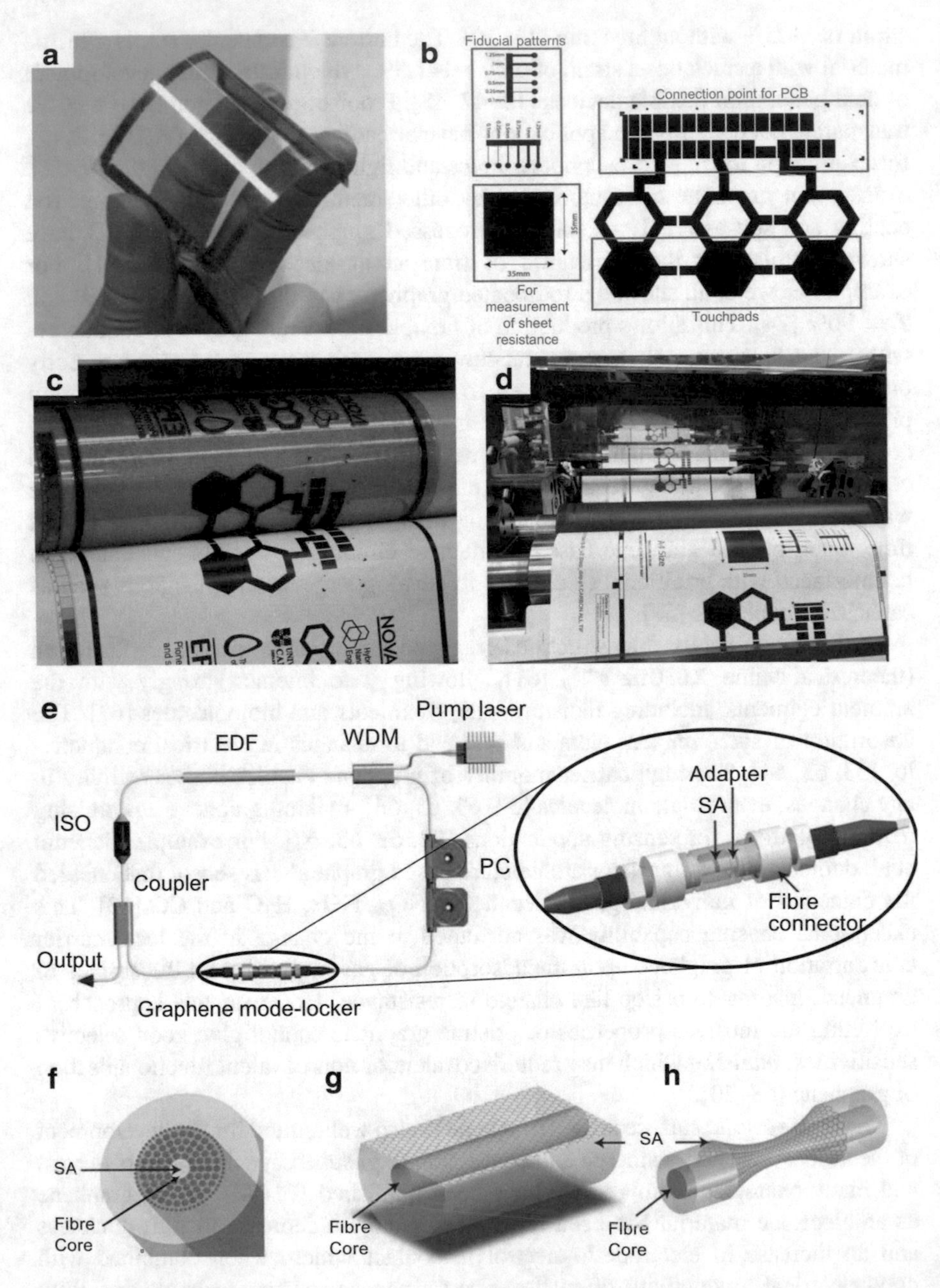

Fig. 2.4 (**a**) Flexible graphene transparent conductors. Adapted with permission from Ref. [49]. Copyright (2010), Nature Publishing Group. (**b**) Printed conductive graphene circuit schematic (the purpose of the different sections are labeled) with the method of fabrication via (**c, d**) flexographic printing. (**e**) Graphene saturable absorbers for ultrafast lasers showing (**f–h**) various device implementation strategies. (**e**) adapted with permission from Ref. [40]. Copyright (2010) American Chemical Society. (**f–h**) Adapted with permission from Ref. [129]. Copyright (2013) Nature Publishing Group

More recently, Liu et al. reported a lithium-air battery using a reduced graphene oxide electrode [73]. The device exhibited high specific capacities, excellent energy efficiency (93.2%) and impressive rechargeability.

Graphene is mechanically strong, with a measured breaking strength of up to 42 $\mathrm{N m^{-1}}$ (100 times that of steel) [50], and a Young's modulus of ~1 TPa (5 times that of steel) [50]. This makes graphene a likely filler material to enhance the strength of polymer-based composites. Polymer-composite technologies date back as far as the 1930s and are typically produced by dispersing stiff fibres into a liquid polymer matrix to create high-performance lightweight composites on curing. Their bulk strength and weight, combined with their ability to easily mould into complex structures find applications across automotive, aerospace and many domestic consumer products. An addition of a mere 0.05 wt% of graphene into a composite material has been reported to offer a significant improvement to the Young's modulus of composite materials [77].

The thermal conductivity of graphene is also exceptional with a measured conductivity value for a suspended single layer sample of up to ~5,000 $\mathrm{W\,m^{-1}\,K^{-1}}$ at room temperature (compared to, for example, silver which has a conductivity of ~406 $\mathrm{W\,m^{-1}\,K^{-1}}$) [78]. As such, graphene-containing composites designed for efficient heat transport have found application opportunities in numerous systems including automotive and aerospace products. Sung et al. successfully achieved a thermal conductivity of 1.53 $\mathrm{W\,m^{-1}\,K^{-1}}$ in epoxy resin with a 10 wt% loading of graphene flakes [79]. This is comparable to the performance of existing nano-fillers. An increase in the loading of graphene in the composite matrix can achieve remarkable results with a thermal conductivity of 6.44 $\mathrm{W\,m^{-1}\,K^{-1}}$ in epoxy resin, with only a 25 vol% incorporation of graphene. This has already surpassed the performance of conventional fillers that require a loading of ~70 vol% in order to achieve these values [80]. This high thermal conductivity, coupled with the limited Seebeck coefficient due to the semi-metallic nature of graphene [81], suggests a modest thermoelectric conversion efficiency for graphene [82]. Utilising nanostructure engineering, (e.g. graphene nanoribbons, heterostructures, nanopore structures and graphene-based organic nanocomposites) for improved Seebeck coefficient, graphene has been demonstrated as a promising candidate for thermoelectric energy conversion [82, 83]. Recently, Juntunen et al. reported high-performance graphene thermoelectrics from an inkjet-printed nanoporous all-graphene structure [82]. The authors demonstrated a room-temperature thermoelectric power factor of 18.7 $\mathrm{\mu W\,m^{-1}\,K^{-2}}$, a threefold improvement over previous solution-processed all-graphene structures. This potentially foresees future flexible, scalable and low-cost thermoelectric applications, such as harvesting energy from body heat in wearable applications [82].

The linear dispersion [25] of Dirac electrons, in addition to nonlinear optical absorption and ultrafast carrier dynamics [84–86], makes graphene particularly suited to operate as a nonlinear optical material for passive optical switches. This has led to the development of graphene-based saturable absorbers (SAs) for ultrafast lasers operating over a wide spectral range [40, 87]. One well-developed technology for graphene SA fabrication is through polymer nanocomposites from solution-processed graphene dispersions [88]. Figure 2.4e–h presents the integration of such graphene nanocomposite SAs into an ultrafast laser cavity for the generation of ultrashort (e.g. ~460 fs) pulses, pioneered by Sun et al. [40, 88].

Recently, graphene has been found to show proven value in water-filtration and biomedical applications due to its anti-bacterial and relatively non-toxic nature [89, 90]. In 2017, Abraham et al. reported the development of a 'tunable' sieve utilising a graphene oxide composite membrane capable of removing 97% of salt from salinated water [91]. Further applications of solution-processed graphene will be discussed in the penultimate chapter on applications of 2D materials.

2.3 Transition Metal Dichalcogenides (TMDs)

TMDs are a group of ~40 compounds sharing a general formula MX_2, where M represents a transition metal (e.g. molybdenum (Mo), tungsten (W), niobium (Nb)) and X is a group VI element (e.g. sulphur (S), selenium (Se), tellurium (Te)) [10, 92, 93]. Like graphite, bulk TMDs are crystals formed with stacked layers. Monolayer TMDs consist of a plane of hexagonally bonded M atoms sandwiched between two layers of X atoms via chemical bonds [94], as shown in top, orthogonal and side view in Fig. 2.5a–c. Depending on the coordination and oxidation states of the M atoms and the chemical structures between M and X atoms, TMDs can be either metallic (e.g. $NbSe_2$, $NiTe_2$, $ZrTe_2$) or semiconducting (e.g. MoS_2, WS_2 and $MoSe_2$) [10, 92, 93].

Exfoliation of bulk TMDs can lead to radical changes in their physical properties [10, 92]. An example of a widely studied TMD is MoS_2 [17]. Monolayer MoS_2 is transparent ($T > 90\%$ at ~652 nm [97]), with a mechanical strength 30 times that of steel [98]. The larger exposed surface area of the exfoliated MoS_2, compared to bulk MoS_2, greatly enhances its chemical and physical reactivities [14]. On its own, bulk MoS_2 is not a very efficient electrocatalyst for hydrogen-evolution reaction. However, exfoliated MoS_2 has been demonstrated to be an efficient electrocatalyst as a result of the significantly increased exposed catalytically active edges [99].

In the case of semiconducting TMDs (s-TMDs), decrease in thickness can cause changes in the electronic band structure [10, 92]. As shown in Fig. 2.5d, with a decrease in the layer numbers of MoS_2, the direct excitonic states near the K point remain relatively unchanged; whilst those at the Γ point shift significantly from indirect to direct such that the indirect bulk bandgap of ~1.3 eV moves towards a direct monolayer bandgap of ~1.9 eV [100]. This bandgap transition allows the observation of novel and interesting optoelectronic properties that are simply not present in bulk MoS_2. This is most evident in the intensity of photoluminescence (PL) of s-TMD materials which depends on the number of layers.

Eda et al. investigated PL from solution-processed MoS_2 flakes with respect to restacking thickness and demonstrated that both the intensity and position of the PL peak were thickness dependent [101]. Splendiani et al. also observed a 10^4 enhancement in PL from monolayer MoS_2 compared to the bulk [29]. Similarly

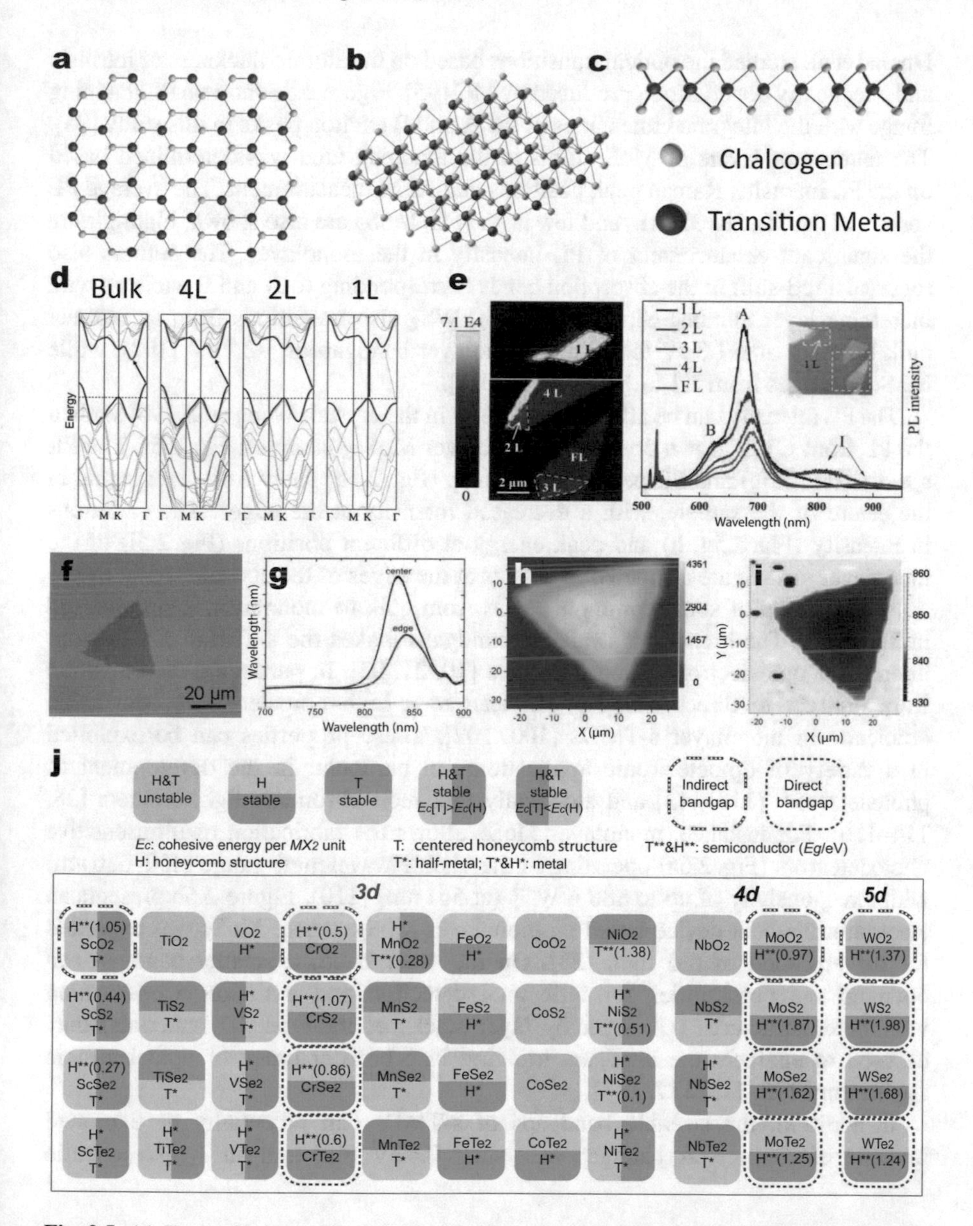

Fig. 2.5 (**a**) Front, (**b**) isometric and (**c**) side-view chemical structure of TMDs. (**d**) Evolution of band structure in bulk, quad-layer, bi-layer and monolayer MoS_2, respectively. Adapted with permission from Ref. [29]. Copyright (2010) American Chemical Society. (**e**) Photoluminescence spectra of MoS_2 with varied thickness. Adapted with permission from Ref. [95]. Copyright (2014) The Royal Society of Chemistry. (**f**) Optical image of small $MoSe_2$ monolayer grown by CVD. (**g**) PL spectra at the centre (black) and edge (red). (**h, i**) PL intensity and peak position maps. Reprinted with permission from Ref. [96]. Copyright (2014) American Chemical Society. (**j**) Summary of structures and properties of TMDs. Adapted with permission from Ref. [29]. Copyright (2010) American Chemical Society

Dhakal et al. studied the optical transitions based on the atomic thickness of intrinsic and chemically doped MoS_2 produced by MC [95]. Figure 2.5e shows a PL mapping image with the integrated intensities of the A and B exciton peaks in this study [95]. The number of layers of MoS_2 films in the mapping area was determined based on the PL intensity, Raman peak positions and AFM measurement. The average PL spectra of the 1L, 2L, 3L, 4L, and few layer (FL) MoS_2 are also shown, highlighting the significant enhancement of PL intensity in the monolayer. The authors also reported a red-shift in the absorption bands corresponding to A and B excitons with increasing layer number. Similar to MoS_2, WS_2 also transitions from an indirect bulk bandgap of ~1.3 eV to a direct monolayer bandgap of ~2.1 eV [102], while $MoSe_2$ changes from 1.1 to 1.6 eV [103, 104].

The PL intensity can be affected by defects in the crystal. Wang et al. [96] studied the PL from CVD grown crystalline monolayer $MoSe_2$ samples; Fig. 2.5f. The PL spectra (Fig. 2.5g) and PL peak intensity map (Fig. 2.5h) show a stronger signal in the centre of the sample, with a decreased intensity at the edges. The variations in intensity (Fig. 2.5g, h) and peak energy at different positions (Fig. 2.5i) of the monolayer sample are attributed to defects at the edges of the crystal.

The bandgap of some common TMDs from bulk to monolayer is summarised in Table 2.1. The breadth of available bandgaps makes the s-TMDs of particular interest for optoelectronics and photonics [10, 21, 92]. In particular, the transition from indirect to direct bandgap can lead to a high optoelectronic conversion efficiency in monolayer s-TMDs [100, 102]. These properties can be exploited in a variety of optoelectronic applications, in particular in the development of photodetectors [110–113] and atomically thin monochromatic light emitters [28, 114–116]. For instance, monolayer MoS_2 allows the fabrication of ultrasensitive photodetectors (Fig. 2.6a) operating in the visible wavelength range (400–680 nm) with a responsivity of up to 880 A W^{-1} (at 561 nm) [110]. Figure 2.6b presents an electroluminescent device based on monolayer MoS_2, where MoS_2 was exploited for the emission of red light [28]. On the other hand, a smaller bandgap can open up the possibilities for NIR photodetection and emission. Constructing vdW heterostructures (i.e. layer by layer stacks of multiple 2D materials) may be used to engineer the bandgap, bringing in a broader range of optoelectronic applications [92, 115, 117, 118].

In addition, the sizeable bandgaps of s-TMDs can potentially be exploited in logic electronics. A bandgap exceeding 0.4 eV [42] and an I_{ON}/I_{OFF} ratio

Table 2.1 Bandgap of some common TMDs. Data collected from Refs. [100, 103, 105–109]

Bandgap (eV)		Mo	W	Ti	Zr	Hf	Pd	Pt
S	Monolayer	1.9 eV	2.1 eV	0.65 eV	1.2 eV	1.3 eV	1.2 eV	1.9 eV
	Bulk	1.3 eV	1.3 eV	0.3 eV	1.6 eV	1.6 eV	1.1 eV	1.8 eV
Se	Monolayer	1.6 eV	1.7 eV	0.51 eV	0.7 eV	0.7 eV	1.1 eV	1.5 eV
	Bulk	1.1 eV	1.2 eV	Metallic	0.8 eV	0.6 eV	1.3 eV	1.4 eV
Te	Monolayer	1.1 eV	1.1 eV	0.1 eV	0.4 eV	0.3 eV	0.3 eV	0.8 eV
	Bulk	1.0 eV	0.7 eV	Metallic	Metallic	Metallic	0.2 eV	0.8 eV

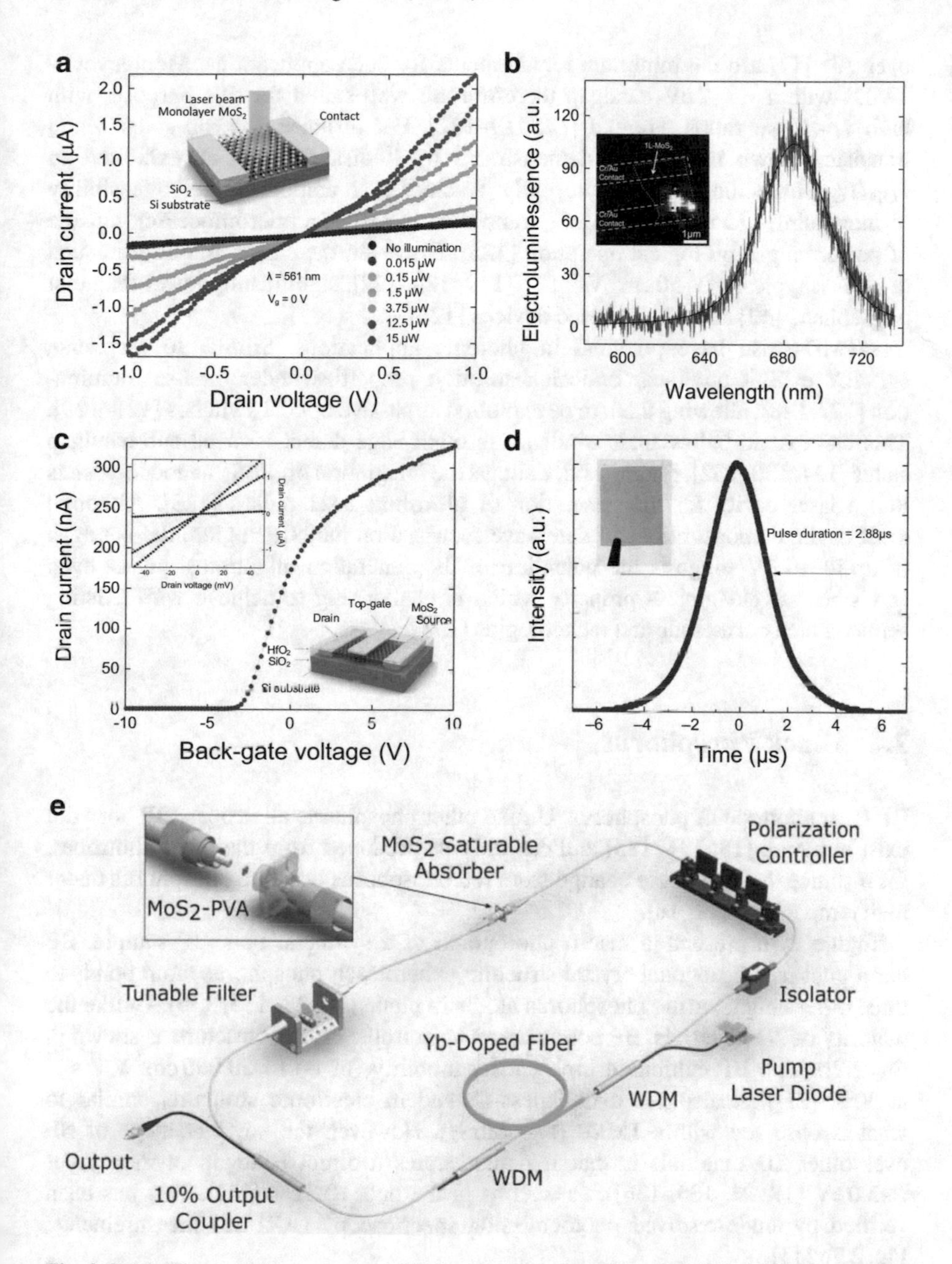

Fig. 2.6 (**a**) Ultrasensitive monolayer MoS_2 photodetector characteristics. Adapted with permission from Ref. [110]. Copyright (2013) Nature Publishing Group. (**b**) Electroluminescence spectrum of a MoS_2 emitter, inset: photograph of MoS_2 emitter with emission, scale bar 1 μm. Adapted with permission from Ref. [28]. Copyright (2013) American Chemical Society. (**c**) MoS_2 transistor transfer characteristics; inset: Schematic figure for top-gated MoS_2 transistor, reprinted with permission from Ref. [17]. Copyright (2010) Nature Publishing Group. (**d**) MoS_2 saturable absorber for ultrafast lasers; inset: image of MoS_2-PVA composite. Adapted with permission from Ref. [123]. Copyright (2015) Springer Publishing. (**e**) Tunable MoS_2 Q-switched fiber laser cavity schematic. Reprinted with permission from Ref. [124]. Copyright (2015) OSA Publishing

over 10^4 [17] are the minimum requirements for such applications. Monolayer s-TMDs with a ~1–2 eV bandgap therefore are well-suited for this purpose, with high I_{ON}/I_{OFF} ratios reported [17, 119–122]. For instance, the top-gated MoS_2 transistor shown in Fig. 2.6c demonstrated by Radisavljevic et al. exhibited an I_{ON}/I_{OFF} exceeding 10^8 [17]. Recently, Wachter et al. demonstrated the feasibility of integrating 115 individual MoS_2 transistors to realise a microprocessor capable of performing 1-bit logical operation [125]. However, the carrier mobility in such devices is typically 1–50 $cm^2\,V^{-1}\,s^{-1}$ [119, 120, 122], significantly lower than that of graphene [42] and silicon based devices [126].

s-TMDs also have potential in photonic applications. Similar to graphene, s-TMDs exhibit nonlinear optical absorption properties under intense illumination [127, 128], allowing them to be exploited as passive optical switches [127–129]. This can extend below their bandgap, through edge-defect induced sub-bandgap states [124, 130–132]. Figure 2.6d, e shows the integration of MoS_2 nanocomposites into a laser cavity for the generation of ultrashort laser pulses [123]. Although most of the demonstrations to date have focussed on the NIR region, the bandgap of up to ~2 eV suggests the potential for the generation of ultrafast pulses even at visible wavelengths; a property which is challenging to achieve with existing semiconductor materials and technologies [133].

2.4 Black Phosphorus

BP is an allotrope of phosphorus. Unlike other phosphorus allotropes, BP does not exist in nature [18, 134, 135] and can only be produced from the other allotropes, for instance through phase change from red phosphorus or white phosphorus under high temperature [11, 18].

Figure 2.7a presents a macro photograph of a synthetic bulk BP sample. BP has a puckered hexagonal crystal structure, where each phosphorus atom bonds to three other neighbouring phosphorus atoms in plane [18, 23, 135, 136]. Unlike the majority of 2D materials, BP possesses an anisotropic crystal structure as shown in Fig. 2.7b [11]. BP exhibits a high carrier mobility (of up to 50,000 $cm^2\,V^{-1}\,s^{-1}$ at 30 K [21]). It also has a thickness-dependent electronic structure, similar to what is observed with s-TMDs (e.g. MoS_2). However, the key advantage of BP over other 2D materials is that it demonstrates a direct bandgap in monolayer (~2.0 eV [18, 23, 135, 136]), as well as in the bulk (0.3 eV [11]). This has been verified by angle-resolved photoemission spectroscopy (ARPES) measurements; Fig. 2.7c [11]

The crystallographic anisotropy allows BP to exhibit in-plane anisotropic opto-electronic [23, 135, 136], thermal [137] and electrical [138] properties and intrinsic strength [139]. For instance, Fig. 2.7d shows that the measured photoluminescence peak intensity from monolayer BP is indeed dependent on the polarisation detection angle [140]. Note that this anisotropic nature is not unique to

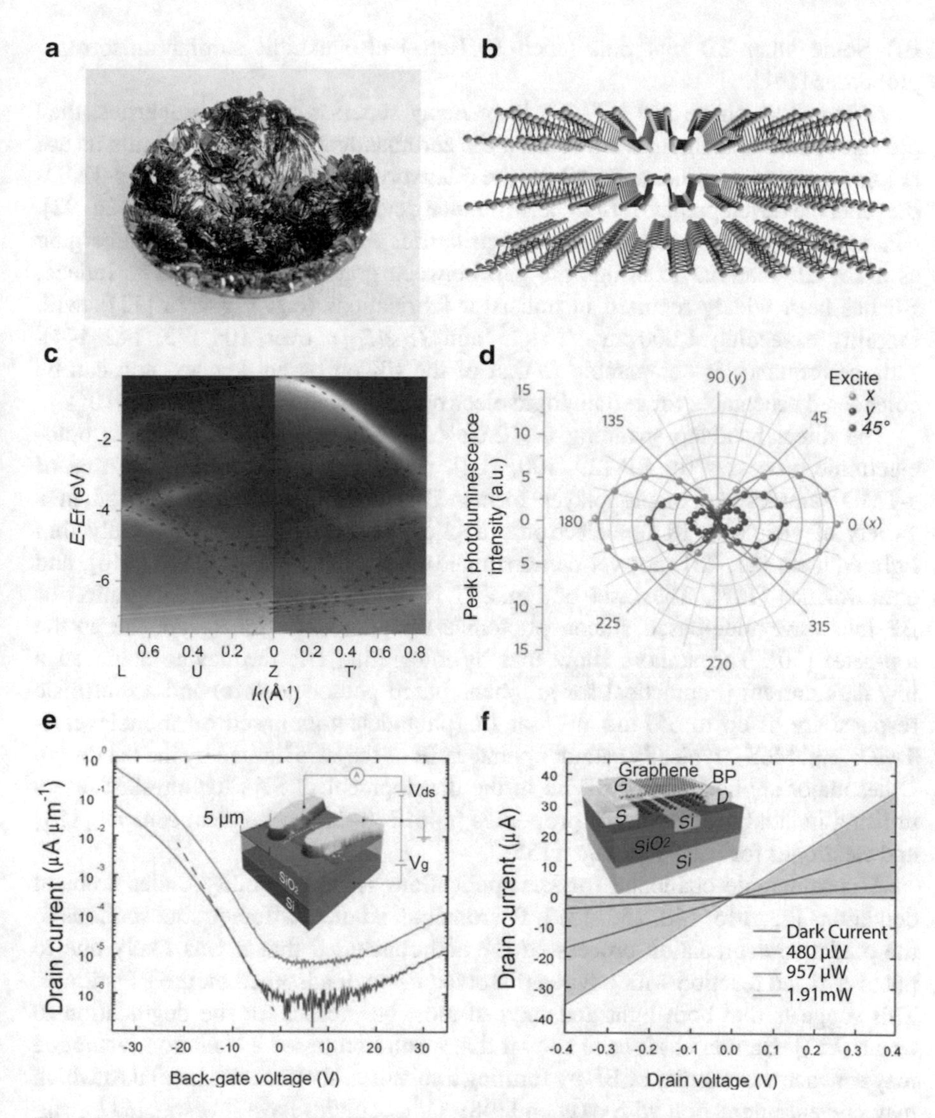

Fig. 2.7 (**a**) Close-up view of bulk BP crystals, source: Smart Elements (www.smart-elements.com). (**b**) Side-view chemical structure of BP. (**c**) Band structure of bulk BP mapped by ARPES. Adapted with permission from Ref. [11]. Copyright (2014) Nature Publishing Group. (**d**) PL peak intensity of a BP crystal as a function of polarisation detection angle. Adapted with permission from Ref. [140]. Copyright (2015) Nature Publishing Group. (**e**) Typical MC produced BP transistor transfer characteristics, inset: Schematic of back-gated BP transistor. Adapted with permission from Ref. [11]. Copyright (2014) Nature Publishing Group. (**f**) Source-drain current versus bias voltage under varied illumination of a waveguide-integrated BP photodetector, inset: Schematic of the photodetector. Adapted with permission from Ref. [30]. Copyright (2015) Nature Publishing Group

BP. Some other 2D materials (such as ReS_2) also exhibit similar anisotropic properties [141].

Although graphene and s-TMDs have many superior material properties, they present certain fundamental drawbacks: the zero bandgap of graphene limits its use in logic electronic applications, whilst the relatively low carrier mobility of s-TMDs prevents the development of high performance device applications [21, 23, 26, 92]. BP, with a sizeable bandgap and a high carrier mobility, is therefore emerging as a key 2D material to bridge the gap between graphene and s-TMDs. Indeed, BP has been widely reported in transistor fabrications (e.g. Fig. 2.7e [11]), with mobility exceeding 1,000 $cm^2\,V^{-1}\,s^{-1}$ and I_{ON}/I_{OFF} over 10^5 [23, 142–144]. This performance is comparable to that of the silicon based devices, and can be considered acceptable for certain logic electronic applications ($I_{ON}/I_{OFF} > 10^4$).

The direct bandgap spanning 0.3–2.0 eV leads to a high efficiency in optoelectronic processes for BP [29, 100, 102], noting a big difference from that of s-TMD samples not in monolayer forms. This can potentially be utilised in a variety of visible to IR optoelectronic applications, for instance atomically thin light emitters [21, 23] (not yet demonstrated), photodetectors [30, 145, 146], and photovoltaics [147]. The inset of Fig. 2.7f is one example of the integration of BP into waveguide-based silicon photonics for photodetection (graphene as the top-gate) [30]. The authors show that by integrating BP, the device achieved a low dark current (impractical for graphene-based photodetectors) and an intrinsic responsivity of up to 657 $mA\,W^{-1}$ at IR (photodetectors based on monolayer s-TMDs, e.g. MoS_2, typically cannot operate at IR as they are limited by the bandgap). Other major applications of BP lie in the development of SAs for ultrafast lasers utilising its nonlinear saturable properties [146, 148–153], chemical sensors [154] and electrodes for energy storage [155].

A considerable challenge for BP applications is its instability under ambient conditions [23, 146, 148, 156–159]. Favron et al. studied different contributions to the oxidative degradation process of BP and concluded that it was likely due to photo-assisted reaction with oxygen dissolved in water adsorbed on the BP surface. This suggests that both light and oxygen must be present for the degradation to occur [157]. Hanlon et al. have shown that a solution based exfoliation technique may enhance the stability of BP by forming a solvation shell around the flakes which may prevent interaction with oxygen [148]. Other studies have investigated encapsulants such as polydimethylsiloxane (PDMS) [158] and parylene C [146, 157]. In particular, using Raman and time-dependent optical absorption spectroscopy, Hu et al. recently demonstrated parylene-C coating to be effective against oxidation for inkjet printed BP thin films. This offers new opportunities for scalable fabrication of chemically unmodified BP based optoelectronic devices [146].

2.5 Hexagonal Boron Nitride (*h*-BN)

Boron nitride is an inorganic compound that can exist in cubic, wurtzite or hexagonal lattices forms [160]. *h*-BN is the hexagonal crystalline form of BN and is usually white in colour. Figure 2.8a shows a photograph of *h*-BN crystals (~5 μm particle size). In *h*-BN, B and N atoms are alternatively covalently bonded in-plane and are weakly stacked out-of-plane via vdW forces [160, 164]; Fig. 2.8b. With such a layered structure similar to graphite, *h*-BN has been widely used as a lubricant in the industry, well known as the 'white graphite' [160, 164]. However, unlike graphite, *h*-BN is an electrically insulating and thermally conductive material [163–166]. *h*-BN therefore exhibits better lubricating properties than graphite, especially when under extreme conditions (e.g. in vacuum, high temperature or oxidising atmosphere) [160, 164]. Figure 2.8c shows the band structure of *h*-BN of (6 eV [21, 160, 167, 168]). This wide bandgap can potentially be exploited in deep ultraviolet light emitters [21, 31]. We note that there has been some debate on this, with widely differing values between 3.6 and 7.1 eV reported [164].

Few- and monolayer *h*-BN has attracted recent attention for high performance devices [21, 160, 168, 169]. In its 2D form, *h*-BN shows many interesting material properties, for instance, an atomically smooth surface, the absence of dangling bonds and charge traps, and a significantly enhanced thermal stability [161]. This combination allows *h*-BN to be used as a dielectric screening substrate for other 2D materials in device fabrication [162, 170–172]. For example, Dean et al. used *h*-BN as a substrate for graphene transistors, replacing SiO_2 (Fig. 2.8d) to demonstrate an increase in the device carrier mobility [162]. As such, *h*-BN has recently emerged as a fundamental building block for vdW heterostructures, heralding a myriad of novel structures and properties [118, 173]. More recently, *h*-BN has been used to encapsulate BP devices to protect against atmospheric degradation [23].

The combination of the dielectric and thermal properties of *h*-BN also allow it to be incorporated into polymer nanocomposites as a dielectric and thermally conductive filler [161, 163, 174]. For example, Li et al. introduced solution-processed *h*-BN flakes into divinyltetramethyldisiloxane-bis(benzocyclobutene) (BCB) and developed a flexible *h*-BN nanocomposite with a high dielectric constant stability ($\epsilon_r >$ 3) against high temperatures (250 °C) and frequency (~10^6 Hz); Fig. 2.8e [163].

2.6 Transition Metal Carbides and/or Carbonitrides (MXenes)

Transition metal carbides and/or carbonitrides are a group of 2D materials with a composition of $M_{n+1}X_n$, where M stands for a transition metal, X being either carbon or nitrogen and n represents the number of X atoms [20]. They are termed as 'MXenes' to emphasise the graphene-like layered structure [20, 175]. MXenes can only be artificially produced from their parent compounds, layered ternary transition

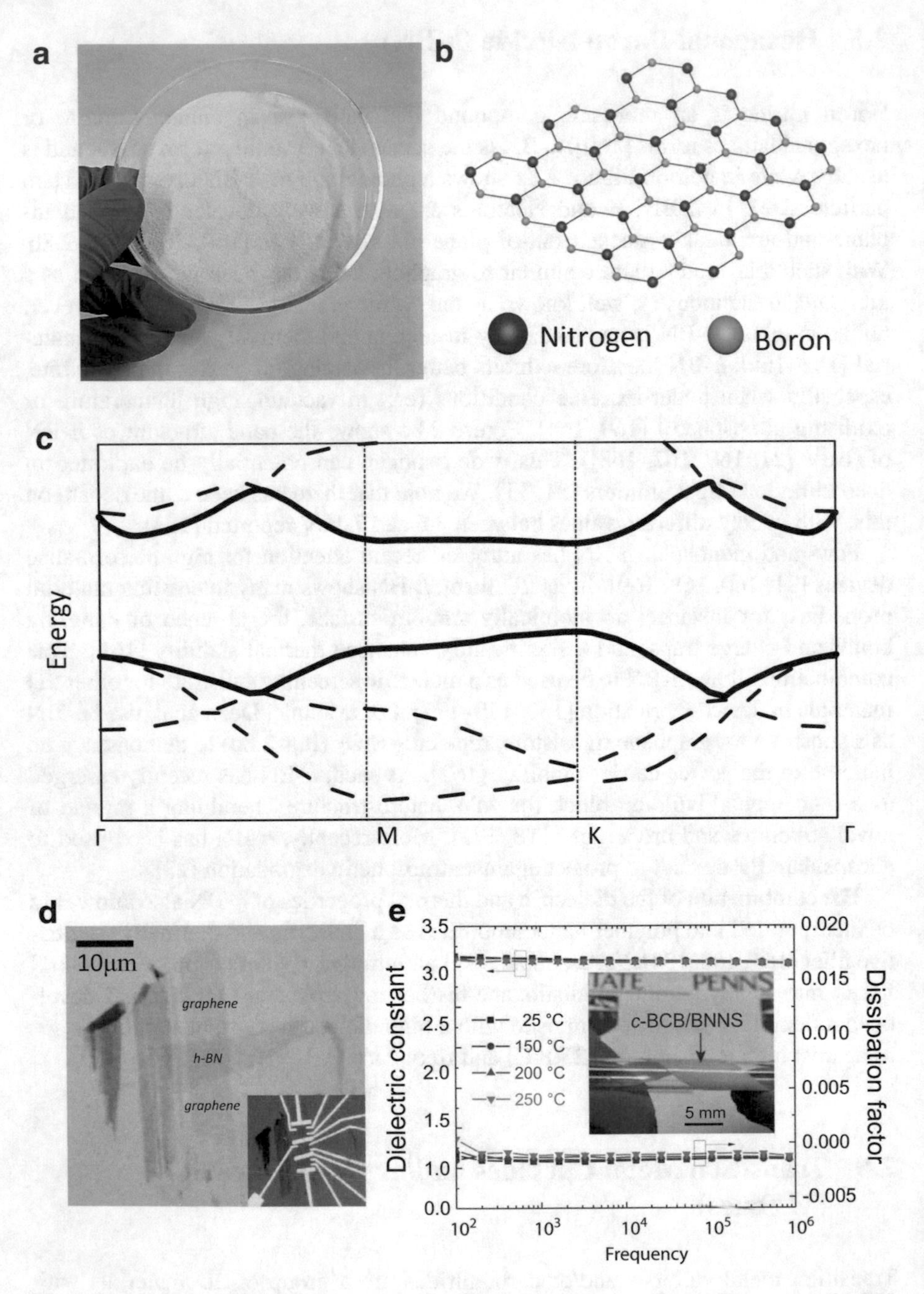

Fig. 2.8 (**a**) Photograph of bulk *h*-BN powder. (**b**) Orthogonal view chemical structure of *h*-BN; (**c**) Band structure of *h*-BN. Adapted with permission from Ref. [161]. Copyright (2014) Royal Society of Chemistry. (**d**) Use of MC *h*-BN as atomically thin dielectric substrate for graphene transistor, scale bar 10 μm. Adapted with permission from Ref. [162]. Copyright (2010) Nature Publishing Group. (**e**) High dielectric stability of flexible *h*-BN nanocomposite against temperature and frequency, inset: *h*-BN nanocomposite. Adapted with permission from Ref. [163]. Copyright (2010) Nature Publishing Group

metal carbides and carbonitrides, which are commonly termed as MAX phases ($M_{n+1}AX_n$) [19, 20].

In the MAX phases, 'A' represents a group IIIA or IVA element on the periodic table such as aluminium or silicon [19, 20]. The MAX phases are arranged in a layered hexagonal lattice, where the M planes are closely packed with the X planes in an octahedral prism. It has a composition of $M_{n+1}X_n$, with A planes interleaved between the X planes. Therefore, the MAX phases are not strictly layered crystals like graphite. Unlike the other 2D materials, the interlayer bonds between A and $M_{n+1}X_n$ are too strong to allow for direct exfoliation [14, 19, 20]. However, the differences between the chemical bonds, M-X and M-A, allow for the selective etching of the A layers away from the structure [19, 20], as shown in Fig. 2.9. MXenes were first experimentally demonstrated by Naguib et al. in 2011 [175]. The authors reported the production of graphene-like Ti_3C_2 thin layers by selectively etching away the aluminium atoms from its parent MAX compound Ti_3AlC_2, where the Al atoms form a plane sandwiched between two Ti_3C_2 layers.

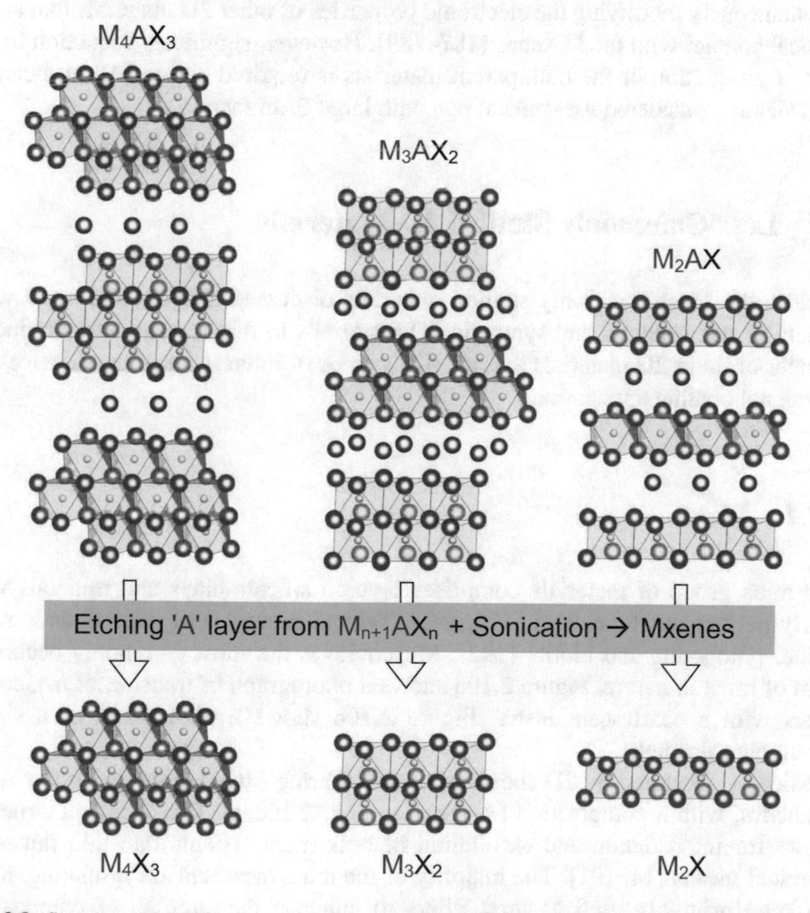

Fig. 2.9 Structure of MAX phases and the corresponding produced MXenes. Adapted from Ref. [20]

Besides Ti_3C_2, many other types of MXenes, such as Ti_2C, Ti_3CN, Nb_2C and V_2C, have thus far been reported [19, 20, 176, 177]. The number of possible combinations for M, A, and X gives >60 known MAX phases, well representing the wide spectrum of properties of MXenes [19, 20]. The electronic properties of MXenes are of special interest, with theoretical studies suggesting that MXenes are either metallic or possess a very small bandgap [20, 175, 177]. However, the bandgap of MXenes can be easily engineered through chemical functionalisation [19, 20].

One major application of MXenes is therefore in electrodes for electrochemical energy storage [19, 20, 178], with many successful demonstrations of high-performance proof-of-concept batteries and supercapacitors already published [179–183]. Potential applications of MXenes also range from sensors, electronics to conductive reinforcement additives [19, 20, 176, 177, 184–187]. MXenes are also emerging as potential building blocks for vdW heterostructures (as-yet largely unexplored) when integrated with other 2D materials [19, 20, 188, 189]. In these heterostructures, MXenes may work as conductive layers whilst simultaneously modifying the electronic properties of other 2D materials that are in vertical contact with the MXenes [187–189]. However, significant reduction in the cost of production of the bulk parent materials is required before MXenes can be realistically considered for applications with large form factors.

2.7 Less Commonly Studied 2D Materials

Besides the most commonly studied materials discussed above, there are a wide variety of other natural and synthetic 2D materials. In this section, we introduce a number of these 2D materials that are likely to be of interest for printed device and functional coating applications.

2.7.1 Mica

The mica group of materials comprises layered silicate clays and minerals with nearly perfect basal cleavage [190, 191]. Typical examples of mica include muscovite, phlogopite and biotite [192]. Muscovite is the most commonly occurring form of mica in nature. Figure 2.10a shows a photograph of translucent muscovite flakes with a pearlescent lustre. Figure 2.10b shows a photograph of a single muscovite mica flake.

Mica is structured by 2D sheets of corner sharing SiO_2 tetrahedra and/or AlO_2 octahedra, with a composition $(Al_nSi)_3O_4$; Fig. 2.10c, d. The chemical structure allows for intercalation and exfoliation of bulk mica crystals into thin flakes by chemical means [14, 191]. The majority of the mica members are insulating. Mica has been primarily used as inert fillers to enhance the strength of composites, as barrier materials for packaging, and as a pigment for special effect inks and

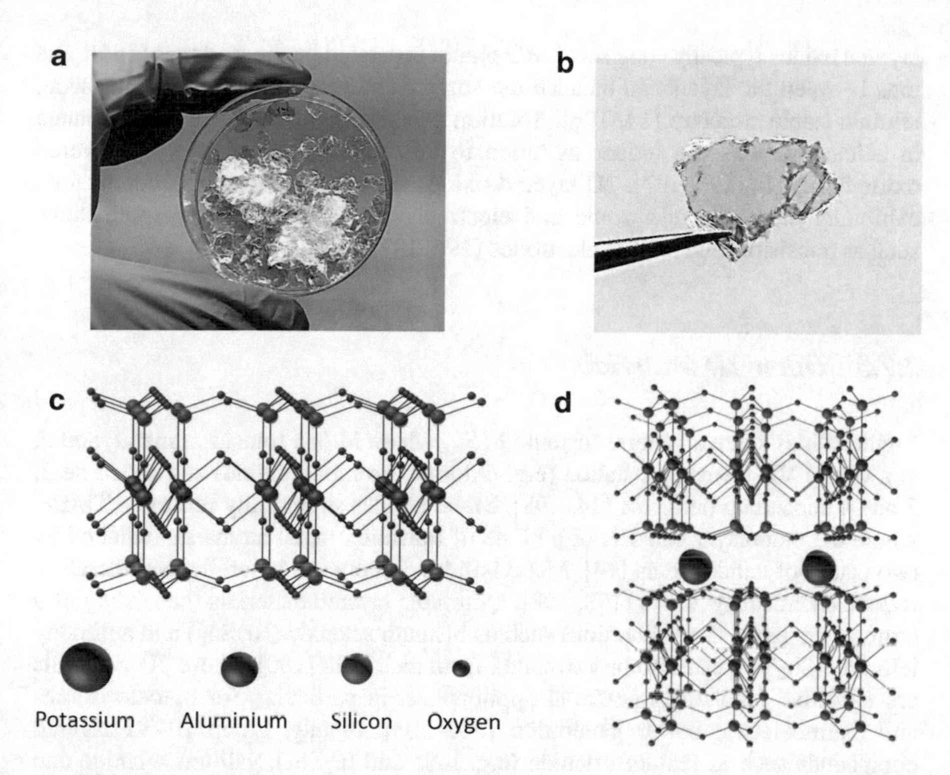

Fig. 2.10 (**a**) Photograph of flakes of muscovite mica showcasing its pearlescent appearance. (**b**) A single flake of muscovite mica. Chemical structure of muscovite mica in (**c**) single layer and (**d**) bi-layer configurations

paints [190–192]. The dielectric properties of mica also allow it to be used in capacitors [192]. More recently, the large surface area and the possibility of intercalation and exfoliation of the mica structure has indicated potential applications in catalysis, energy storage and sensing [193]. At present, very little work has been done on the 2D form of mica and its exfoliation. This is largely due to the challenges involved in the exfoliation of mica, since its charged state in 2D form requires constant presence of a charge balancing counterion. Only recently, Harvey et al. demonstrated successful liquid phase exfoliation of mica sheets as charged layered crystals [194].

2.7.2 *Metal Oxides*

Layered oxides are a large group of materials with a general formula, A_nMO_2, where A is an alkali metal (e.g. lithium, potassium), M is transition metal, O is oxygen, and $0.5 \leq n \leq 1$ [195]. Layered oxides typically occur as, for example, metal dioxides and trioxides, perovskites and niobates. The chemical structure of layered oxides typically consists of MO planes separated by the A planes [195]. The ions between the layers can balance the surface charge of the layers and as such, retain a stable structure [14, 196]. Solution processing methods via ion exchange in acidic solutions can induce agitation to this stable structure to yield layered oxide flakes [14, 196, 197]. 2D layered oxides are wide bandgap semiconductors, exhibiting interesting electronic and electrochemical properties for applications such as transistors and battery electrodes [196, 197].

2.7.3 Other 2D Materials

Metal halides share a general formula MX_n, where M is a transition metal, and X is a Group VII element or halide (e.g. chlorine, bromine, iodine) and n can be 2, 3 and 4 depending on the M [14, 198]. Metal halides structurally resemble TMDs, where the monolayer consists of a plane of transition metal atoms sandwiched by two planes of halide atoms [14]. Metal halides do not exist naturally, and therefore require chemical synthesis [198, 199]. Quintuple layered materials (consisting of 5 atom layers along the z-direction) such as bismuth selenide (Bi_2Se_3) and antimony telluride (Sb_2Te_3) can also be exfoliated from their bulk [200]. These 2D materials are attractive for various potential applications, in particular, for optoelectronics and thermoelectric power generation [201–203]. Finally, group III–VI layered compounds such as indium selenide (e.g. InSe and In_2Se_3), gallium selenide and copper indium selenide have also recently emerged as another promising family of 2D materials for (opto)electronic applications [204–207].

2.8 Conclusion

In this chapter, we have discussed the structure, properties and applications of the most widely studied 2D materials. The unique properties of 2D materials can potentially enable a broad range of high-performance, novel applications.

Of all these 2D materials, graphene and graphene variants like graphene oxide have attracted the most research attention. This understandably has led to a far more developed range of applications when compared with TMDs, BP, *h*-BN or other 2D materials. Indeed, many laboratory scale prototype applications of graphene or graphene variants such as transistors, sensors, electrodes and water filtration have already been widely reported.

Although 2D materials offer outstanding physical properties, the examples outlined within this chapter have been usually based on highly pristine samples produced via MC. In practice, however, MC is an unrealistic process to scale up and control on a practical industrial scale. The last decade therefore has seen the development of alternative production methods for 2D materials. For example, CVD, a high temperature growth technique on catalyst substrates, may be suitable for large-scale device fabrication. However, the production temperature and the complexities of post-growth processing and handling present a practical obstacle to cost-effective volume manufacturing. Solution processing methods are capable of delivering mass production of 2D materials at a low-cost. The quality of these materials however is significantly different to those produced through MC. This can adversely affect performances in many applications.

The balance between cost and performance should be carefully considered and the selection of the scale-up production method for 2D materials is therefore dependent on the final application. The next chapter will consider the practicalities of production methods for 2D materials and discuss their key advantages and drawbacks.

References

1. M. Garašanin, The Eneolithic period in the Central Balkan Area, in *The Cambridge Ancient History*, ed. by J. Boardman, I.E.S. Edwards, N.G.L. Hammond, E. Sollberger (Cambridge University Press, Cambridge, 1982), pp. 136–162
2. R. Mas-Ballesté, C. Gómez-Navarro, J. Gómez-Herrero, F. Zamora, 2D materials: to graphene and beyond. Nanoscale 3(1), 20–30 (2011)
3. B.C. Brodie, On the atomic weight of graphite. Philos. Trans. R. Soc. Lond. **149**, 249–259 (1859)
4. H.P. Boehm, A. Clauss, G.O. Fischer, U. Hofmann, Dunnste Kohlenstoff-Folien. Z. fur Naturforsch. Sect. B J. Chem. Sci. **17**(3), 150–153 (1962)
5. A.K. Bodenmann, A.H. MacDonald, Graphene: exploring carbon flatland. Phys. Today **60**(8), 35–41 (2007)
6. G. Ruess, F. Vogt, Hochstlamellarer kohlenstoff aus graphitoxyhydroxyd. Monatsh. Chem. **78**(3-4), 222–242 (1948)
7. S. Mouras, A. Hamm, D. Djurado, J.-C. Cousseing, Synthesis of first stage graphite intercalation compounds with fluorides. Rev. Chim. minér. **24**(5), 572–582 (1987)
8. K.S. Novoselov, D. Jiang, F. Schedin, T.J. Booth, V.V. Khotkevich, S.V. Morozov, A.K. Geim, Two-dimensional atomic crystals. Proc. Natl. Acad. Sci. **102**(30), 10451–10453 (2005)
9. K.S. Novoselov, A.K. Geim, S.V. Morozov, D. Jiang, Y. Zhang, S.V. Dubonos, I.V. Grigorieva, A.A. Firsov, Electric field effect in atomically thin carbon films. Science **306**(5696), 666–669 (2004)
10. M. Xu, T. Liang, M. Shi, H. Chen, Graphene-like two-dimensional materials. Chem. Rev. **113**(5), 3766–3798 (2013)
11. L. Li, Y. Yu, G.J. Ye, Q. Ge, X. Ou, H. Wu, D. Feng, X.H. Chen, Y. Zhang, Black phosphorus field-effect transistors. Nat. Nanotechnol. **9**(5), 372–377 (2014)
12. A. Lipp, K.A. Schwetz, K. Hunold, Hexagonal boron nitride: fabrication, properties and applications. J. Eur. Ceram. Soc. **5**(1), 3–9 (1989)
13. F. Bonaccorso, Z. Sun, T. Hasan, A.C. Ferrari, Graphene photonics and optoelectronics. Nat. Photonics **4**(9), 611–622 (2010)
14. V. Nicolosi, M. Chhowalla, M.G. Kanatzidis, M.S. Strano, J.N. Coleman, Liquid exfoliation of layered materials. Science **340**(6139), 1226419–1226437 (2013)
15. R.F. Frindt, A.D. Yoffe, Physical properties of layer structures: optical properties and photoconductivity of thin crystals of molybdenum disulphide. Proc. R. Soc. A Math. Phys. Eng. Sci. **273**(1352), 69–83 (1963)
16. P. Joensen, R.F. Frindt, S.R. Morrison, Single-layer MoS_2. Mater. Res. Bull. **21**(4), 457–461 (1986)

17. B. Radisavljevic, A. Radenovic, J. Brivio, V. Giacometti, A. Kis, Single-layer MoS_2 transistors. Nat. Nanotechnol. **6**(3), 147–150 (2011)
18. P.W. Bridgman, Two new modifications of phosphorus. J. Am. Chem. Soc. **36**(7), 1344–1363 (1914)
19. B. Anasori, Y. Xie, M. Beidaghi, J. Lu, B.C. Hosler, L. Hultman, P.R.C. Kent, Y. Gogotsi, M.W. Barsoum, Two-dimensional, ordered, double transition metals carbides (MXenes). ACS Nano **9**(10), 9507–9516 (2015)
20. M. Naguib, V.N. Mochalin, M.W. Barsoum, Y. Gogotsi, MXenes: a new family of two-dimensional materials. Adv. Mater. **26**(7), 992–1005 (2014)
21. F. Xia, H. Wang, D. Xiao, M. Dubey, A. Ramasubramaniam, Two-dimensional material nanophotonics. Nat. Photonics **8**(12), 899–907 (2014)
22. A.C. Ferrari, F. Bonaccorso, V. Fal'ko, K.S. Novoselov, S. Roche, P. Bøggild, S. Borini, F.H.L. Koppens, V. Palermo, N. Pugno, J.A. Garrido, R. Sordan, A. Bianco, L. Ballerini, M. Prato, E. Lidorikis, J. Kivioja, C. Marinelli, T. Ryhänen, A. Morpurgo, J.N. Coleman, V. Nicolosi, L. Colombo, A. Fert, M. Garcia-Hernandez, A. Bachtold, G.F. Schneider, F. Guinea, C. Dekker, M. Barbone, Z. Sun, C. Galiotis, A.N. Grigorenko, G. Konstantatos, A. Kis, M. Katsnelson, L. Vandersypen, A. Loiseau, V. Morandi, D. Neumaier, E. Treossi, V. Pellegrini, M. Polini, A. Tredicucci, G.M. Williams, B.H. Hong, J.-H. Ahn, J.M. Kim, H. Zirath, B.J. van Wees, H. van der Zant, L. Occhipinti, A. Di Matteo, I.A. Kinloch, T. Seyller, E. Quesnel, X. Feng, K. Teo, N. Rupesinghe, P. Hakonen, S.R.T. Neil, Q. Tannock, T. Löfwander, J. Kinaret, Science and technology roadmap for graphene, related two-dimensional crystals, and hybrid systems. Nanoscale **7**(11), 4598–4810 (2015)
23. A. Castellanos-Gomez, Black phosphorus: narrow gap, wide applications. J. Phys. Chem. Lett. **6**(21), 4280–4291 (2015)
24. R.C.T. Howe, G. Hu, Z. Yang, T. Hasan, Functional inks of graphene, metal dichalcogenides and black phosphorus for photonics and (opto)electronics. Proc. SPIE **9553**, 95530R (2015)
25. A.K. Geim, K.S. Novoselov, The rise of graphene. Nat. Mater. **6**(3), 183–191 (2007)
26. A.K. Geim, Graphene: status and prospects. Science **324**(5934), 1530–1534 (2009)
27. A.G. Kelly, T. Hallam, C. Backes, A. Harvey, A.S. Esmaeily, I. Godwin, J. Coelho, V. Nicolosi, J. Lauth, A. Kulkarni, S. Kinge, L.D.A. Siebbeles, G.S. Duesberg, J.N. Coleman, All-printed thin-film transistors from networks of liquid-exfoliated nanosheets. Science **356**(6333), 69–73 (2017)
28. R.S. Sundaram, M. Engel, A. Lombardo, R. Krupke, A.C. Ferrari, P. Avouris, M. Steiner, Electroluminescence in single layer MoS_2. Nano Lett. **13**(4), 1416–1421 (2013)
29. A. Splendiani, L. Sun, Y. Zhang, T. Li, J. Kim, C.-Y. Chim, G. Galli, F. Wang, Emerging photoluminescence in monolayer MoS_2. Nano Lett. **10**(4), 1271–1275 (2010)
30. N. Youngblood, C. Chen, S.J. Koester, M. Li, Waveguide-integrated black phosphorus photodetector with high responsivity and low dark current. Nat. Photonics **9**(4), 247 (2015)
31. Y. Kubota, K. Watanabe, O. Tsuda, T. Taniguchi, Deep ultraviolet light-emitting hexagonal boron nitride synthesized at atmospheric pressure. Science **317**(5840), 932–934 (2007)
32. F. Hui, C. Pan, Y. Shi, Y. Ji, E. Grustan-Gutierrez, M. Lanza, On the use of two dimensional hexagonal boron nitride as dielectric. Microelectron. Eng. **163**, 119–133 (2016)
33. A.H. Castro Neto, F. Guinea, N.M.R. Peres, K.S. Novoselov, A.K. Geim, The electronic properties of graphene. Rev. Mod. Phys. **81**(1), 109–162 (2009)
34. J. Ye, M.F Craciun, M. Koshino, S. Russo, S. Inoue, H. Yuan, H. Shimotani, A.F. Morpurgo, Y. Iwasa, Accessing the transport properties of graphene and its multilayers at high carrier density. Proc. Natl. Acad. Sci. **108**(32), 13002–13006 (2011)
35. Y. Huang, E. Sutter, N.N. Shi, J. Zheng, T. Yang, D. Englund, H.J. Gao, P. Sutter, Reliable exfoliation of large-area high-quality flakes of graphene and other two-dimensional materials. ACS Nano **9**(11), 10612–10620 (2015)
36. A.F. Morpurgo, The ABC of 2D materials. Nat. Phys. **11**(2), 99–100 (2015)
37. Y.-J. Kim, Y. Kim, K. Novoselov, B.H. Hong, Engineering electrical properties of graphene: chemical approaches. 2D Mater. **2**(4), 042001 (2015)

38. A. Das, S. Pisana, B. Chakraborty, S. Piscanec, S.K. Saha, U.V. Waghmare, K.S. Novoselov, H.R. Krishnamurthy, A.K. Geim, A.C. Ferrari, A.K. Sood, Monitoring dopants by Raman scattering in an electrochemically top-gated graphene transistor. Nat. Nanotechnol. **3**(4), 210–215 (2008)
39. R.R. Nair, P. Blake, A.N. Grigorenko, K.S. Novoselov, T.J. Booth, T. Stauber, N.M.R. Peres, A.K. Geim, Fine structure constant defines visual transparency of graphene. Science **320**(5881), 1308–1308 (2008)
40. Z. Sun, T. Hasan, F. Torrisi, D. Popa, G. Privitera, F. Wang, F. Bonaccorso, D.M. Basko, A.C. Ferrari, Graphene mode-locked ultrafast laser. ACS Nano **4**(2), 803–810 (2010)
41. U. Keller, Recent developments in compact ultrafast lasers. Nature **424**(6950), 831–838 (2003)
42. F. Schwierz, Graphene transistors. Nat. Nanotechnol. **5**(7), 487–496 (2010)
43. L. Liao, Y.-C. Lin, M. Bao, R. Cheng, J. Bai, Y. Liu, Y. Qu, K.L. Wang, Y. Huang, X. Duan, High-speed graphene transistors with a self-aligned nanowire gate. Nature **467**(7313), 305–308 (2010)
44. F. Torrisi, T. Hasan, W. Wu, Z. Sun, A. Lombardo, T.S. Kulmala, G.-W. Hsieh, S. Jung, F. Bonaccorso, P.J. Paul, D. Chu, A.C. Ferrari, Inkjet-printed graphene electronics. ACS Nano **6**(4), 2992–3006 (2012)
45. L. Liao, J. Bai, R. Cheng, Y.-C. Lin, S. Jiang, Y. Huang, X. Duan, Top-gated graphene nanoribbon transistors with ultrathin high-*k* dielectrics. Nano Lett. **10**(5), 1917–1921 (2010)
46. X. Huang, Z. Zeng, Z. Fan, J. Liu, H. Zhang, Graphene-based electrodes. Adv. Mater. **24**(45), 5979–6004 (2012)
47. S. De, J.N. Coleman, Are there fundamental limitations on the sheet resistance and transmittance of thin graphene films? ACS Nano **4**(5), 2713–2720 (2010)
48. M.F. Craciun, T.H. Bointon, S. Russo, Is graphene a good transparent electrode for photovoltaics and display applications? IET Circuits Devices Syst. **9**(6), 403–412 (2015)
49. S. Bae, H. Kim, Y. Lee, X. Xu, J.-S. Park, Y. Zheng, J. Balakrishnan, T. Lei, H.R. Kim, Y.I. Song, Y.-J. Kim, K.S. Kim, B. Ozyilmaz, J.-H. Ahn, B.H. Hong, S. Iijima. Roll-to-roll production of 30 inch graphene films for transparent electrodes. Nat. Nanotechnol. **5**(8), 574–578 (2010)
50. C. Lee, X. Wei, J.W. Kysar, J. Hone, Measurement of the elastic properties and intrinsic strength of monolayer graphene. Science **321**(5887), 385–388 (2008)
51. S. De, P.J. King, M. Lotya, A. O'Neill, E.M. Doherty, Y. Hernandez, G.S. Duesberg, J.N. Coleman, Flexible, transparent, conducting films of randomly stacked graphene from surfactant-stabilized, oxide-free graphene dispersions. Small **6**(3), 458–464 (2010)
52. X. Li, G. Zhang, X. Bai, X. Sun, X. Wang, E. Wang, H. Dai, Highly conducting graphene sheets and Langmuir-Blodgett films. Nat. Nanotechnol. **3**(9), 538–542 (2008)
53. J. Shim, J.M. Yun, T. Yun, P. Kim, K.E. Lee, W.J. Lee, R. Ryoo, D.J. Pine, G.-R. Yi, S.O. Kim, Two-minute assembly of pristine large-area graphene based films. Nano Lett. **14**(3), 1388–1393 (2014)
54. M. Hempel, D. Nezich, J. Kong, M. Hofmann, A novel class of strain gauges based on layered percolative films of 2D materials. Nano Lett. **12**(11), 5714–5718 (2012)
55. D. Dodoo-Arhin, R.C.T. Howe, G. Hu, Y. Zhang, P. Hiralal, A. Bello, G. Amaratunga, T. Hasan, Inkjet-printed graphene electrodes for dye-sensitized solar cells. Carbon **105**, 33–41 (2016)
56. E.B. Secor, T.Z. Gao, A.E. Islam, R. Rao, S.G. Wallace, J. Zhu, K.W. Putz, B. Maruyama, M.C. Hersam, Enhanced conductivity, adhesion, and environmental stability of printed graphene inks with nitrocellulose. Chem. Mater. **29**, 2332–2340 (2017)
57. E.B. Secor, M.C. Hersam, Graphene inks for printed electronics, http://www.sigmaaldrich.com/technical-documents/articles/technology-spotlights/graphene-inks-for-printed-electronics.html. Accessed 24 May 2015
58. New graphene based inks for high-speed manufacturing of printed electronics, University of Cambridge, http://www.cam.ac.uk/research/news/new-graphene-based-inks-for-high-speed-manufacturing-of-printed-electronics

59. P.G. Karagiannidis, S.A. Hodge, L. Lombardi, F. Tomarchio, N. Decorde, S. Milana, I. Goykhman, Y. Su, S.V. Mesite, D.N. Johnstone, R.K. Leary, P.A. Midgley, N.M. Pugno, F. Torrisi, A.C. Ferrari, Microfluidization of graphite and formulation of graphene-based conductive inks. ACS Nano **11**(3), 2742–2755 (2017)
60. Y. Liu, Z. Xu, J. Zhan, P. Li, C. Gao, Superb electrically conductive graphene fibers via doping strategy. Adv. Mater. **28**(36), 7941–7947 (2016)
61. A. Peigney, C. Laurent, E. Flahaut, R.R. Bacsa, A. Rousset, Specific surface area of carbon nanotubes and bundles of carbon nanotubes. Carbon **39**(4), 507–514 (2001)
62. F. Yavari, N. Koratkar, Graphene-based chemical sensors. J. Phys. Chem. Lett. **3**(13), 1746–1753 (2012)
63. F. Schedin, A.K. Geim, S.V. Morozov, E.W. Hill, P. Blake, M.I. Katsnelson, K.S. Novoselov, Detection of individual gas molecules adsorbed on graphene. Nat. Mater. **6**(9), 652–655 (2007)
64. P. Martin, Electrochemistry of graphene: new horizons for sensing and energy storage. Chem. Rec. **9**(4), 211–223 (2009)
65. S. Rumyantsev, G. Liu, M.S. Shur, R.A. Potyrailo, A.A. Balandin, Selective gas sensing with a single pristine graphene transistor. Nano Lett. **12**(5), 2294–2298 (2012)
66. W. Yuan, G. Shi, Graphene-based gas sensors. J. Mater. Chem. A **1**(35), 10078 (2013)
67. K. Shehzad, T. Shi, A. Qadir, X. Wan, H. Guo, A. Ali, W. Xuan, H. Xu, Z. Gu, X. Peng, J. Xie, L. Sun, Q. He, Z. Xu, C. Gao, Y.-S. Rim, Y. Dan, T. Hasan, P. Tan, E. Li, W. Yin, Z. Cheng, B. Yu, Y. Xu, J. Luo, X. Duan, Designing an efficient multimode environmental sensor based on graphene-silicon heterojunction. Adv. Mater. Technol. **2**(4), 1600262 (2017)
68. Y. Shao, J. Wang, H. Wu, J. Liu, I.A. Aksay, Y. Lin, Graphene based electrochemical sensors and biosensors: a review. Electroanalysis **22**(10), 1027–1036 (2010)
69. V. Georgakilas, M. Otyepka, A.B. Bourlinos, V. Chandra, N. Kim, K.C. Kemp, P. Hobza, R. Zboril, K.S. Kim, Functionalization of graphene: covalent and non-covalent approaches, derivatives and applications. Chem. Rev. **112**(11), 6156–6214 (2012)
70. G. Hu, J. Kang, L.W.T. Ng, X. Zhu, R.C.T. Howe, C. Jones, M.C. Hersam, T. Hasan, Functional inks and printing of two-dimensional materials. Chem. Soc. Rev. **47**(9), 3265–3300 (2018)
71. Y. Shao, M.F. El-Kady, L.J. Wang, Q. Zhang, Y. Li, H. Wang, M.F. Mousavi, R.B. Kaner, Graphene-based materials for flexible supercapacitors. Chem. Soc. Rev. **44**(11), 3639–3665 (2015)
72. M.F. El-Kady, V. Strong, S. Dubin, R.B. Kaner, Laser scribing of high-performance and flexible graphene-based electrochemical capacitors. Science **335**(6074), 1326–1330 (2012)
73. T. Liu, M. Leskes, W. Yu, A.J. Moore, L. Zhou, P.M. Bayley, G. Kim, C.P. Grey, Cycling Li-O_2 batteries via LiOH formation and decomposition. Science **350**(6260), 530–533 (2015)
74. Y. Xie, Y. Liu, Y. Zhao, Y.H. Tsang, S.P. Lau, H. Huang, Y. Chai, Stretchable all-solid-state supercapacitor with wavy shaped polyaniline/graphene electrode. J. Mater. Chem. A **2**(24), 9142–9149 (2014)
75. J. Cao, Y. Wang, Y. Zhou, J.-H. Ouyang, D. Jia, L. Guo, High voltage asymmetric supercapacitor based on MnO_2 and graphene electrodes. J. Electroanal. Chem. **689**, 201–206 (2013)
76. L.T. Le, M.H. Ervin, H. Qiu, B.E. Fuchs, W.Y. Lee, Graphene supercapacitor electrodes fabricated by inkjet printing and thermal reduction of graphene oxide. Electrochem. Commun. **13**(4), 355–358 (2011)
77. T. Ramanathan, A. Abdala, S. Stankovich, D.A. Dikin, M. Herrera-Alonso, R.D. Piner, D.H. Adamson, H.C. Schniepp, X. Chen, R.S. Ruoff, S.T. Nguyen, I.A. Aksay, R.K. Prud'Homme, L.C. Brinson, Functionalized graphene sheets for polymer nanocomposites. Nat. Nanotechnol. **3**, 327–331 (2008)
78. A.A. Balandin, S. Ghosh, W. Bao, I. Calizo, D. Teweldebrhan, F. Miao, C.N. Lau, Superior thermal conductivity of single-layer graphene. Nano Lett. **8**(3), 902–907 (2008)

79. S.H. Song K.H. Park, B.H. Kim, Y.W. Choi, G.H. Jun, D.J. Lee, B.-S. Kong, K.-W. Paik, S. Jeon, Enhanced thermal conductivity of epoxy–graphene composites by using non-oxidized graphene flakes with non-covalent functionalization. Adv. Mater. **25**(5), 732–737 (2013)
80. A. Yu, P. Ramesh, M.E. Itkis, E. Bekyarova, R.C. Haddon, Graphite nanoplatelet - epoxy composite thermal interface materials. J. Phys. Chem. C **111**, 7565–7569 (2007)
81. P. Dollfus, V.H. Nguyen, Thermoelectric effects in graphene nanostructures. J. Phys. Condens. Matter **27**, 133204 (2015)
82. T. Juntunen, H. Jussila, M. Ruoho, S. Liu, G. Hu, T. Albrow-Owen, L.W.T. Ng, R.C.T. Howe, T. Hasan, Z. Sun, I. Tittonen, Inkjet printed large-area flexible graphene thermoelectrics. Adv. Funct. Mater. **28**(22), 1800480 (2018)
83. J. Saint-Martin, V.H. Nguyen, P. Dollfus, M.C. Nguyen, High thermoelectric figure of merit in devices made of vertically stacked graphene layers, in *2015 International Conference on Simulation of Semiconductor Processes and Devices (SISPAD)* (2015), pp. 169–172
84. M. Breusing, C. Ropers, T. Elsaesser, Ultrafast carrier dynamics in graphite. Phys. Rev. Lett. **102**(8), 086809 (2009)
85. D. Sun, Z.-K. Wu, C. Divin, X. Li, C. Berger, W.A. de Heer, P.N. First, T.B. Norris, Ultrafast relaxation of excited dirac fermions in epitaxial graphene using optical differential transmission spectroscopy. Phys. Rev. Lett. **101**(15), 157402 (2008)
86. K. Seibert, G.C. Cho, W. Kütt, H. Kurz, D.H. Reitze, J.I. Dadap, H. Ahn, M.C. Downer, A.M. Malvezzi, Femtosecond carrier dynamics in graphite. Phys. Rev. B **42**(5), 2842–2851 (1990)
87. T. Hasan, F. Torrisi, Z. Sun, D. Popa, V. Nicolosi, G. Privitera, F. Bonaccorso, A.C. Ferrari, Solution-phase exfoliation of graphite for ultrafast photonics. Phys. Status Solidi B **247**(11–12), 2953–2957 (2010)
88. T. Hasan, Z. Sun, F. Wang, F. Bonaccorso, P.H. Tan, A.G. Rozhin, A.C. Ferrari, Nanotube-polymer composites for ultrafast photonics. Adv. Mater. **21**(38–39), 3874–3899 (2009)
89. R.R. Nair, H.A. Wu, P.N. Jayaram, I.V. Grigorieva, A.K. Geim, Unimpeded permeation of water through helium-leak-tight graphene-based membranes. Science **335**(6067), 442–444 (2012)
90. R.K. Joshi, P. Carbone, F.C. Wang, V.G. Kravets, Y. Su, I.V. Grigorieva, H.A. Wu, A.K. Geim, R.R. Nair, Precise and ultrafast molecular sieving through graphene oxide membranes. Science **343**(6172), 752–754 (2014)
91. J. Abraham, K.S. Vasu, C.D. Williams, K. Gopinadhan, Y. Su, C.T. Cherian, J. Dix, E. Prestat, S.J. Haigh, I.V. Grigorieva, P. Carbone, A.K. Geim, R.R. Nair, Tunable sieving of ions using graphene oxide membranes. Nat. Nanotechnol. **12**(6), 546–550 (2017)
92. Q.H. Wang, K. Kalantar-Zadeh, A. Kis, J.N. Coleman, M.S. Strano, Electronics and optoelectronics of two-dimensional transition metal dichalcogenides. Nat. Nanotechnol. **7**(11), 699–712 (2012)
93. M. Chhowalla, H.S. Shin, G. Eda, L.-J. Li, K.P. Loh, H. Zhang, The chemistry of two-dimensional layered transition metal dichalcogenide nanosheets. Nat. Chem. **5**(4), 263–275 (2013)
94. E.A. Marseglia, Transition metal dichalcogenides and their intercalates. Int. Rev. Phys. Chem. **3**(2), 177–216 (1983)
95. K.P. Dhakal, D.L. Duong, J. Lee, H. Nam, M. Kim, M. Kan, Y.H. Lee, J. Kim, Confocal absorption spectral imaging of MoS_2: optical transitions depending on the atomic thickness of intrinsic and chemically doped MoS_2. Nanoscale **6**(21), 13028–13035 (2014)
96. X. Wang, Y. Gong, G. Shi, W.L. Chow, K. Keyshar, G. Ye, R. Vajtai, J. Lou, Z. Liu, E. Ringe, B.K. Tay, P.M. Ajayan, Chemical vapor deposition growth of crystalline monolayer $MoSe_2$. ACS Nano **8**(5), 5125–5131 (2014)
97. M. Amani, D.-H. Lien, D. Kiriya, J. Xiao, A. Azcatl, J. Noh, S.R. Madhvapathy, R. Addou, S. KC, M. Dubey, K. Cho, R.M. Wallace, S.-C. Lee, J.-H. He, J.W. Ager III, X. Zhang, E. Yablonovitch, A. Javey, Near-unity photoluminescence quantum yield in MoS_2. Science **350**, 1065–1068 (2015)
98. S. Bertolazzi, J. Brivio, A. Kis, Stretching and breaking of ultrathin MoS_2. ACS Nano **5**(12), 9703–9709 (2011)

99. A.B. Laursen, S. Kegnæs, S. Dahl, I. Chorkendorff, Molybdenum sulfides - efficient and viable materials for electro - and photoelectrocatalytic hydrogen evolution. Energy Environ. Sci. **5**(2), 5577 (2012)
100. K.F. Mak, C. Lee, J. Hone, J. Shan, T.F. Heinz, Atomically thin MoS_2: a new direct-gap semiconductor. Phys. Rev. Lett. **105**(13), 136805 (2010)
101. G. Eda, H. Yamaguchi, D. Voiry, T. Fujita, M. Chen, M. Chhowalla, Photoluminescence from chemically exfoliated MoS_2. Nano Lett. **11**(12), 5111–5116 (2011)
102. A. Kuc, N. Zibouche, T. Heine, Influence of quantum confinement on the electronic structure of the transition metal sulfide TS_2. Phys. Rev. B **83**(24), 245213 (2011)
103. Y. Zhang, T.-R. Chang, B. Zhou, Y.-T. Cui, H. Yan, Z. Liu, F. Schmitt, J. Lee, R. Moore, Y. Chen, H. Lin, H.-T. Jeng, S.-K. Mo, Z. Hussain, A. Bansil, Z.-X. Shen. Direct observation of the transition from indirect to direct bandgap in atomically thin epitaxial $MoSe_2$. Nat. Nanotechnol. **9**(2), 111–115 (2013)
104. W.S. Yun, S.W. Han, S.C. Hong, I.G. Kim, J.D. Lee, Thickness and strain effects on electronic structures of transition metal dichalcogenides: 2H-MX_2 semiconductors (M = Mo, W; X = S, Se, Te). Phys. Rev. B **85**(3), 033305 (2012)
105. X. Duan, C. Wang, A. Pan, R. Yu, X. Duan, Two-dimensional transition metal dichalcogenides as atomically thin semiconductors: opportunities and challenges. Chem. Soc. Rev. **44**(24), 8859–8876 (2015)
106. K.K. Kam, B.A. Parkinson, Detailed photocurrent spectroscopy of the semiconducting group VIB transition metal dichalcogenides. J. Phys. Chem. **86**(4), 463–467 (1982)
107. S. Tongay, J. Zhou, C. Ataca, K. Lo, T.S. Matthews, J. Li, J.C. Grossman, J. Wu, Thermally driven crossover from indirect toward direct bandgap in 2D semiconductors: $MoSe_2$ versus MoS_2. Nano Lett. **12**(11), 5576–5580 (2012)
108. C. Ruppert, O.B. Aslan, T.F. Heinz, Optical properties and band gap of single- and few-layer $MoTe_2$ crystals. Nano Lett. **14**(11), 6231–6236 (2014)
109. W. Zhao, Z. Ghorannevis, L. Chu, M. Toh, C. Kloc, P.-H. Tan, G. Eda, Evolution of electronic structure in atomically thin sheets of WS_2 and WSe_2. ACS Nano **7**(1), 791–797 (2013)
110. O. Lopez-Sanchez, D. Lembke, M. Kayci, A. Radenovic, A. Kis, Ultrasensitive photodetectors based on monolayer MoS_2. Nat. Nanotechnol. **8**(7), 497–501 (2013)
111. D. Kufer, G. Konstantatos, Highly sensitive, encapsulated MoS_2 photodetector with gate controllable gain and speed. Nano Lett. **15**(11), 7307–7313 (2015)
112. N. Perea-López, A.L. Elías, A. Berkdemir, A. Castro-Beltran, H.R. Gutiérrez, S. Feng, R. Lv, T. Hayashi, F. López-Urías, S. Ghosh, B. Muchharla, S. Talapatra, H. Terrones, M. Terrones, Photosensor device based on few-layered WS_2 films. Adv. Funct. Mater. **23**(44), 5511–5517 (2013)
113. Y. Xue, Y. Zhang, Y. Liu, H. Liu, J. Song, J. Sophia, J. Liu, Z. Xu, Q. Xu, Z. Wang, J. Zheng, Y. Liu, S. Li, Q. Bao, Scalable production of a few-layer MoS_2/WS_2 vertical heterojunction array and its application for photodetectors. ACS Nano **10**(1), 573–580 (2016)
114. C. Palacios-Berraquero, M. Barbone, D.M. Kara, X. Chen, I. Goykhman, D. Yoon, A.K. Ott, J. Beitner, K. Watanabe, T. Taniguchi, A.C. Ferrari, M. Atatüre, Atomically thin quantum light-emitting diodes. Nat. Commun. **7**, 12978 (2016)
115. F. Withers, O. Del Pozo-Zamudio, A. Mishchenko, A.P. Rooney, A. Gholinia, K. Watanabe, T. Taniguchi, S.J. Haigh, A.K. Geim, A.I. Tartakovskii, K.S. Novoselov, Light-emitting diodes by band-structure engineering in van der Waals heterostructures. Nat. Mater. **14**(3), 301–306 (2015)
116. S. Jo, N. Ubrig, H. Berger, A.B. Kuzmenko, A.F. Morpurgo, Mono- and bilayer WS_2 light-emitting transistors. Nano Lett. **14**(4), 2019–2025 (2014)
117. M. Shanmugam, T. Bansal, C.A. Durcan, B. Yu, Molybdenum disulphide/titanium dioxide nanocomposite-poly 3-hexylthiophene bulk heterojunction solar cell. Appl. Phys. Lett. **100**(15), 153901 (2012)
118. A.K. Geim, I.V. Grigorieva, Van der Waals heterostructures. Nature **499**(7459), 419–425 (2013)

119. M.S. Fuhrer, J. Hone, Measurement of mobility in dual-gated MoS_2 transistors. Nat. Nanotechnol. **8**(3), 146–147 (2013)
120. K. Kang, S. Xie, L. Huang, Y. Han, P.Y. Huang, K.F. Mak, C.-J. Kim, D. Muller, J. Park, High-mobility three-atom-thick semiconducting films with wafer-scale homogeneity. Nature **520**(7549), 656–660 (2015)
121. T. Georgiou, R. Jalil, B.D. Belle, L. Britnell, R.V. Gorbachev, S.V. Morozov, Y.-J. Kim, A. Gholinia, S.J. Haigh, O. Makarovsky, L. Eaves, L.A. Ponomarenko, A.K. Geim, K.S. Novoselov, A. Mishchenko, Vertical field-effect transistor based on graphene-WS_2 heterostructures for flexible and transparent electronics. Nat. Nanotechnol. **8**(2), 100–103 (2012)
122. B. Radisavljevic, A. Kis, Mobility engineering and a metal-insulator transition in monolayer MoS_2. Nat. Mater. **12**(9), 815–820 (2013)
123. M. Zhang, R.C.T. Howe, R.I. Woodward, E.J.R. Kelleher, F. Torrisi, G. Hu, S.V. Popov, J.R. Taylor, T. Hasan, Solution processed MoS_2-PVA composite for sub-bandgap mode-locking of a wideband tunable ultrafast Er:fiber laser. Nano Res. **8**(5), 1522–1534 (2015)
124. R.I. Woodward, R.C.T. Howe, T.H. Runcorn, G. Hu, F. Torrisi, E.J.R. Kelleher, T. Hasan, Wideband saturable absorption in few-layer molybdenum diselenide ($MoSe_2$) for Q-switching Yb-, Er- and Tm-doped fiber lasers. Opt. Express **23**(15), 20051 (2015)
125. S. Wachter, D.K. Polyushkin, O. Bethge, T. Mueller, A microprocessor based on a two-dimensional semiconductor. Nat. Commun. **8**, 14948 (2017)
126. Electrical properties of silicon, http://www.ioffe.ru/SVA/NSM/Semicond/Si/electric.html
127. K. Wang, J. Wang, J. Fan, M. Lotya, A. O'Neill, D. Fox, Y. Feng, X. Zhang, B. Jiang, Q. Zhao, H. Zhang, J.N. Coleman, L. Zhang, W.J. Blau, Ultrafast saturable absorption of two-dimensional MoS_2 nanosheets. ACS Nano **7**(10), 9260–9267 (2013)
128. K. Wu, X. Zhang, J. Wang, X. Li, J. Chen, WS_2 as a saturable absorber for ultrafast photonic applications of mode-locked and Q-switched lasers. Opt. Express **23**(9), 11453 (2015)
129. Z. Sun, A. Martinez, F. Wang, Optical modulators with 2D layered materials. Nat. Photonics **10**(4), 227–238 (2016)
130. L.A.B. Marçal, M.S.C. Mazzoni, L.N. Coelho, E. Marega, G.J. Salamo, R. Magalhães-Paniago, A. Malachias, Quantitative measurement of manganese incorporation into (In,Mn)As islands by resonant x-ray scattering. Phys. Rev. B **96**(24), 245301 (2017)
131. R.I. Woodward, E.J.R. Kelleher, R.C.T. Howe, G. Hu, F. Torrisi, T. Hasan, S.V. Popov, J.R. Taylor, Tunable Q-switched fiber laser based on saturable edge-state absorption in few-layer molybdenum disulfide (MoS_2). Opt. Express **22**(25), 31113 (2014)
132. M. Zhang, G. Hu, G. Hu, R.C.T. Howe, L. Chen, Z. Zheng, T. Hasan, Yb- and Er-doped fiber laser Q-switched with an optically uniform, broadband WS_2 saturable absorber. Sci. Rep. **5**, 17482 (2015)
133. R.I. Woodward, R.C.T. Howe, G. Hu, F. Torrisi, M. Zhang, T. Hasan, E.J.R. Kelleher, Few-layer MoS_2 saturable absorbers for short-pulse laser technology: current status and future perspectives. Photonics Res. **3**, A30–A42 (2015)
134. S. Liu, N. Huo, S. Gan, Y. Li, Z. Wei, B. Huang, J. Liu, J. Li, H. Chen, Thickness-dependent Raman spectra, transport properties and infrared photoresponse of few-layer black phosphorus. J. Mater. Chem. C **3**(42), 10974–10980 (2015)
135. L. Kou, C. Chen, S.C. Smith, Phosphorene: fabrication, properties, and applications. J. Phys. Chem. Lett. **6**(14), 2794–2805 (2015)
136. X. Ling, H. Wang, S. Huang, F. Xia, M.S. Dresselhaus, The renaissance of black phosphorus. Proc. Natl. Acad. Sci. **112**(15), 4523–4530 (2015)
137. A. Jain, A.J.H. McGaughey, Strongly anisotropic in-plane thermal transport in single-layer black phosphorene. Sci. Rep. **5**, 8501 (2015)
138. R. Fei, L. Yang, Strain-engineering the anisotropic electrical conductance of few-layer black phosphorus. Nano Lett. **14**(5), 2884–2889 (2014)
139. Q. Wei, X. Peng, Superior mechanical flexibility of phosphorene and few-layer black phosphorus. Appl. Phys. Lett. **104**(25), 251915 (2014)

140. X. Wang, A.M. Jones, K.L. Seyler, V. Tran, Y. Jia, H. Zhao, H. Wang, L. Yang, X. Xu, F. Xia, Highly anisotropic and robust excitons in monolayer black phosphorus. Nat. Nanotechnol. **10**(6), 517–521 (2015)
141. H. Yang, H. Jussila, A. Autere, H.-P. Komsa, G. Ye, X. Chen, T. Hasan, Z. Sun, Optical waveplates based on birefringence of anisotropic two-dimensional layered materials. ACS Photonics **4**, 3023–3030 (2017)
142. J. Li, M.M. Naiini, S. Vaziri, M.C. Lemme, M. Östling, Inkjet printing of MoS_2. Adv. Funct. Mater. **24**(41), 6524–6531 (2014)
143. H. Liu, A.T. Neal, Z. Zhu, Z. Luo, X. Xu, D. Tomanek, P.D. Ye, Phosphorene: an unexplored 2D semiconductor with a high hole mobility. ACS Nano **8**(4), 4033–4041 (2014)
144. J. Kang, J. Wood, S. Wells, J.-H. Lee, X. Liu, K.-S. Chen, M. Hersam, Solvent exfoliation of electronic-grade, two-dimensional black phosphorus. ACS Nano **9**(4), 3596–3604 (2015)
145. M. Engel, M. Steiner, P. Avouris, Black phosphorus photodetector for multispectral, high-resolution imaging. Nano Lett. **14**(11), 6414–6417 (2014)
146. G. Hu, T. Albrow-Owen, X. Jin, A. Ali, G. Hu, C.T. Richard, Z. Yang, X. Zhu, R. Woodward, T.-C. Wu, H. Jussila, P. Tan, Z. Sun, E. Kelleher, Y. Xu, M. Zhang, Black phosphorus ink formulation for inkjet printing of optoelectronics and photonics. Nat. Commun. **8**, 278 (2017)
147. M. Buscema, D.J. Groenendijk, G.A. Steele, H.S.J. van der Zant, A. Castellanos-Gomez, Photovoltaic effect in few-layer black phosphorus PN junctions defined by local electrostatic gating. Nat. Commun. **5**, 4651 (2014)
148. D. Hanlon, C. Backes, E. Doherty, C. Cucinotta, N. Berner, C. Boland, K. Lee, A. Harvey, P. Lynch, Z. Gholamvand, S. Zhang, K. Wang, G. Moynihan, A. Pokle, Q. Ramasse, N. McEvoy, W. Blau, J. Wang, G. Abellan, F. Hauke, A. Hirsch, S. Sanvito, D. O'Regan, G.S. Duesberg, V. Nicolosi, J. Coleman, Liquid exfoliation of solvent-stabilized few-layer black phosphorus for applications beyond electronics. Nat. Commun. **6**, 8563 (2015)
149. Y. Chen, G. Jiang, S. Chen, Z. Guo, X. Yu, C. Zhao, H. Zhang, Q. Bao, S. Wen, D. Tang, D. Fan, Mechanically exfoliated black phosphorus as a new saturable absorber for both Q-switching and Mode-locking laser operation. Opt. Express **23**(10), 12823 (2015)
150. J. Sotor, G. Sobon, W. Macherzynski, P. Paletko, K.M. Abramski, Black phosphorus saturable absorber for ultrashort pulse generation. Appl. Phys. Lett. **107**(5), 051108 (2015)
151. H. Mu, S. Lin, Z. Wang, S. Xiao, P. Li, Y. Chen, H. Zhang, H. Bao, S.P. Lau, C. Pan, D. Fan, Q. Bao, Black phosphorus-polymer composites for pulsed lasers. Adv. Opt. Mater. **10**, 1446 (2015)
152. D. Li, R. Cheng, H. Zhou, C. Wang, A. Yin, Y. Chen, N.O. Weiss, Y. Huang, X. Duan, Electric-field-induced strong enhancement of electroluminescence in multilayer molybdenum disulfide. Nat. Commun. **6**, 7509 (2015)
153. Z. Guo, H. Zhang, S. Lu, Z. Wang, S. Tang, J. Shao, Z. Sun, H. Xie, H. Wang, X.-F. Yu, P.K. Chu, From black phosphorus to phosphorene: basic solvent exfoliation, evolution of Raman scattering, and applications to ultrafast photonics. Adv. Funct. Mater. **25**(45), 6996–7002 (2015)
154. S.-Y. Cho, Y. Lee, H.-J. Koh, H. Jung, J.-S. Kim, H.-W. Yoo, J. Kim, H.-T. Jung, Superior chemical sensing performance of black phosphorus: comparison with MoS_2 and graphene. Adv. Mater. **28**(32), 7020–7028 (2016)
155. J. Sun, G. Zheng, H.-W. Lee, N. Liu, H. Wang, H. Yao, W. Yang, Y. Cui, Formation of stable phosphorus-carbon bond for enhanced performance in black phosphorus nanoparticle-graphite composite battery anodes. Nano Lett. **14**(8), 4573–4580 (2014)
156. A. Castellanos-Gomez, L. Vicarelli, E. Prada, J.O. Island, K.L. Narasimha-Acharya, S.I. Blanter, D.J. Groenendijk, M. Buscema, G.A. Steele, J.V. Alvarez, H.W. Zandbergen, J.J. Palacios, H.S.J. van der Zant, Isolation and characterization of few-layer black phosphorus. 2D Mater. **1**(2), 025001 (2014)
157. A. Favron, E. Gaufres, F. Fossard, A.-L. Phaneuf-L'Heureux, N.Y.-W. Tang, P.L. Levesque, A. Loiseau, R. Leonelli, S. Francoeur, R. Martel, Photooxidation and quantum confinement effects in exfoliated black phosphorus. Nat. Mater. **14**(8), 826–832 (2015)

158. J.O. Island, G.A. Steele, H.S.J. van der Zant, A. Castellanos-Gomez, Environmental instability of few-layer black phosphorus. 2D Mater. **2**(1), 011002 (2015)
159. S.-Z. Huang, J. Jin, Y. Cai, Y. Li, Z. Deng, J.-Y. Zeng, J. Liu, C. Wang, T. Hasan, B.-L. Su, Three-dimensional (3D) bicontinuous hierarchically porous Mn_2O_3 single crystals for high performance lithium-ion batteries. Sci. Rep. **5**, 14686 (2015)
160. X.-F. Jiang, Q. Weng, X.-B. Wang, X. Li, J. Zhang, D. Golberg, Y. Bando, Recent progress on fabrications and applications of boron nitride nanomaterials: a review. J. Mater. Sci. Technol. **31**(6), 589–598 (2015)
161. P. Miró, M. Audiffred, T. Heine, An atlas of two-dimensional materials. Chem. Soc. Rev. **43**(18), 6537 (2014)
162. C.R. Dean, A.F. Young, I. Meric, C. Lee, L. Wang, S. Sorgenfrei, K. Watanabe, T. Taniguchi, P. Kim, K.L. Shepard, J. Hone, Boron nitride substrates for high-quality graphene electronics. Nat. Nanotechnol. **5**(10), 722–726 (2010)
163. Q. Li, L. Chen, M.R. Gadinski, S. Zhang, G. Zhang, H. Li, A. Haque, L.-Q. Chen, T. Jackson, Q. Wang, Flexible high-temperature dielectric materials from polymer nanocomposites. Nature **523**(7562), 576–579 (2015)
164. L. Liu, Y.P. Feng, Z.X. Shen, Structural and electronic properties of *h*-BN. Phys. Rev. B **68**(10), 104102 (2003)
165. I. Jo, M.T. Pettes, J. Kim, K. Watanabe, T. Taniguchi, Z. Yao, L. Shi, Thermal conductivity and phonon transport in suspended few-layer hexagonal boron nitride. Nano Lett. **13**(2), 550–554 (2013)
166. H. Liem, H.S. Choy, Superior thermal conductivity of polymer nanocomposites by using graphene and boron nitride as fillers. Solid State Commun. **163**, 41–45 (2013)
167. K. Watanabe, T. Taniguchi, H. Kanda, Direct-bandgap properties and evidence for ultraviolet lasing of hexagonal boron nitride single crystal. Nat. Mater. **3**(6), 404–409 (2004)
168. A. Pakdel, Y. Bando, D. Golberg, Nano boron nitride flatland. Chem. Soc. Rev. **43**(3), 934–959 (2014)
169. D. Golberg, Y. Bando, Y. Huang, T. Terao, M. Mitome, C. Tang, C. Zhi, Boron nitride nanotubes and nanosheets. ACS Nano **4**(6), 2979–2993 (2010)
170. I. Meric, C. Dean, A. Young, J. Hone, P. Kim, K.L. Shepard, Graphene field-effect transistors based on boron nitride gate dielectrics, in *2010 International Electron Devices Meeting* (IEEE, New York, 2010), pp. 2321–2324
171. I. Meric, C.R. Dean, N. Petrone, L. Wang, J. Hone, P. Kim, K.L. Shepard, Graphene field-effect transistors based on boron-nitride dielectrics. Proc. IEEE **101**(7), 1609–1619 (2013)
172. L.H. Li, E.J.G. Santos, T. Xing, E. Cappelluti, R. Roldán, Y. Chen, K. Watanabe, T. Taniguchi, Dielectric screening in atomically thin boron bitride nanosheets. Nano Lett. **15**(1), 218–223 (2015)
173. F. Withers, H. Yang, L. Britnell, A.P. Rooney, E. Lewis, A. Felten, C.R. Woods, V. Sanchez Romaguera, T. Georgiou, A. Eckmann, Y.J. Kim, S.G. Yeates, S.J. Haigh, A.K. Geim, K.S. Novoselov, C. Casiraghi, Heterostructures produced from nanosheet-based inks. Nano Lett. **14**(7), 3987–3992 (2014)
174. X. Wang, A. Pakdel, J. Zhang, Q. Weng, T. Zhai, C. Zhi, D. Golberg, Y. Bando, Large-surface-area BN nanosheets and their utilization in polymeric composites with improved thermal and dielectric properties. Nanoscale Res. Lett. **7**(1), 662 (2012)
175. M. Naguib, M. Kurtoglu, V. Presser, J. Lu, J. Niu, M. Heon, L. Hultman, Y. Gogotsi, M.W. Barsoum, Two-dimensional nanocrystals produced by exfoliation of Ti_3AlC_2. Adv. Mater. **23**(37), 4248–4253 (2011)
176. Z. Ling, C.E. Ren, M.-Q. Zhao, J. Yang, J.M. Giammarco, J. Qiu, M.W. Barsoum, Y. Gogotsi, Flexible and conductive MXene films and nanocomposites with high capacitance. Proc. Natl. Acad. Sci. **111**(47), 16676–16681 (2014)
177. M. Khazaei, M. Arai, T. Sasaki, C.-Y. Chung, N.S. Venkataramanan, M. Estili, Y. Sakka, Y. Kawazoe, Novel electronic and magnetic properties of two-dimensional transition metal carbides and nitrides. Adv. Funct. Mater. **23**(17), 2185–2192 (2013)

178. Q. Tang, Z. Zhou, P. Shen, Are MXenes promising anode materials for Li ion batteries? Computational studies on electronic properties and Li storage capability of Ti_3C_2 and $Ti_3C_2X_2$ (X = F, OH) monolayer. J. Am. Chem. Soc. **134**(40), 16909–16916(2012)
179. Y. Xie, Y. Dall'Agnese, M. Naguib, Y. Gogotsi, M.W. Barsoum, H.L. Zhuang, P.R.C. Kent, Prediction and characterization of MXene nanosheet anodes for non-lithium-ion batteries. ACS Nano **8**(9), 9606–9615 (2014)
180. D. Sun, M. Wang, Z. Li, G. Fan, L.-Z. Fan, A. Zhou, Two-dimensional Ti_3C_2 as anode material for Li-ion batteries. Electrochem. Commun. **47**, 80–83 (2014)
181. D. Er, J. Li, M. Naguib, Y. Gogotsi, V.B. Shenoy, Ti_3C_2 MXene as a high capacity electrode material for metal (Li, Na, K, Ca) ion batteries. ACS Appl. Mater. Interfaces **6**(14), 11173–11179 (2014)
182. X. Wang, S. Kajiyama, H. Iinuma, E. Hosono, S. Oro, I. Moriguchi, M. Okubo, A. Yamada. Pseudocapacitance of MXene nanosheets for high-power sodium-ion hybrid capacitors. Nat. Commun. **6**, 6544 (2015)
183. Y. Dall'Agnese, P.-L. Taberna, Y. Gogotsi, P. Simon, Two-dimensional vanadium carbide (MXene) as positive electrode for sodium-ion capacitors. J. Phys. Chem. Lett. **6**(12), 2305–2309 (2015)
184. R.B. Rakhi, B. Ahmed, M.N. Hedhili, D.H. Anjum, H.N. Alshareef, Effect of postetch annealing gas composition on the structural and electrochemical properties of Ti_2CT_x MXene electrodes for supercapacitor applications. Chem. Mater. **27**(15), 5314–5323 (2015)
185. B. Xu, M. Zhu, W. Zhang, X. Zhen, Z. Pei, Q. Xue, C. Zhi, P. Shi, Ultrathin MXene-micropattern-based field-effect transistor for probing neural activity. Adv. Mater. **28**(17), 3333–3339 (2016)
186. F. Wang, C. Yang, C. Duan, D. Xiao, Y. Tang, J. Zhu, An organ-like titanium carbide material (MXene) with multilayer structure encapsulating hemoglobin for a mediator-free biosensor. J. Electrochem. Soc. **162**(1), B16–B21 (2014)
187. J. Xu, J. Shim, J.-H. Park, S. Lee, MXene electrode for the integration of WSe_2 and MoS_2 field effect transistors. Adv. Funct. Mater. **26**(29), 5328–5334 (2016)
188. Z. Ma, Z. Hu, X. Zhao, Q. Tang, D. Wu, Z. Zhou, L. Zhang, Tunable band structures of heterostructured bilayers with transition-metal dichalcogenide and MXene monolayer. J. Phys. Chem. C **118**(10), 5593–5599 (2014)
189. Z. Guo, N. Miao, J. Zhou, B. Sa, Z. Sun, Strain-mediated type-I/type-II transition in MXene/blue phosphorene van der Waals heterostructures for flexible optical/electronic devices. J. Mater. Chem. C **5**(4), 978–984 (2017)
190. F. Annabi-Bergaya, Layered clay minerals. Basic research and innovative composite applications. Microporous Mesoporous Mater. **107**(1–2), 141–148 (2008)
191. B. Chen, J.R.G. Evans, H.C. Greenwell, P. Boulet, P.V. Coveney, A.A. Bowden, A. Whiting, A critical appraisal of polymer clay nanocomposites. Chem. Soc. Rev. **37**(3), 568–594 (2008)
192. T.P. Dolley, *2008 Minerals Yearbook: Mica* (National Minerals Information Center, Reston, 2008)
193. W.-G. Kim, S. Nair, Membranes from nanoporous 1D and 2D materials: a review of opportunities, developments, and challenges. Chem. Eng. Sci. **104**, 908–924 (2013)
194. A. Harvey, J.B. Boland, I. Godwin, A.G. Kelly, B.M. Szydłowska, G. Murtaza, A. Thomas, D.J. Lewis, P. O'Brien, J.N. Coleman, Exploring the versatility of liquid phase exfoliation: producing 2D nanosheets from talcum powder, cat litter and beach sand. 2D Mater. **4**(2), 25054 (2017)
195. C. Delmas, C. Fouassier, P. Hagenmuller, Structural classification and properties of the layered oxides. Physica B+C **99**(1–4), 81–85 (1980)
196. R. Ma, T. Sasaki, Nanosheets of oxides and hydroxides: ultimate 2D charge-bearing functional crystallites. Adv. Mater. **22**(45), 5082–5104 (2010)
197. M. Osada, T. Sasaki, Exfoliated oxide nanosheets: new solution to nanoelectronics. J. Mater. Chem. **19**(17), 2503 (2009)

198. A.R. Pray, R.F. Heitmiller, S. Strycker, V.D. Aftandilian, T. Muniyappan, D. Choudhury, M. Tamres, Anhydrous metal chlorides, in *Inorganic Syntheses: Reagents for Transition Metal Complex and Organometallic Syntheses*, ed. by R.J. Angelici (Wiley, New York, 1990), pp. 321–323
199. N.N. Greenwood, A. Earnshaw, *Chemistry of the Elements* (Butterworth-Heinemann, Oxford, 1997)
200. E. Carroll, D. Buckley, N.V.V. Mogili, D. McNulty, M.S. Moreno, C. Glynn, G. Collins, J.D. Holmes, K.M. Razeeb, C. O'Dwyer. 2D nanosheet paint from solvent-exfoliated Bi_2Te_3 ink. Chem. Mater. **29**, 7390–7400 (2017)
201. W. Zheng, T. Xie, Y. Zhou, Y.L. Chen, W. Jiang, S. Zhao, J. Wu, Y. Jing, Y. Wu, G. Chen, Y. Guo, J. Yin, S. Huang, H.Q. Xu, Z. Liu, H. Peng, Patterning two-dimensional chalcogenide crystals of Bi_2Se_3 and In_2Se_3 and efficient photodetectors. Nat. Commun. **6**, 6972 (2015)
202. V.D. Das, N. Soundararajan, Thermoelectric power and electrical resistivity of crystalline antimony telluride (Sb_2Te_3) thin films: temperature and size effects. J. Appl. Phys. **65**, 2332–2341 (1989)
203. S.K. Mishra, S. Satpathy, O. Jepsen, Electronic structure and thermoelectric properties of bismuth telluride and bismuth selenide. J. Phys. Condens. Matter **9**, 461–470 (1997)
204. R.B. Jacobs-Gedrim, M. Shanmugam, N. Jain, C.A. Durcan, M.T. Murphy, T.M. Murray, R.J. Matyi, R.L. Moore, B. Yu, Extraordinary photoresponse in two-dimensional In_2Se_3 nanosheets. ACS Nano **8**, 514–521 (2014)
205. D.A. Bandurin, A.V. Tyurnina, G.L. Yu, A. Mishchenko, V. Zólyomi, S.V. Morozov, R.K. Kumar, R.V. Gorbachev, Z.R. Kudrynskyi, S. Pezzini, Z.D. Kovalyuk, U. Zeitler, K.S. Novoselov, A. Patanè, L. Eaves, I.V. Grigorieva, V.I. Fal'ko, A.K. Geim, Y. Cao, High electron mobility, quantum Hall effect and anomalous optical response in atomically thin InSe. Nat. Nanotechnol. **12**, 223–227 (2017)
206. S. Ghosh, P.D. Patil, M. Wasala, S. Lei, A. Nolander, P. Sivakumar, R. Vajtai, P. Ajayan, S. Talapatra, Fast photoresponse and high detectivity in copper indium selenide ($CuIn_7Se_{11}$) phototransistors. 2D Mater. **5**, 015001 (2017)
207. J.O. Island, S.I. Blanter, M. Buscema, H.S.J. van der Zant, A. Castellanos-Gomez, Gate controlled photocurrent generation mechanisms in high-gain In_2Se_3 phototransistors. Nano Lett. **15**, 7853–7858 (2015)

Chapter 3
2D Material Production Methods

Abstract The widespread use of printing of 2D materials in their relevant applications is highly dependent on the cost and scalability of their methods of production. This chapter serves as an introduction to the key methods for 2D material production and characterisation. Methods such as chemical vapour deposition, plasma cracking of hydrocarbons, intercalation, chemical exfoliation and liquid phase exfoliation are described and their relative merits are discussed. Particular emphasis is given to the 2D materials relevant to ink production. The latter half of the chapter discusses commonly used processing steps and characterisation methods for the 2D materials and their respective roles in qualifying and quantifying the material produced.

A fundamental element of research into 2D materials has been in the production methods, with a particular emphasis on scalability and cost-effectiveness. In general, the two key approaches to 2D material production are via either bottom-up (i.e. synthesis or growth) or top-down (i.e. exfoliation) methods. Both of these processes can be further divided into dry and solution-based methods. We will devote more attention to the latter, since these are ultimately more relevant to ink formulation, although non-solution-based techniques are nevertheless used to produce 2D material flakes and films, especially for lab use. A more comprehensive summary of 2D material production may be found in a review article by Bonaccorso et al. [1]. We note that most of the techniques discussed in this chapter are focussed around graphene, since this has dominated the research landscape.

3.1 Non-Solution-Based Methods for 2D Material Production

These methods involve the exfoliation of bulk crystals or growth of atomically thin 2D materials, the key differentiator from the solution-based methods being that this approach does not require a liquid phase. There are some advantages to following a non-solution processing route. Depending on the production strategies, these include the ability to produce high quality pristine samples, the ability to

L. W. T. Ng et al., *Printing of Graphene and Related 2D Materials*,
https://doi.org/10.1007/978-3-319-91572-2_3

produce large quantities of dry nanoplatelets in large quantity quickly and the ability to incorporate 2D material growth directly into a continuous roll-to-roll process. We will discuss three distinct methods in this section. The first being mechanical cleavage (MC), which was the method initially used by Novosolev et al. in 2004 to exfoliate graphene and other 2D materials [2, 3]. Another prevalently used method is chemical vapour deposition (CVD), which has proven to be effective in the growth of large-area 2D material films [1, 4, 5]. The third method, demonstrated only for graphene, is through plasma cracking of hydrocarbons, a method that is gaining popularity and perhaps has the best potential among these non-solution methods for economically producing material suitable for formulating inks and coatings [6, 7].

3.1.1 Mechanical Cleavage

For fundamental research purposes, one of the most widely used techniques for producing 2D materials is mechanical exfoliation from bulk crystals. In this method, atomically thin flakes are exfoliated from the counterpart bulk crystal by 'rubbing' the face of the crystal against a solid surface such as a silicon wafer, or by cleaving with adhesive tape, as shown in Fig. 3.1a, b. Exfoliation is possible due to the relatively weak interlayer van der Waals (vdW) forces compared with the strong covalent intra-layer bonds. There are several key advantages to this method. As a purely physical process, mechanical cleavage causes no chemical alteration of the exfoliated material, producing exceptionally high-quality flakes. This approach has been used for decades by crystal growers and cryptographers [1] and can be used to produce atomically smooth surfaces for techniques such as crystallography. Indeed, reports of its use to exfoliate few-layer 2D material flakes date back to the 1960s [9–11]. In 1999, Lu et al. indicated that it was possible to use this technique to produce graphene [12]. This was first achieved in 2004 [2], with other 2D materials following in 2005 [3]. This method has since been used to produce pristine samples of 2D materials such as graphene; Fig. 3.1c, MoS_2; Fig. 3.1d, and black phosphorus (BP); Fig. 3.1e. It should be noted that from the fabrication point of view, the key advance in 2004 was as much the actual identification of the monolayer flakes as successful exfoliation [1–3, 13]. Mechanical cleavage produces flakes of many different thicknesses and, prior to that time, no method existed for accurately distinguishing the monolayer flakes [1–3, 13]. Whilst the process has subsequently been optimised to exfoliate monolayer flakes of up to several mm in size (usually limited by the grain size of the bulk crystal), for scale-up purposes it is impractical and the yield remains extremely low, rendering it unsuitable for large-scale device manufacture [1, 13].

3.1.2 Substrate-Bound High Temperature Growth

To produce large area films, the leading technique to emerge is chemical vapour deposition (CVD); Fig. 3.2a [1, 4, 5]. CVD has been the workhorse for depositing

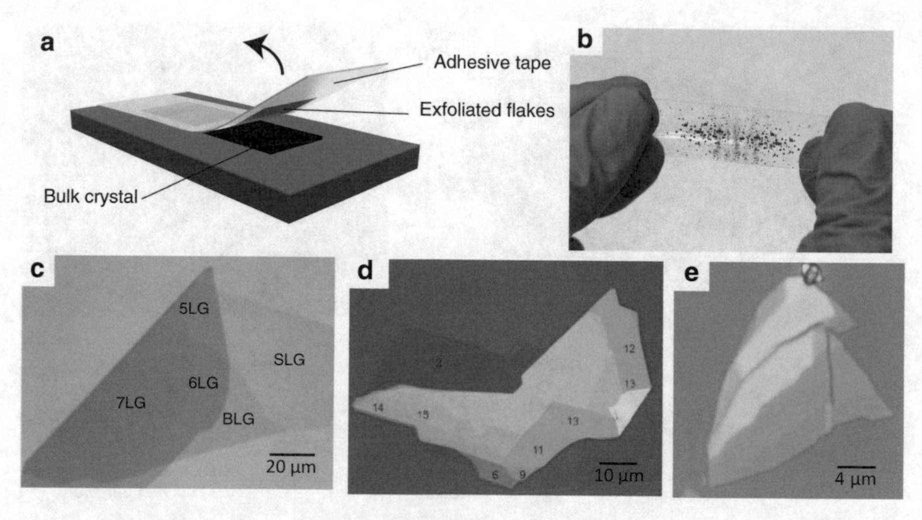

Fig. 3.1 Mechanical cleavage of 2D materials. (**a**) Schematic of the cleaving process. (**b**) Partially cleaved graphite flakes using a scotch tape. (**c**) Optical micrograph of a mechanically cleaved flake of mono- and few-layer graphene, consisting of regions of different thicknesses. Adapted with permission from Ref. [8]. Copyright (2012) Materials Today. (**d**) Example of cleaved flake of MoS_2 (numbers indicate layer count) Adapted with permission from Ref. [1]. Copyright (2013) American Chemical Society. (**e**) Mechanically cleaved BP

many materials used in semiconductor devices for several decades [1]. Figure 3.2b depicts a typical quartz tube furnace used for this purpose. This technique involves flowing of one or more vaporised or gaseous precursors over a substrate at high temperature (typically 600–1,000 °C) [1, 4, 5]. The substrate catalyses reactions between the precursors, leading to the growth of a mono- or few-layer film of 2D materials [1, 4, 5]; Fig. 3.2c. Selection of the actual growth substrate for synthesising graphene has a great influence on the final properties of the graphene itself [15]. Pristine graphene is often fabricated using CVD on both conducting and dielectric substrates. CVD graphene has been successfully grown on many metal substrates including copper, nickel, gold, platinum and iron [16]. Typically, CVD graphene on metal produces graphene of high quality with few defects [16]. A field-effect electron mobility of ~27,000–45,000 $cm^2 V^{-1} s^{-1}$ was reported by Petrone et al. [17]. However, any CVD growth on metal does require significant additional processing and post growth transfer to target substrates for device fabrication. An example of a typical transfer process is depicted in Fig. 3.2c. We note that this is equally applicable to other other CVD grown 2D materials, and that such transfer processes are prone to damaging or introducing unwanted defects.

Hence, research efforts have been directed at growth onto device-specific dielectric substrates to eliminate the transfer process. This has already been demonstrated on substrates such as glass, quartz, hexagonal boron nitride (*h*-BN), and silicon dioxide [15]. The performance of the grown graphene has been as effective as those on metal substrates with the highest reported field-effect mobility value of

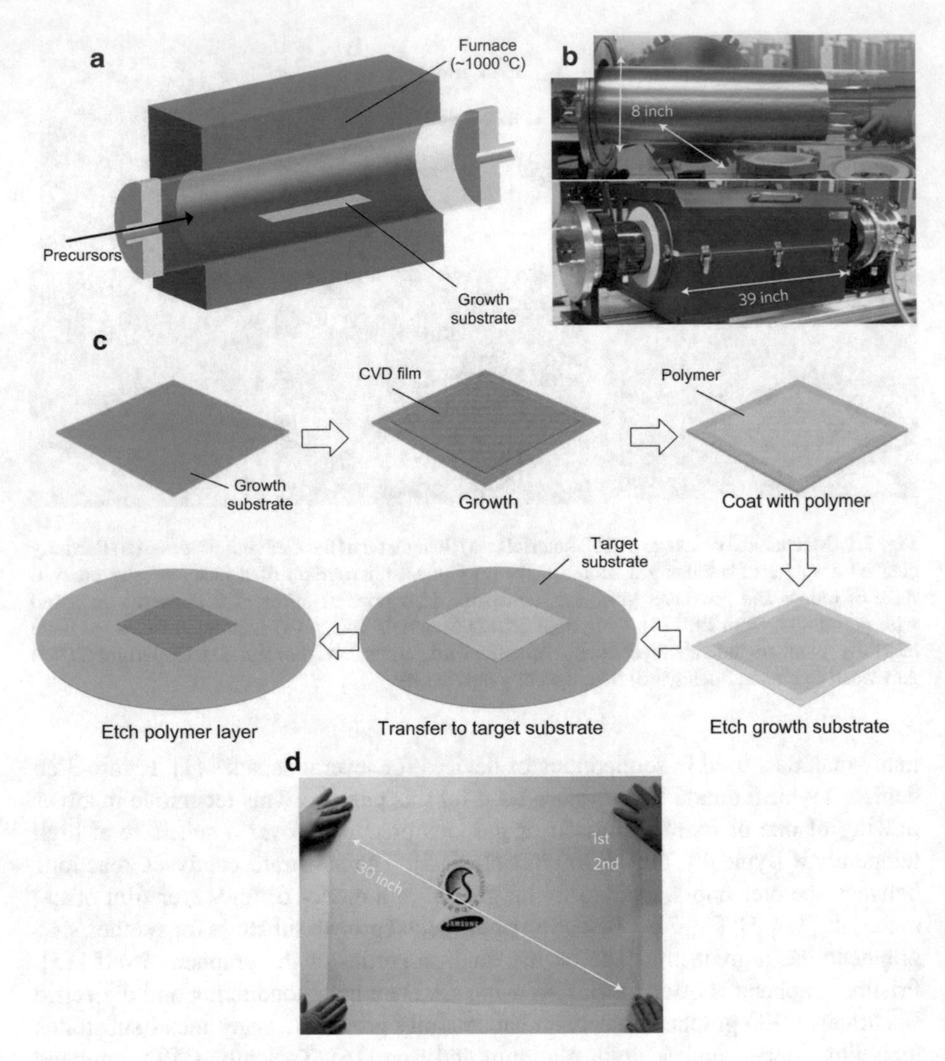

Fig. 3.2 Growth of 2D materials via CVD. (**a**) Schematic and (**b**) experimental set-up for CVD growth. (**c**) Typical transfer process for CVD films. (**d**) Large area CVD growth and transfer of a 30 inch graphene film. (**b**) and (**d**) adapted with permission from Ref. [14]. Copyright (2010) Nature Publishing Group

20,000 $cm^2 V^{-1} s^{-1}$, grown on *h*-BN via a silane-catalysed CVD process [16]. Two fundamental issues remain barriers for effective growth on dielectric substrates. Firstly, the effective control of the nucleation density of graphene, which leads to uneven film growth [15]. Secondly, it has proven difficult to control the growth rate of graphene [15]. Further information on substrates used can be found in the 2016 review paper by Chen et al. [15].

Precursor selection is usually steered by a balance of material availability, desired film properties, and what is cost effective for the specific application [1]. The precursors used for CVD growth of graphene are predominantly hydrocarbon gases such as methane and benzene [1, 4, 5]. Some research, however, has been directed at exploring alternative precursors. Ruan et al. demonstrated the use of exotic sources of carbon, including food, waste and insects via pyrolysis [18]. This was achieved by passing heated hydrogen/argon gas over a carbon-based compound on a copper substrate, such as a cockroach leg [18].

Other types of 2D materials such as transition metal dichalcogenides (TMDs) and *h*-BN require combinations of precursors to produce the compounds, and are primarily grown onto substrates such as silicon [19–25] or sapphire [20, 26]. For instance, Lee et al. demonstrated large-area MoS_2 growth on SiO_2 using molybdenum trioxide (MoO_3) and sulphur (S) as reactants [19]. Also, *h*-BN has also been successfully grown via CVD using Cu or Fe as a substrate and ammonia borane (NH_3–BH_3) or borazine ($B_3H_6N_3$) as a primary precursor [23, 25, 27–31]. Additionally, multi-material CVD such as the growth of graphene following the growth of single-layer *h*-BN has also been demonstrated by, for instance, Han et al. and Wang et al. [27, 28]. This provides advantages in electrical carrier transport abilities as compared to graphene transferred onto *h*-BN film or SiO_2 substrates [28].

Industrial use of CVD has only been realised for graphene, where it has been used to produce transparent electrical conductors such as the one shown in Fig. 3.2d. Kobayashi et al.[32] of the Sony Corporation in 2013 reported successful integration of CVD in a continuous roll-to-roll growth and transfer process, producing large-area graphene films of up to $100 \times 0.23\,\mathrm{m}^2$. In this process, Kobayashi et al. heated large rolls of copper wire using a Joule heating process as it was exposed to hydrocarbon gases for graphene growth; Fig. 3.3a. This copper coil was then transferred and bonded to a PET substrate in a reverse gravure coating process shown in Fig. 3.3b which then enabled the resultant copper to be etched away as in Fig. 3.3c. This produced CVD graphene bonded onto PET film; Fig. 3.3d, e.

The key role that the substrates play in any growth process, as well as the requirement for the high growth temperatures, means that the choice of substrates for CVD is relatively limited. In turn this leads to a need for complex and multi-step processes in order to transfer the as-grown film to the target substrate [1, 16, 33]. In spite of the advancements made in the above examples, the complexity in the transfer process increases the possibility for damage to the grown film, arising either from physical damage during the transfer (e.g. cracking) or from chemical residues (e.g. polymers, adhesives) [16, 33]. There has been recent progress in reducing these effects [16], including the development of dry transfer techniques to avoid the damage from etchants [1, 34, 35], and improved understanding of polymer removal [16]. However, rapid and residue free transfer remains elusive.

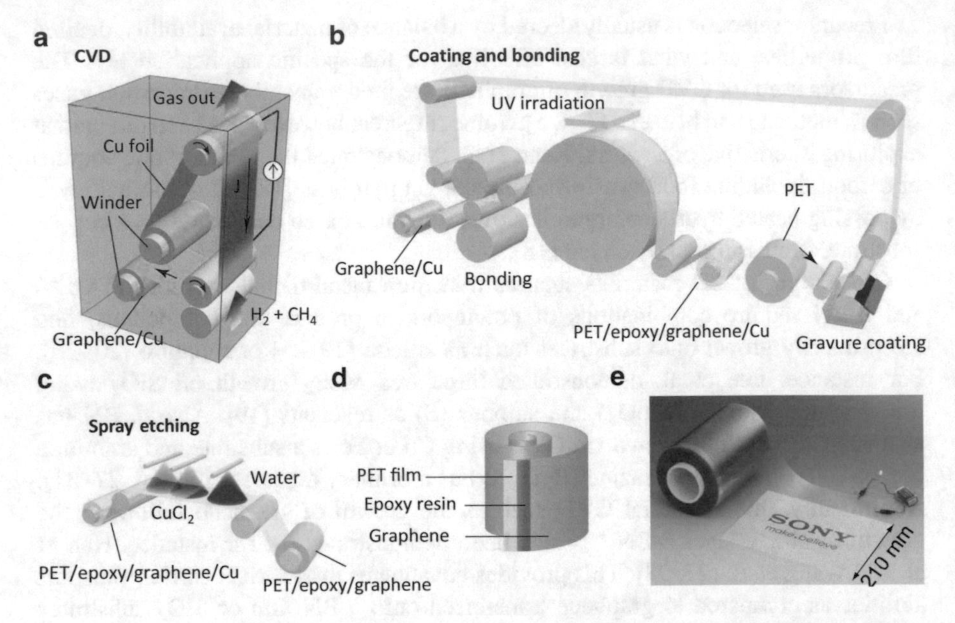

Fig. 3.3 (**a**) Continuous roll-to-roll CVD system using selective Joule heating to heat a copper foil suspended between two current-feeding electrode rollers for graphene growth. (**b**) Reverse gravure coating of a photocurable epoxy resin onto PET film and bonding to graphene/copper foil, followed by the curing of the epoxy resin. (**c**) Spray etching of the copper foil. (**d**) Structure of the fabricated graphene/epoxy/PET film. (**e**) Photograph of the graphene/epoxy/PET roll. Adapted with permission from Ref. [32]. Copyright (2013) AIP Publishing LLC

3.1.3 Plasma Cracking of Hydrocarbons

A related technique to CVD is plasma cracking of hydrocarbons such as methane and ethanol. This technique can be used to synthesise a variety of nanostructured carbon materials including graphene flakes [36–40]; Fig. 3.4a. Instead of using substrates for graphene growth, this technique uses thermally or microwave generated plasma to decompose the hydrocarbon precursor directly into hydrogen and carbon which then forms mono- and few-layer graphene; Fig. 3.4b–d [36–40]. The plasma utilised in this method also effectively removes amorphous impurities and provides an enhanced heat transfer effect that produces better bonding.

The introduction of additional gases into the plasma reactor during the synthesis allows the flakes to be functionalised (to improve their conductivity, for example) [39]. As plasma cracking is a flow process, there is scope to produce large quantities of graphene flakes. However, careful optimisation and control of the plasma and flow conditions is required to ensure that graphene is synthesised as opposed to other carbonaceous materials. Graphene grown via plasma cracking represents a promising starting point for high-concentration ink formulation, as the graphene flakes can be directly dispersed into solvents or binders adopting similar protocols to those discussed in the following section.

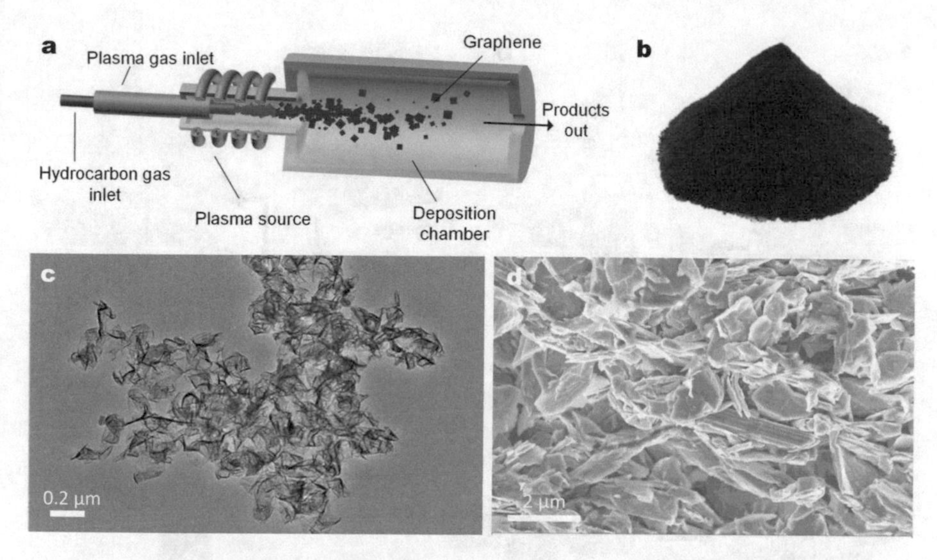

Fig. 3.4 (**a**) Schematic of plasma cracking of hydrocarbons for substrate-free synthesis of graphene flakes. (**b**) Photograph of plasma-cracked graphene nanoplatelets. (**c**) TEM image of graphene nanoplatelets produced by plasma cracking. Adapted with permission from Ref. [36]. Copyright (2015) Elsevier B.V. (**d**) SEM image of few-layer graphene obtained through thermal plasma cracking of methane

3.2 Solution-Based Methods for 2D Material Production

Solution processing offers some particular advantages for applications such as ink and paint formulation and composite manufacture. These include the comparatively low cost of both equipment and raw materials and the relative simplicity in scaling-up for large formulation/manufacture volumes. In addition, the process does not require demanding production conditions such as high temperature, or the complications of materials transfer. The resultant materials in liquid form are also easy to handle. The approaches we focus on here are all methods of exfoliation from the bulk. It should be noted that whilst attempts have been made to synthesise 2D materials in solution, the protocols are lengthy and complex, requiring elevated temperatures and/or pressures, and the produced flakes are typically too small and defective to be of practical usage to date. Solution-based exfoliation is based on the principle of either chemically altering the 2D material or tuning the solvent environment in order to overcome the weak interlayer forces. As a top-down exfoliation strategy, it is capable of isolating large quantities of mono- and few-layer flakes from the bulk in a liquid medium mainly via intercalation, ion exchange or liquid phase exfoliation (LPE) [1, 41], as schematically illustrated in Fig. 3.5. Among these approaches, intercalation and LPE are widely applicable to the 2D materials, while ion exchange is generally limited to certain 2D materials (e.g. layered oxides) that contain an interlayer of cations which can be exchanged by, for

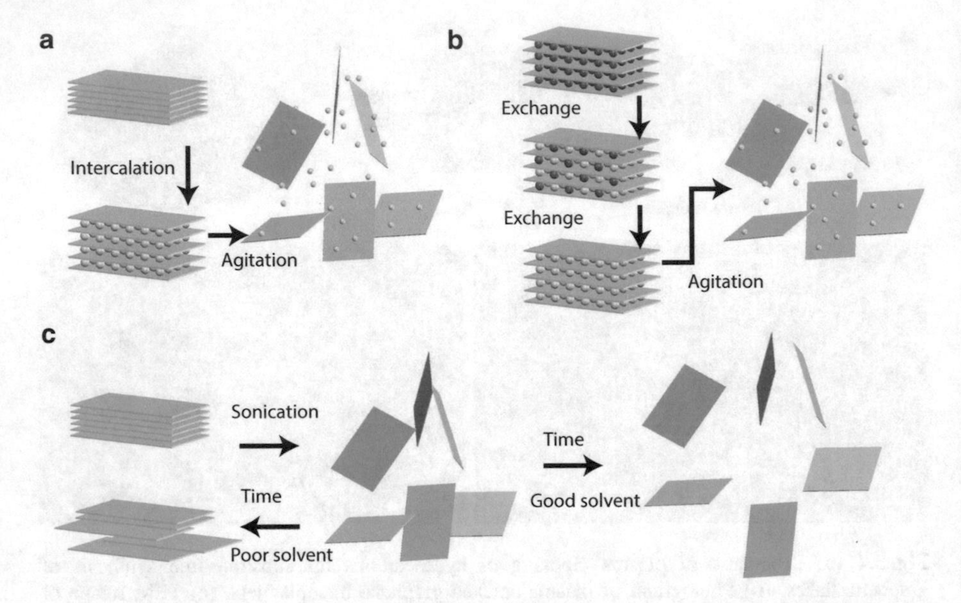

Fig. 3.5 Solution processing via **(a)** intercalation, **(b)** ion exchange, and **(c)** LPE, showing that in 'good solvents' the exfoliated flakes are stabilised against reaggregation, while in 'bad solvents' they reaggregate and sediment

example, organic ions [1, 41]. In the following sections, we will discuss intercalation and LPE in detail.

3.2.1 Exfoliation through Intercalation

Ion Intercalation

Ion intercalation dates back to 1841 when Schaffautl et al. reported a graphite intercalation compound [42]. Ion intercalation as a method of producing monolayer MoS_2 was reported as early as in 1986 [43]. It was later successfully exploited in obtaining monolayer graphene [44, 45] as well as inorganic 2D materials [46]. Figure 3.6a, b are representative photographs of graphene and MoS_2 dispersions produced via ion intercalation [44, 47].

In an ion intercalation process, small molecules like ionic species are used to intercalate between the crystal layers. For instance, the intercalants containing alkalis such as lithium or potassium are particularly well-suited to exfoliation in water [42, 49]. Figure 3.6c, d are schematic figures presenting the use of alkali metals (e.g. Li^+, Na^+ and K^+) to intercalate graphite and MoS_2, respectively [45, 47]. The increased interlayer distance created by intercalation weakens interlayer vdW forces, leading to the separation of the layers. Subsequent sonication or stirring further increases the interlayer distance and the separated mono- and few-

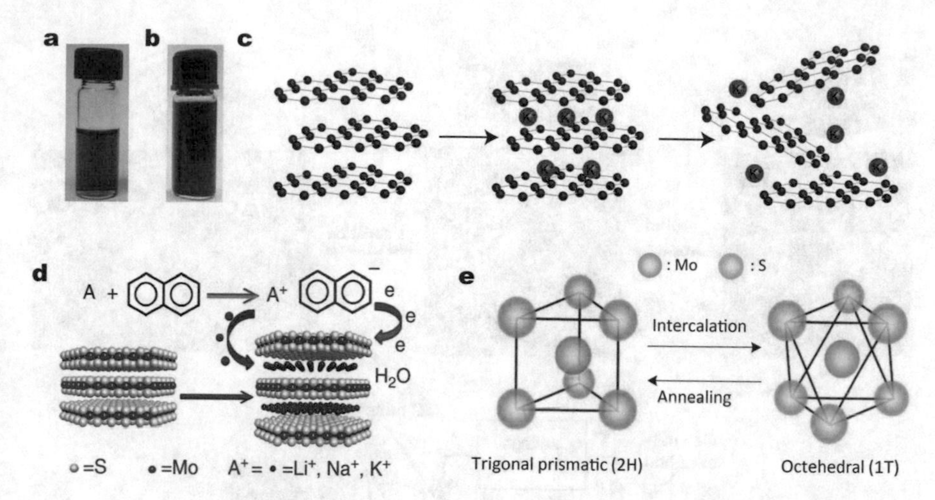

Fig. 3.6 Dispersions of (**a**) graphene and (**b**) MoS_2 produced via ion intercalation. (**a**) Reprinted with permission from Ref. [44]. Copyright (2008) Nature Publishing Group. (**b**) Reprinted with permission from Ref. [47]. Copyright (2014) American Chemical Society. (**c**) Production of graphene flakes using potassium ions as the intercalation agent. Reprinted with permission from Ref. [45]. Copyright (2008) American Chemical Society. (**d**) Production of MoS_2 flakes using alkali ions as the intercalation agents. Reprinted with permission from Ref. [47]. Copyright (2014) American Chemical Society. (**e**) Schematic depicting the phase change of MoS_2 induced by ion intercalation. Reprinted with permission from Ref. [48]. Copyright (2011) American Chemical Society

layer fakes are exfoliated from each other and dispersed in the liquid medium. Ion intercalation allows production of relatively large flakes with a high yield of monolayer flakes [44–48]. However, this approach may introduce defects, impurities and structural changes to the 2D material lattices which can lead to alteration of resultant 2D material properties. Eda et al. demonstrated that Li^+ intercalation of MoS_2 caused a lattice change from trigonal prismatic phase (semiconducting) to metastable octahedral phase (metallic). To make use of the optoelectronic properties of MoS_2, the metallic phase would require a high temperature annealing to transform back to the semiconducting phase (Fig. 3.6e) [48]. Furthermore, ion intercalation requires stringent production conditions, which increases the production cost [1, 41, 47, 48].

Electrochemical Intercalation

A variant of the intercalation process is electrochemical exfoliation; Fig. 3.7. In this process, a bulk 2D material (such as a rod of graphite) is used as an electrode in an electrochemical cell [49, 51–57]; Fig3.7a. This allows rapid intercalation, as well as increased control over the process when compared with simpler solution-based intercalation [51, 53]. The bulk 2D material is immersed (along with a counter electrode: typically platinum) in an electrolyte containing the intercalation compound. When a positive voltage is applied to the bulk material electrode, the potential difference causes intercalants to move from the electrolyte and between the

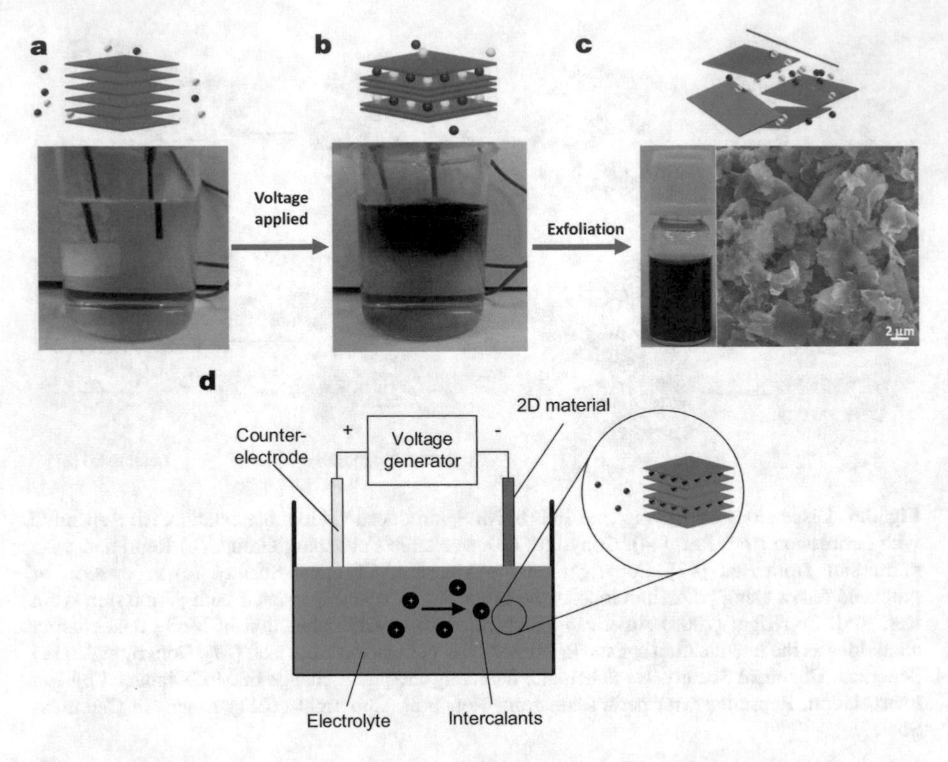

Fig. 3.7 Electrochemical intercalation and exfoliation starting with (**a**) immersion of 2D material electrode and counter electrode into intercalant-containing electrolyte solution, (**b**) subsequent application of voltage, and (**c**) exfoliation of intercalated material in solvent by stirring. Reprinted with permission from Ref. [50]. Copyright (2013) The Royal Society of Chemistry. (**d**) Schematic of the electrochemical intercalation set-up

2D layers; Fig3.7b. A simple stirring or sonication process is then applied as a post processing technique that dislodges the 2D material flakes; Fig.3.7c. A complete schematic is shown in Fig. 3.7d.

Chemical Intercalation

Besides the above intercalation approaches, a widely adopted process for exfoliating of 2D materials, generally graphene, is through chemical intercalation [1, 51, 58–64]. The processes used for chemical intercalation are based on work carried out on the production of graphite oxide in 1958 reported by Hummers et al., a process now known as the Hummers method [65]. This has now evolved to become perhaps the most widely used method for the production of graphene oxide (GO) and reduced graphene oxide (rGO). Chemical intercalation has also been used as a method for producing other 2D materials such as monolayer MoS_2 [43]. Chemical intercalation takes advantage of the layered structure of 2D materials whereby various processes can be used to introduce intercalants (i.e. molecules or compounds) between the atomic layers to increase the interlayer distance (i.e.

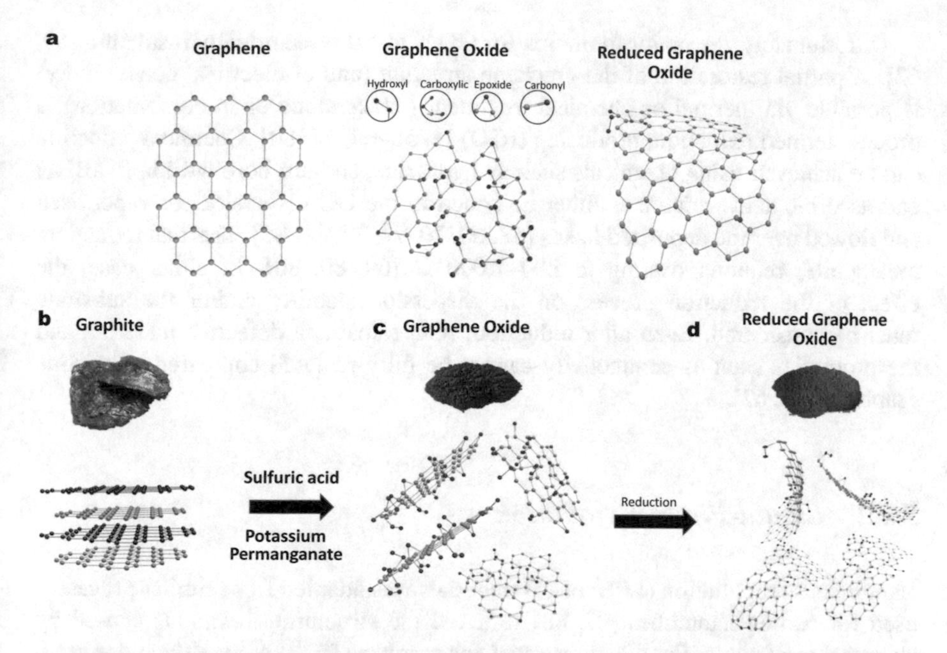

Fig. 3.8 (**a**) Structures of graphene, GO and rGO. Chemical intercalation starts with (**b**) graphite in a stacked form that can be (**c**) oxidised with an acid and oxidising agent. This is then exfoliated via a chemical oxidation process followed by (**d**) reduction either by thermal, photocatalytic or chemical means to partially restore the graphene structure

reducing the interlayer vdW forces) therefore allowing individual layers to be more readily separated from the bulk [1, 42, 49].

Figure 3.8 provides a complete schema for the Hummers method [65]. Figure 3.8a shows the difference in chemical structures between the two products formed by this process, GO and rGO as compared to pristine graphene. Graphite can be oxidised by exposure to a concentrated acid in the presence of an oxidising agent (e.g. via the Hummers method, which uses sulphuric acid and potassium permanganate); Fig. 3.8b, c [58, 59, 64, 65]. The oxidation process introduces additional functional groups (primarily composed of oxygen and hydrogen atoms) onto the 'face', of the graphene sheets that disrupt the interlayer bonding and increases the interlayer spacing; Fig. 3.8c [59, 66]. This allows graphite oxide to be exfoliated to form a dispersion of GO through processes such as stirring or mild ultrasonication; Fig. 3.8d [51, 58, 59, 61, 62]. The functional groups mean that, unlike pristine graphene, GO is hydrophilic, allowing it to readily interact with water [51, 58, 59, 61, 64]. GO exfoliation into water can produce relatively high concentration (1–4 $g\,L^{-1}$) dispersions, with close to 100% monolayer yield [60, 62, 64, 67]. This makes it an attractive process for ink and composite production [68–76].

Unfortunately, the major limitation for GO is that it is electrically insulating [59, 62]. A partial restoration of the graphene structure (and of electrical conductivity) is possible via thermal or chemical treatment (either alone or in combination), a process termed reduction, producing (rGO) [1, 58–62, 77–80]. Chemical reduction can be achieved using chemicals such as hydrazine, sodium borohydride ($NaBH_4$) and ascorbic acid which can either be added to the GO dispersion, or vapourised and flowed over the deposited flakes [58, 60, 70, 72, 77, 81–85]. Thermal reduction, meanwhile, requires heating to 200–1,000 °C [69, 80, 86]. In either case, the effect of the reduction process on the dispersion stability and/or the substrate must be considered. Even after reduction, rGO remains a defective material, and the properties such as conductivity cannot be fully restored compared to pristine graphene [51, 62].

3.2.2 *Liquid-Phase Exfoliation*

Liquid phase exfoliation (LPE) of 2D materials was adapted from similar processes used for carbon nanotubes [87, 88] to avoid the structural alterations caused by chemical treatments. First demonstrated for graphene in 2008 by Hernandez et al. [89], it has since been extended to a wide range of other 2D materials [1, 90–99]. The LPE process begins with bulk crystals and achieves exfoliation and dispersion of atomically thin flakes through shear forces in a liquid medium, for example, a solvent or solvent mixture, sometimes with the addition of dispersants such as polymers and surfactants.

In general, the exfoliation and dispersion process in LPE is dependent on the intermolecular interactions between the solvent and the 2D material, and requires the ‘matching’ of the solvent and the 2D material [90, 92, 100]. The characteristics that define whether a solvent is suitable for LPE, and the role of dispersants are discussed later in the chapter. It should be noted here that the same principles can be applied when forming dispersions of few-layer 2D materials (as opposed to starting from bulk material) such as those produced by plasma cracking. This therefore allows dispersions to be made from these industrially produced materials, enabling high throughput and high concentration formulations.

To exfoliate the 2D material a shear force sufficient to overcome the interlayer vdW forces must be applied to the mixture. A number of processes such as ultrasound assisted liquid phase exfoliation (UALPE); Fig. 3.9, high-shear mixing [101–103]; Fig. 3.10, high pressure mixing [104–107]; Fig. 3.11 and ball-milling [108–111]; Fig 3.12 have emerged for this purpose.

Ultrasound Assisted Liquid Phase Exfoliation (UALPE)

UALPE, first demonstrated for graphene in 2008, is widely used to overcome the inter-layer vdW forces of a wide range of layered materials [1, 90–92, 97];

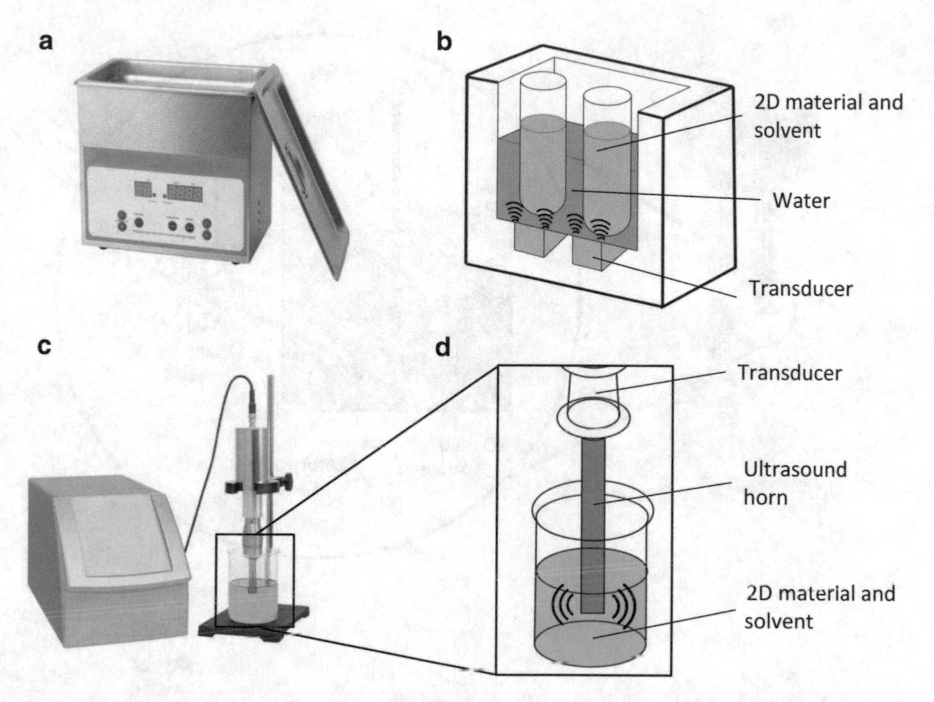

Fig. 3.9 LPE by ultrasound (UALPE) using (**a**, **b**) bath sonication and (**c**, **d**) tip sonication

Fig. 3.9. Ultrasound, produced either from a vibrating tip immersed in the solvent (tip or horn sonication), or using an ultrasonic bath (bath sonication), generates rapid and localised pressure variations in the exfoliation liquid [112]. These lead to the formation of unstable microbubbles in the liquid [112]. When these bubbles collapse, they produce high shear forces that are capable of breaking the weak vdW interactions between the layered materials [113]. Due to the simplicity and low-cost of equipment, UALPE is a popular laboratory-based method for the exfoliation of layered materials including graphene [90], *h*-BN [91], BP [97], TMDs [97] and even materials as diverse as talcum powder, sand and cat litter [99]. While effective, UALPE is usually time consuming (∼2–12 h) and requires further post processing. This will be discussed in Sect. 3.2.3.

High-Shear Mixing

High-shear mixing mechanically generates shear forces through liquid turbulence by the rotation of a mixing blade in a rotor-stator configuration [102, 103]; Fig. 3.10a. This is further assisted by homogenisation by the flow of the liquid through a meshed screen; Fig. 3.10b (e.g. impeller mixing) [101]. This type of high-shear mixing is a mature, scalable technology that has extensive use across a wide range of industries from pharmaceuticals to paint manufacture. Paton et al.

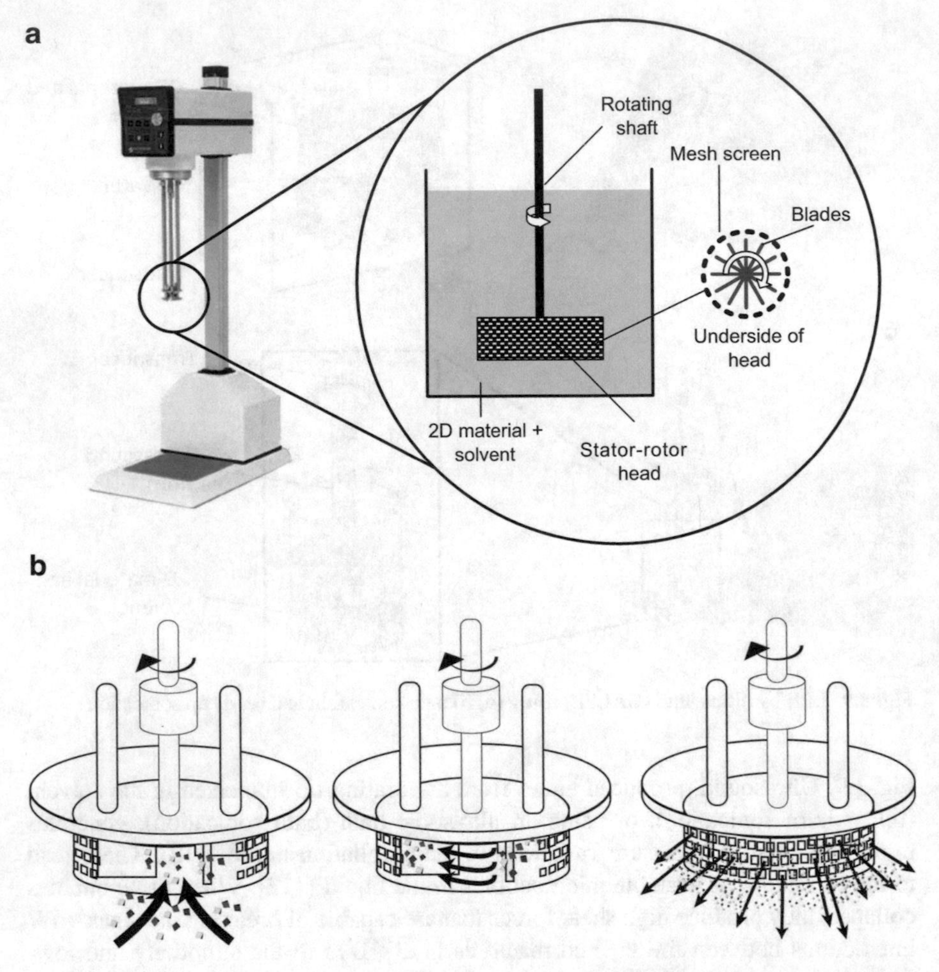

Fig. 3.10 High-shear mixing using a rotor-stator mill. (**a**) Schematic of a rotor-stator mill. (**b**) The flow of liquid and the breaking up of particles from forcing large particle through meshed holes in the rotor-stator mixer

demonstrated the effectiveness of this method for the production for few-layer graphene sheets without any defects [101].

High-shear exfoliation occurs whenever the liquid shear rate exceeds $10^4\ s^{-1}$ [101]. Such shear rates can be achieved in a range of mixers including household kitchen blenders. Varrla et al. simulated the turbulent effects of high-shear mixing by exfoliating MoS_2 nanosheets [102] and graphene [103] using a kitchen blender and detergent as a surfactant. The graphene produced in this way is virtually indistinguishable from that produced by sonication of graphite in solvents or surfactants. Importantly, high-shear mixing can also be used to exfoliate a range of other layered materials.

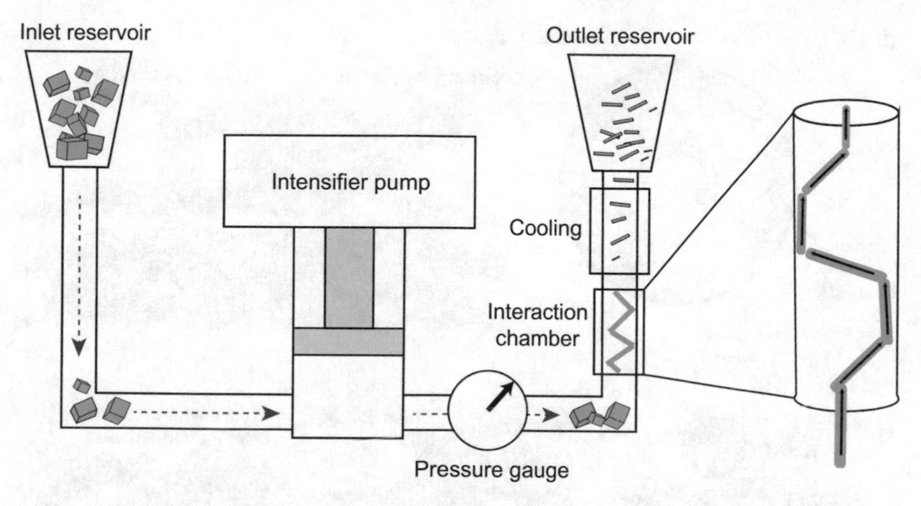

Fig. 3.11 Schematic of the microfluidisation process

Microfluidisation

Microfluidisation is a homogenisation method that applies high pressure (from tens to several hundreds of MPa) [114] to a fluid, forcing it to pass through a microchannel (typical diameter $d < 100\,\mu m$) [104–107]; Fig 3.11. The main advantage of this method over sonication and shear-mixing is that high pressure can be applied to the whole fluid volume, and not just locally [107]. Microfluidisation is a well-established technique which has been used for the production of polymer nanosuspensions [115], in pharmaceuticals, cosmetics and food industries [113, 116, 117].

Microfluidisation has been demonstrated to be a promising method of exfoliation for graphene by a number of research groups including Shang et al. and more recently Karagiannidis et al. [104, 107]. In addition, microfluidisation of exfoliated MoS_2 was reported recently by Xue et al. [105]. This technique can also be used for the exfoliation of other 2D materials in compatible solvents.

Ball Milling

Ball milling makes use of a cylindrical 'jar' containing a large number of small balls/beads (typically steel or zirconia) which act as the grinding media. Under the correct rotation conditions (i.e. fast enough to generate movement, but not so fast as to cause tumbling) the balls generate shear forces perpendicular to the walls of the jar [108–111]; Fig 3.12a. Exfoliation takes place by shear forces caused by the individual grinding media generating two forces on layered materials, shear force and compression force, which can cleave layered materials into 2D nanosheets from the top/bottom surfaces, and the edge of layered materials [110].

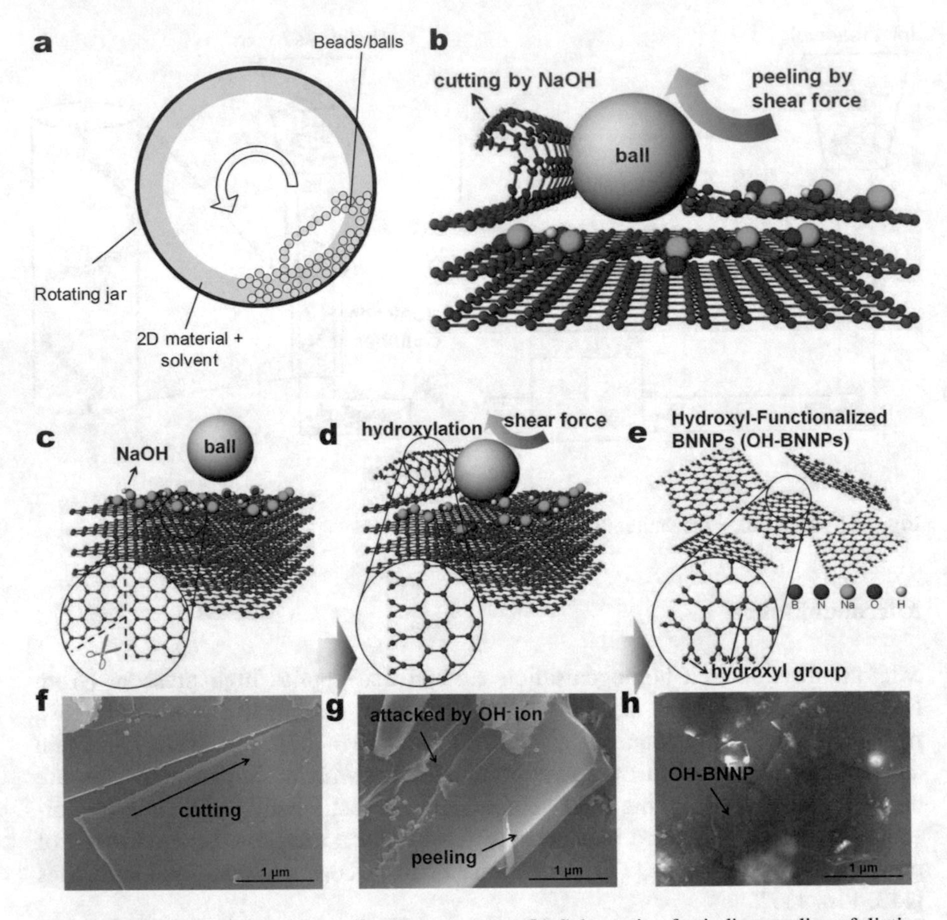

Fig. 3.12 (**a**) Schematic of the ball milling process. (**b**) Schematic of grinding media exfoliation of *h*-BN. (**c–h**) Schematic diagrams and corresponding SEM images of the exfoliation mechanism of *h*-BN: (**c**, **f**) Cutting of large *h*-BN sheets; (**d**, **g**) Thin, curled sheets peeling off the top surface of an *h*-BN particle in response to shear forces created by the milling balls; (**e**, **h**) Exfoliated hydroxyl functionalised *h*-BN nanoparticles (OH-BNNPs). Adapted with permission from Ref. [118]. Copyright (2015) American Chemical Society

Ball milling is widely used to break down the agglomerates and aggregates into primary particles in industries such as ink and powder production. However, ball milling exfoliation is a violent process that can introduce a large number of defects in the 2D materials [118]. The inclusion of the potentially erodible grinding media can also result in contamination of the final 2D material dispersions [111]. Hence any use of ball milling must be carefully controlled to reduce damage to in-plane crystal structure and milling is done in a liquid phase to reduce impact forces [110, 118]. Figure 3.12b–h depicts the cutting and shearing action of *h*-BN in a sodium hydroxide solution into individual 2D sheets. Ball milling has been used to produce graphene [109], *h*-BN [110, 111, 118] and MoS_2 [110], and can be extended to other 2D materials.

Exfoliation in Pure Solvents

The LPE process differs to intercalation and chemical exfoliation as it requires no pre-treatment of the bulk material. Instead, successful exfoliation is reliant on optimising the intermolecular interactions between the solvent and the exfoliated flakes [90, 92, 100]. Exfoliating material exposes additional surface area to the solvent, which has an associated energy 'cost' dependent on the interactions between the solvent and the 2D material. This energy cost (termed the Gibbs free energy of mixing: ΔG_{mix}) can be estimated from:

$$\Delta G_{mix} = \Delta H_{mix} - T \Delta S_{mix} \tag{3.1}$$

where T is the absolute temperature, and ΔS_{mix} and ΔH_{mix} are, respectively, the change in entropy and enthalpy associated with the exfoliation process [90, 100, 119]. If $\Delta G_{mix} < 0$, exfoliation will occur with no energy input, while $\Delta G_{mix} > 0$ indicates additional energy input will be required. This indicates that ΔH_{mix} must be minimised to ensure stable dispersion.

ΔH_{mix} can be estimated through various approaches, the simplest of which is in terms of surface energy of the solvent and 2D material [7, 89, 90, 119]:

$$\Delta H \propto (\sqrt{\gamma_S} - \sqrt{\gamma_{2D}})^2 \frac{\varphi_{\text{Max, 2D}}}{T_{2D}} \tag{3.2}$$

where γ_S and γ_{2D} are the solvent and 2D material surface energies, $\varphi_{\text{Max, 2D}}$ is the maximum dispersed volume fraction, and T_{2D} is the flake thickness, respectively. Equation (3.2) indicates that the most stable dispersion will occur when $\gamma_S \approx \gamma_{2D}$. Early studies correlated the concentration of dispersed 2D materials for ~40 solvents under the same exfoliation conditions, and found that when the concentration is plotted against surface tension, there is a broad peak at ~35–45 mN m^{-1}. However, the surface energy provides only a very rough indication, with many solvents deviating from their expected concentrations [7, 120]. This suggests that the model is incomplete, and that a more refined set of parameters is required to understand the solvent–2D material interactions.

When characterising the interactions between different solvents, and between solvents and molecules, a commonly used approach is to work in terms of Hansen solubility parameters (HSPs), which provide a more detailed breakdown of the intermolecular interactions [121]. As with surface energy, ΔH_{mix} will be minimised when the solvent and 2D material have matching HSPs.

HSPs, subdivide the intermolecular interactions into three components to better reflect their different contributions. The three HSP components, δ_D, δ_P and δ_H, relate to dispersive (D) interactions (i.e. vdW interactions between atoms), dipole–dipole (P) interactions between molecules, and hydrogen bonding (H) interactions [121]. As with the surface energy, HSPs for 2D materials can be empirically derived by measuring the concentration of dispersed material in a wide range of solvents [7, 120]. The dispersed concentration is maximised when all thee

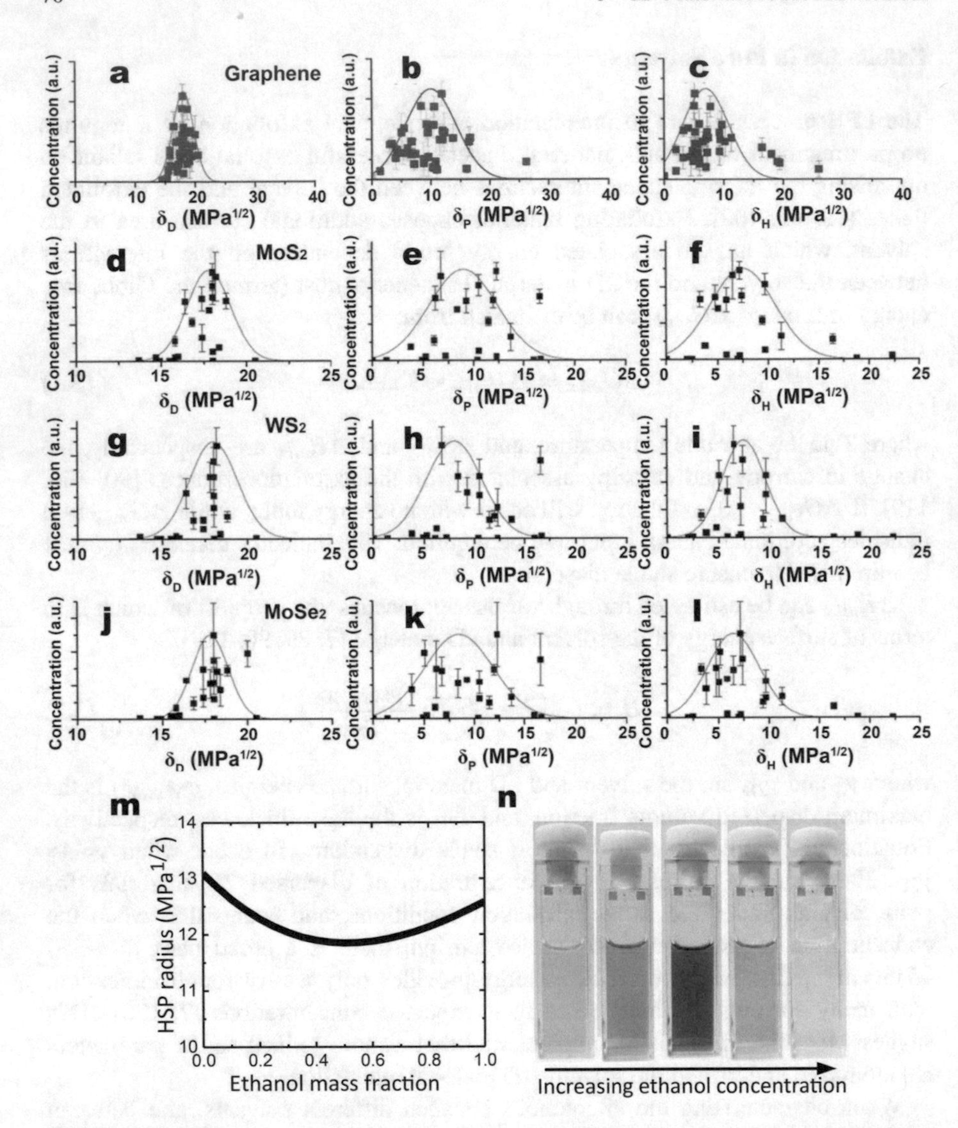

Fig. 3.13 Experimental estimation of the Hansen solubility parameters for (**a–c**) graphene. Adapted with permission from Ref. [92]. Copyright (2012) American Chemical Society. (**d–f**) MoS_2, (**g–i**) WS_2, and (**j–l**) $MoSe_2$. Adapted with permission from Ref. [7]. Copyright (2012) American Chemical Society. Example of using mixture of solvents which are individually incompatible for 2D material exfoliation. (**m**) As the ethanol content is varied, the HSP radius (R_H) between the solvent and MoS_2 reaches a minimum value. This is reflected in (**n**) the concentration of exfoliated material following UALPE. Adapted from Ref. [122]

components for both the solvent and the 2D material are well matched. For example, the empirically derived HSP δ_D, δ_P and δ_H values respectively (in $MPa^{1/2}$) for commonly exfoliated 2D materials are: graphene: 18, 10, 7; MoS_2: 17.9, 9, 7.5; WS_2: 18, 8, 7.5; $MoSe_2$: 17.8, 8.5, 6.5; Fig. 3.13a–l; Ref. [92].

The level of matching between a solvent and the 2D material can then be assessed by calculating the HSP radius, R_H, (i.e. the distance between the two in the 3D HSP space):

$$R_H \propto \sqrt{4(\delta_{D,S} - \delta_{D,N})^2 + (\delta_{P,S} - \delta_{P,N})^2 + (\delta_{H,S} - \delta_{H,N})^2} \qquad (3.3)$$

where $\delta_{D,S}$, $\delta_{P,S}$ and $\delta_{H,S}$, and $\delta_{D,N}$, $\delta_{P,N}$ and $\delta_{H,N}$ are the HSPs of the solvent and 2D material, respectively [7, 93, 120, 121].

Exfoliation in Solvent Mixtures

Many of the solvents best suited to 2D material dispersion (e.g. N-Methyl-2-pyrrolidone (NMP), cyclohexanone, N-cyclohexyl-2-pyrrolidone (CHP)) [7, 90, 92–94, 120] have strong limitations for their use in industrial processes, including high boiling points (NMP ~ 200 °C, cyclohexanone ~ 155 °C, CHP ~ 284 °C), and possible toxicity, leading to challenges in printing and processing [123]. For many applications, therefore, it is preferable to exfoliate in solvents such as water and alcohols. This is possible for some alcohols, although the resulting dispersions typically remain stable only for a brief period (minutes-hours). In general, the mismatch in HSPs is too great to achieve stable dispersions in the solvents alone.

An advantage of working in terms of HSPs is that they can be calculated for solvent mixtures as well as individual solvents which allows combinations of solvents to be formulated to produce stable dispersions [124–126]. The HSPs of a solvent mixture can be calculated from:

$$\delta^2_{i,mix} = \frac{\left(\frac{\phi_1}{\rho_1}\right)\delta_{i,1} + \left(\frac{\phi_2}{\rho_2}\right)\delta_{i,2} + \cdots + \left(\frac{\phi_n}{\rho_n}\right)\delta_{i,n}}{\left(\frac{\phi_1}{\rho_1}\right) + \left(\frac{\phi_2}{\rho_2}\right) + \cdots + \left(\frac{\phi_n}{\rho_n}\right)} \qquad (3.4)$$

where ϕ_n is the volume fraction of each solvent, and δ_i is the HSP (δ_D, δ_P or δ_H) [121, 124, 125]. For example, materials such as *h*-BN, graphene, and MoS_2 can be exfoliated and dispersed in mixtures of water and ethanol or water and isopropanol at far higher concentration than in the individual solvents. An example of this is shown in Fig. 3.13m, n.

Exfoliation Using Dispersants

An alternative to the mixed solvent approach (for example to exfoliate into pure water) is to use a dispersant [1, 90–92, 122]. The most widely used dispersants are polymers such as sodium carboxymethylcellulose (Na-CMC), ethyl cellulose, and polyvinylpyrrolidone (PVP), and surfactants such as sodium cholate (SC), sodium deoxycholate (SDC), sodium dodecyl sulfate (SDS), and sodium dodecylbenzene sulfonate (SDBS) [1, 90–92, 122] (Fig. 3.14).

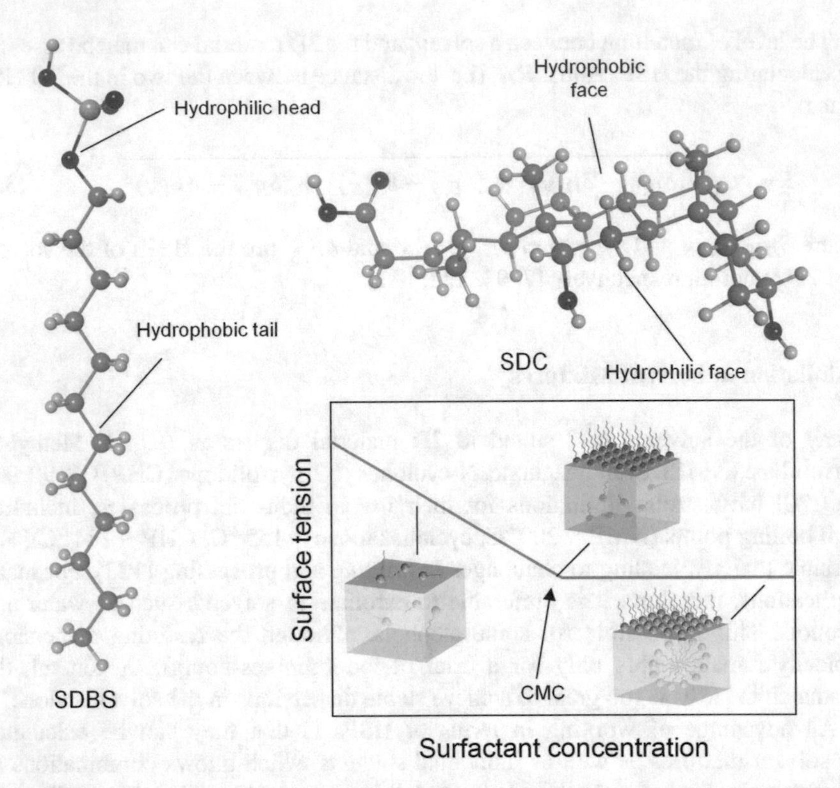

Fig. 3.14 Structure of sodium dodecylbenzene sulfonate (SDBS) and sodium deoxycholate (SDC) surfactants. The surface tension of a surfactant solution changes according to the surfactant concentration but stabilises above the critical micelle concentration (CMC)

The polymeric dispersants effectively coat the faces of the exfoliated 2D material flakes, providing a physical repulsion between the flakes [88, 90, 122]. They are of interest for ink formulation, as polymers can allow tuning of the fluidic properties (e.g. viscosity) of dispersions [96, 97, 127–129]. Some polymers used within composites can also act as dispersants, allowing fabrication of devices and coatings directly from the dispersions [88, 130].

Surfactants are chemicals that contain both hydrophobic (i.e. 'water-hating') and hydrophilic ('water-loving') regions in their molecule, either at opposite ends of a chain, or on opposite faces of a 2D molecule; Fig. 3.14 [131, 132]. As such, they can aid dissolution or exfoliation of hydrophobic materials (such as graphene) by 'bridging' the differences in surface energy. For this reason, surfactants are commonly used in soaps and detergents, where they can allow hydrophobic materials such as oils to be washed away in water. When dissolved in water at low concentration, surfactants self-assemble at the liquid surface to minimise contact between the hydrophobic portions of the molecule and the water. As shown in Fig. 3.14, at a higher surfactant concentration (i.e. greater than the critical micelle concentration, CMC), the surface is saturated, and any further surfactant instead assembles into micelles within the bulk of the liquid. These micelles can then encapsulate and stabilise the exfoliated 2D material flakes [90, 119, 133]. It has been shown that the best-suited surfactants for 2D material exfoliation are quasi-2D molecules such as the bile salts SC and SDC, which have hydrophobic and hydrophilic faces [119, 133–136].

Depending on the application, it may be necessary to remove the polymer or surfactant following deposition (for example in conductive coatings, where interflake contact is required to allow an acceptable level of electrical conductivity). In some instances this can be achieved through rinsing or heating of the deposited film, removing the dispersant while leaving the 2D materials largely unaffected [127, 128, 137].

3.2.3 *Post-Processing*

After solution processing, the exfoliated 2D materials are not uniform in lateral size and thickness and hence, material properties. It is therefore required to apply post-processing of the 2D materials. These include processes to increase the few-layer/monolayer content of the dispersion, to sort and extract uniform 2D material flakes, or to remove undesirable solvents, in addition to the chemical and thermal annealing processes described above.

Thick flakes can be separated and removed by different strategies based on ultracentrifugation in a uniform medium [1], or in a density gradient medium (DGM) [1]. The first is called differential ultracentrifugation (sedimentation based-separation, SBS), while the second is called density gradient ultracentrifugation (DGU) [1]. The SBS process separates various particles on the basis of their sedimentation rate in response to a centrifugal force acting on them.

SBS is the most common separation technique to date, successfully producing flakes ranging from few nanometers to a few microns, with concentrations up to a few gL^{-1} [1]. SBS is particularly useful for solution processing techniques like LPE. While producing a dispersion enriched in mono- and few-layer flakes, LPE still has relatively low yield, and initially contains a high proportion of under-exfoliated flakes. The distribution of flake sizes and thicknesses can be altered by using SBS [1, 90, 92]. In this process, larger and thicker flakes in the dispersion sediment more rapidly, allowing them to be removed.

DGU as a post-processing method of 2D materials was first reported with graphene by Green and Hersam [138]; Fig. 3.15a–c. The authors demonstrated the sorting of graphene flakes of different thickness. As shown in Fig. 3.15d–g, in a DGU process, the flakes are ultracentrifuged in a preformed DGM where they move along the cuvette until they reach the corresponding isopycnic point (i.e. the point where their buoyant density equals that of the surrounding DGM) [1, 139]. The buoyant density is defined as the density (ρ) of the medium at the corresponding isopycnic point. However, the 2D material flakes generally have the same density,

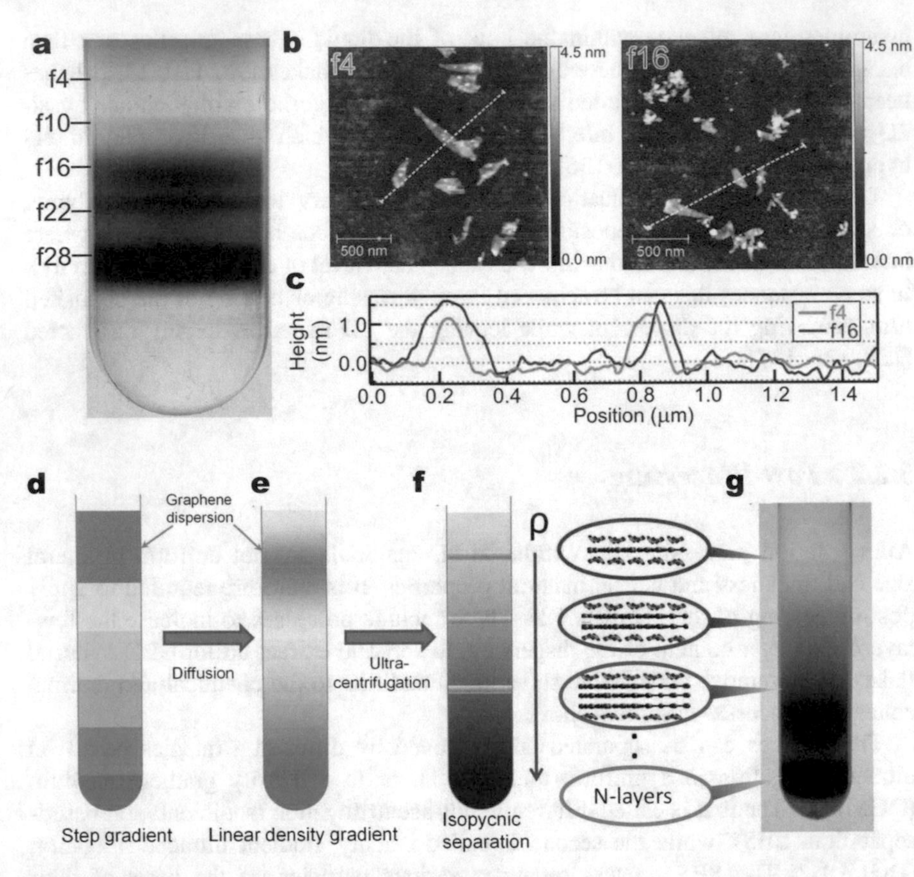

Fig. 3.15 (**a**) Photograph of a centrifuge tube following the first iteration of DGU. Lines mark the position of sorted graphene fractions within the centrifuge tube. (**b**) Representative AFM images of graphene deposited using fractions f4 and f16 onto SiO_2. (**c**) Height profile of regions marked in panels B (blue curve) and C (red curve) demonstrating the different thicknesses of graphene flakes obtained from different DGU fractions. Reprinted with permission from Ref. [138]. Copyright (2009) American Chemical Society. Sorting of graphite flakes via isopycnic separation starting with the (**d**) formation of step gradient by placing a density gradient medium with decreasing concentration; (**e**) linear density gradient formed via diffusion; (**f**) during isopycnic separation the graphite flake-surfactant complexes move along the cuvette, dragged by the centrifugal force, until they reach their isopycnic points. The buoyant density of the flake-surfactant complexes increases with number of layers. (**g**) Photograph of cuvette containing sorted flakes. Adapted with permission from Ref. [1]. Copyright (2012) Materials Today

irrespective of the number of layers. Hence, to assist in increasing buoyant density and the formation of a density gradient, a surfactant is usually used to aid in the process.

Centrifugation can also be used to exchange the exfoliation solvent for one with more desirable properties (e.g. lower boiling point) [140]. Another approach to achieve this is to use filtration, using a filter with a sufficiently small pore size to retain the flakes [127, 141]. Both of these methods allow the flakes to be redispersed into the new solvent (sometimes at increased concentration) via mild sonication or stirring.

3.3 Material and Dispersion Characterisation Methods

Whether produced by growth or by exfoliation it is important to understand the techniques used to characterise 2D materials and their dispersions. A number of techniques have emerged as useful for this purpose. They can be broadly classified into optical absorption, microscopy, spectroscopy and diffraction methods. This section briefly describes the most widely exploited characterisation methods for 2D materials and examples of measurements from published literature.

3.3.1 Optical Absorption Spectroscopy

It is challenging to measure the extent of exfoliation of the starting layered crystals and therefore to directly quantify the concentration of a produced 2D materials in a dispersion. It is also impractical and potentially inaccurate to directly measure the mass of the 2D materials with a typical micro-balance for every single sample due to their low concentration (typically $<1\,\mathrm{g\,L^{-1}}$ for the UALPE dispersions [1, 41, 89, 93, 94, 142–144]).

Therefore a method of indirect estimation using optical absorption spectroscopy following Beer–Lambert law is a convenient tool [89, 93]:

$$A = \alpha_\lambda C l \tag{3.5}$$

where A is the absorbance at wavelength λ, α_λ is the absorption coefficient at the same wavelength, C is concentration, and l is optical path of incident light through the dispersion. α_λ relates absorbance to concentration, and once α_λ is known, C can be calculated from A.

A can be measured using a UV-Vis-NIR spectrophotometer, as schematically illustrated in Fig. 3.16a. UV-Vis-NIR spectrophotometer is designed to measure the optical absorbance of a compound in solution or as a solid (e.g. free standing thin films). The apparatus usually consists of the light source (the lamps), a monochromator, sample and reference holders, and detectors. As shown, a filtered monochrome beam is split and passed through a sample and a reference, respectively, allowing analysis of the absorbance of the sample at a wavelength at a time. This allows collection of an absorbance spectrum spanning a wide wavelength region.

To estimate the concentration from the spectroscopy, an accurate determination of α_λ is crucial. In a commonly adopted method, proposed by Hernandez

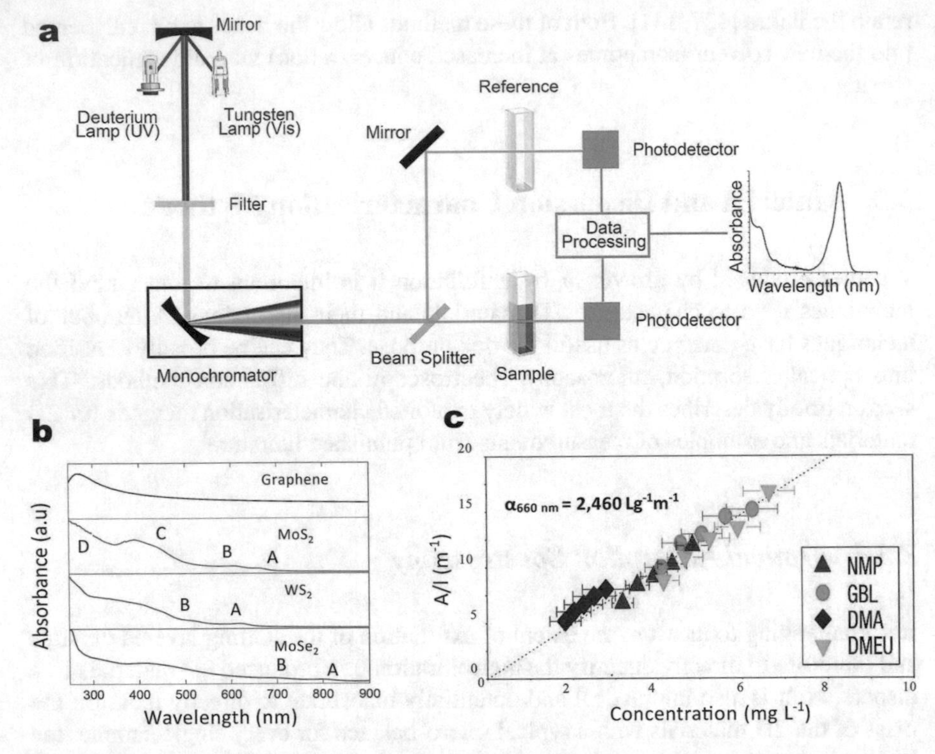

Fig. 3.16 (**a**) Schematic of UV-Vis-NIR spectrophotometer. (**b**) Optical extinction spectra for Graphene, MoS_2, WS_2, $MoSe_2$. (**c**) The absorbance divided by the cell length with respect to the concentration showing that the absorption coefficient at 660 nm is 2,460 L g^{-1} m^{-1}. Reproduced with permission from Ref. [89]. Copyright (2008) Nature Publishing Group

et al. [89, 93], a certain volume of dispersion is filtered through a membrane of known mass such that the flakes are left on the membrane. The residual solvent is then removed through annealing (e.g. 400 °C). The mass of the extracted flakes is therefore the mass difference of the membrane. The α_λ is determined by correlating the absorbance of the dispersion with the mass. This α_λ then allows estimation of the concentration. Since α_λ is a study of the optical absorption of the compound, it should not depend on the solvents where the compound is dispersed. This is demonstrated by Ref. [89], where the authors Hernandez et al. showed that the α_λ of graphene at 660 nm is 2,460 L g^{-1} m^{-1} in different solvents, as shown in Fig. 3.16c [89].

We note that in the preceding studies of UALPE production of 2D materials the optical extinction (E, i.e. the sum of the absorbance and scattering [145]) is actually used to estimate the concentration for the convenience of measurement. For example, Fig. 3.16b presents the extinction spectra at 350–800 nm of different UALPE produced materials. For accurate measurement, an integrating sphere should be used to allow the scattering component to be discounted from the extinction measurements.

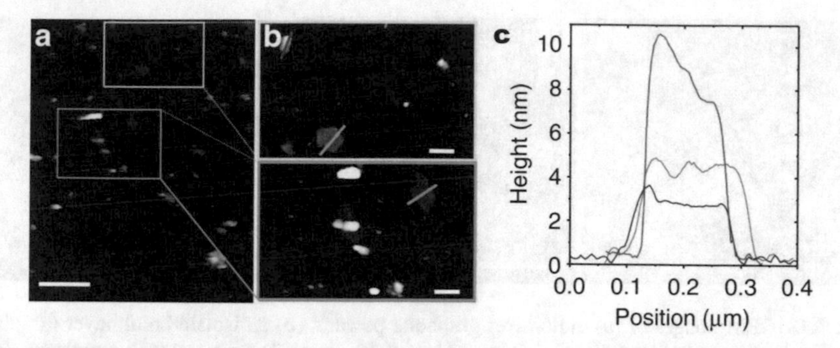

Fig. 3.17 (**a, b**) Representative AFM images of exfoliated UALPE MoS_2 flakes deposited on Si/SiO_2, allowing (**c**) measurement of thickness of individual flakes. Reprinted with permission from Ref. [147]. Copyright (2014) Nature Publishing Group

3.3.2 Characterisation via Microscopy

Atomic Force Microscopy

Atomic force microscopy (AFM) is a type of scanning probe microscopy used to study the thickness and morphology of samples such as nanomaterials [146]. An AFM consists of a cantilever ending in a sharp tip, typically nanometers in radius, scanning across the surface of the sample. The scanning mode of AFM may be either static, whereby the tip and the surface contact, or dynamic, whereby the cantilever vibrates. During operation, the tip is brought into direct contact with, or driven close enough to the surface, with the force between the tip and the sample being collected to represent the thickness and the surface morphology. AFM offers a resolution on the order of nanometres to micrometres, making it a very effective tool to study the morphology and thickness of the 2D materials deposited onto a flat substrate.

Figure 3.17a, b are representative AFM images of deposited UALPE MoS_2 flakes [147]. As shown, the flakes are randomly distributed over the substrate, with varied thickness and lateral size. In the above example, the blue, red and green lines on the flakes in Fig. 3.17a, b show thicknesses of ~3 nm, ~10 nm and ~4 nm, respectively; Fig. 3.17c. Statistical analysis of large quantities of flakes can further allow an estimation of flake size distribution.

Scanning Electron Microscopy

Scanning electron microscopy (SEM) makes use of accelerated electrons as the source of illumination to investigate the morphology and the structure of an object [146]. In operation, the electron beam is fired from an electron gun and subsequently accelerated by a strong electrostatic field. The highly accelerated electron beam is then focused by electrostatic and electromagnetic lenses, and finally projected on to the surface of the object. The electron beam probes over the surface and is subsequently reflected to produce an image of the surface.

Fig. 3.18 SEM images of (**a**) multi-layer graphene powder, (**b**) an isolated multilayer (graphite) flake, and (**c**) a partially exfoliated, thin multilayer flake. (**b**, **c**) Reproduced with permission from Ref. [148]. Copyright (2015) The Royal Society of Chemistry

SEM is a useful tool to study the flake morphology and also the cross-sectional stacking of solution-processed 2D materials. Figure 3.18a shows an SEM image of multi-layer graphene (graphite) flakes. Figure 3.18b depicts an SEM image of an isolated multi-layer flake as is apparent from its 'layered' appearance, and Fig. 3.18c presents graphene that has been exfoliated as a few-layer flake [148]. These images clearly illustrate the flake morphologies, allowing the measurement of the lateral size of the investigated flakes. Unlike AFM, SEM cannot be used to directly measure the thickness of the investigated flakes.

Transmission Electron Microscopy

Transmission electron microscopy (TEM) investigates an object with an accelerated electron beam [146], similar to SEM. However, unlike SEM, the electron beam transmits through the object with some parts of the electron beam scattered, allowing collection of the information of the object. Because the electron beam needs to transmit through and emerge from the opposite side, the object needs to be extremely thin, typically <100 nm. This makes TEM well-suited for the characterisation of solution-processed 2D material flakes.

TEM is capable of producing high quality images of individual flakes of 2D materials. It can typically achieve a resolution at least ten times higher than SEM, down to 50 pm [149]. Figure 3.19a–h depicts TEM images of LPE graphene, MoS_2, WS_2, and *h*-BN, while Fig. 3.19i–l show their corresponding high resolution TEM (HRTEM) images.

As well as offering high resolution imagery of flake morphology, TEM can also be used to distinguish between monolayer and multilayer of 2D materials. Figure 3.20a, b depict graphene monolayer and bilayer, respectively, with Fig. 3.20c showing the normal-incidence electron diffraction pattern on the flake in Fig. 3.20a. This pattern shows a typical sixfold symmetry as expected for graphene [89, 152, 153] which can be labelled with Miller-Bravais (*hkil*) indices [89]. Figure 3.20d, e show normal-incidence selected-area diffraction patterns for the flake

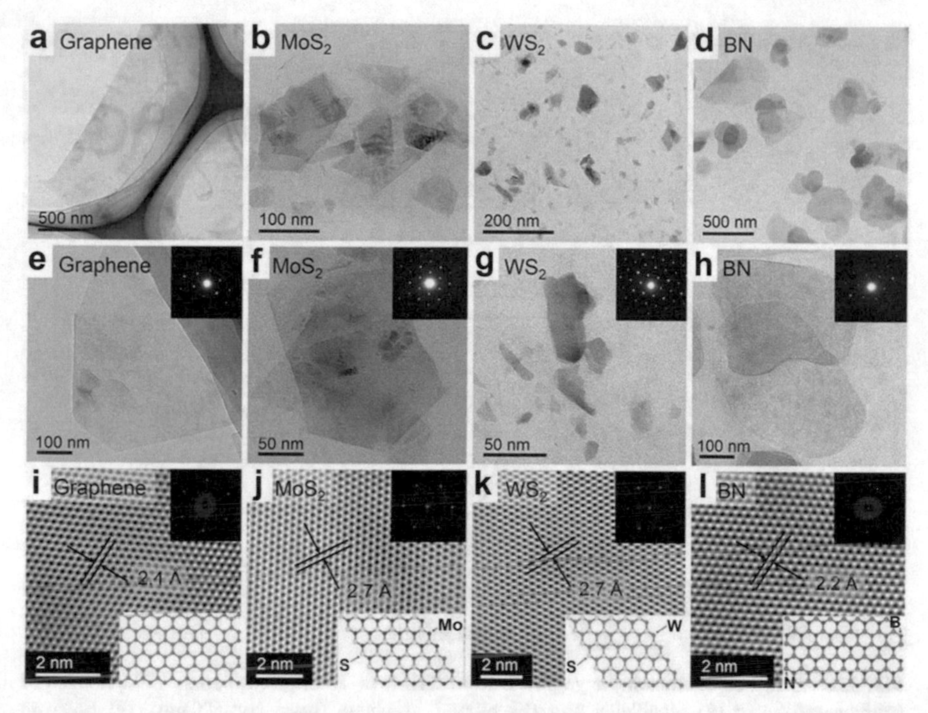

Fig. 3.19 TEM characterisation of 2D materials. (**a–d**) Low-magnification TEM images of flakes of graphene, MoS_2, WS_2, and *h*-BN. (**e–h**) High-magnification TEM images of flakes of graphene, MoS_2, WS_2, and *h*-BN. (**i–l**) HRTEM images of flakes of graphene, MoS_2, WS_2, and *h*-BN nanosheets. (Insets) Top: Fast Fourier transforms of the images; Bottom: Schematic drawing of the atomic structure of graphene, MoS_2, WS_2, and *h*-BN, respectively. Reproduced with permission from Ref. [150]. Copyright (2015) Nature Publishing Group

in Fig. 3.20b with beam positions taken at the black and white dots, respectively. Hence, Fig. 3.20d reflects a hexagonal pattern consistent with monolayer graphene while Fig. 3.20e reflects a pattern consistent with multilayer graphene. The {2110} spots appear to be more intense relative to the {1100} spots [89]. This is important as for multilayers with Bernal (AB) stacking, computational studies have shown that the intensity ratio, $I_{\{1100\}}/I_{\{2110\}}$ is < 1 [89, 152, 153]. Conversely, monolayers have $I_{\{1100\}}/I_{\{2110\}} > 1$. The identification of AB stacking in multilayers can lead to the differentiation between monolayers, bilayers and multilayers by inspection of diffraction pattern intensity ratios, confirming that both the flake in Fig. 3.20a and the region marked by the black dot in Fig. 3.20b are monolayers. Conversely, Fig. 3.20h indicates that the area around the white dot in Fig. 3.20b consists of graphene flakes of more than one layer.

Further analysis can also be carried out in addition to diffraction pattern. As TEM provides the resolution to study individual flakes, it is also an effective tool to measure the lateral size of the investigated individual flakes and hence provide statistical analysis. For example, Fig. 3.20j–l present lateral size histograms of

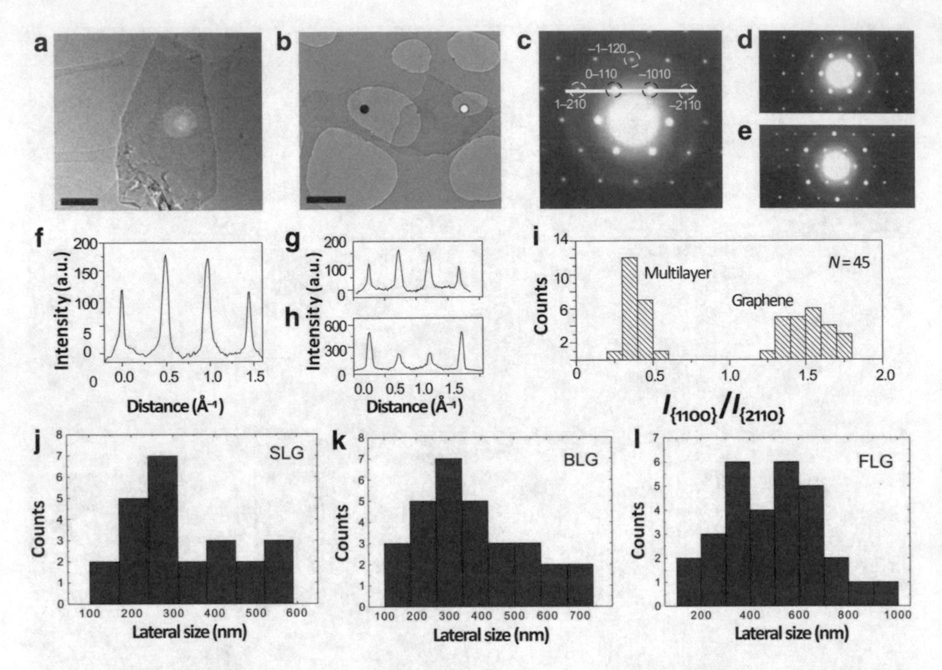

Fig. 3.20 Evidence of monolayer graphene from TEM. (**a**–**b**) High-resolution TEM images of solution-exfoliated (**a**) monolayer and (**b**) bilayer graphene (scale bar 500 nm). (**c**) Electron diffraction pattern of a single monolayer as in (**a**), with the peaks labelled by Miller-Bravais indices. Electron diffraction patterns taken from the positions of the (**d**) black and (**e**) white spots of the sheet shown in (**b**), using the same labels as in (**c**). The graphene is clearly one layer thick in (**d**) and two layers thick in (**e**). (**f**–**h**) Diffracted intensity taken along the 1-210 to -2110 axis for the patterns shown in (**c**–**e**). (**i**) Histogram of the ratios of the intensity of the {1100} and {2110} diffraction peaks for all the diffraction patterns collected. A ratio > 1 is a signature of graphene. Reproduced with permission from Ref. [89]. Copyright (2008) Nature Publishing Group. Statistical analysis of lateral size for (**j**) mono-, (**k**) bi- and (**l**) few-layer graphene flakes. Reproduced with permission from Ref. [151]. Copyright (2012) American Chemical Society

monolayer, bilayer and few-layer graphene flakes, which were measured by Torrisi et al. in determining the lateral size distribution of LPE graphene flakes used for graphene inkjet ink [151]. TEM is hence a highly versatile tool in characterising flake distribution and flake layers.

3.3.3 Spectroscopic Characterisation of 2D Material Dispersions

Various spectroscopy techniques can also be used to study the atomic and electronic structure of the exfoliated 2D material flakes. These include Fourier transform infrared spectroscopy (FTIR) and Raman spectroscopy, which probe interatomic vibrations, X-ray photoelectron spectroscopy (XPS), which probes chemical composition, and photoluminescence spectroscopy (PL), which probes material bandgaps.

Raman Spectroscopy

Raman spectroscopy is one of the most widely used non-destructive techniques for studying the interatomic vibrations (phonons) of 2D materials [154–160]. Interlayer and intralayer bonds in 2D materials can be visualised as springs with a mass at each end. These springs have resonant vibrational modes, and it is these that Raman spectroscopy probes. The sample is typically deposited on a substrate such as silicon, but measurements can also be made in dispersions. In a typical measurement set-up, the sample is illuminated with a monochromatic laser, and the reflected light is collected. Photons in the laser beam scatter inelastically from phonons in the sample (i.e. they gain or lose energy on interacting with the resonant modes), leading to small shifts in the wavelength of the reflected light. The reflected spectrum therefore contains characteristic peaks corresponding to the vibrational modes of the materials being investigated. The position and intensity ratios of the peaks can then indicate parameters such as the number of layers in the sample [154, 156, 161, 162], the density of defects [163, 164], or the level of doping or functionalisation [160, 165]. The following subsections will briefly discuss the unique Raman spectroscopy signatures of common 2D materials.

Raman Spectroscopy of Graphene

The Raman spectrum of a defect free graphene monolayer has two main peaks: the G peak (at $\sim$1,580 cm^{-1}), a primary in-plane vibrational mode arising from the stretching of the C-C bond, and the 2D peak (at $\sim$2,700 cm^{-1}), a second-order overtone of another in-plane vibration mode, the D peak (at $\sim$1,350 cm^{-1}) [168]; Fig. 3.21a. The D peak is the first-order breathing mode for the aromatic carbon rings, but is only observed when flake edges or in defective samples are probed [163, 164]. The G, D, and 2D peaks are also present in few-layer graphene and bulk graphite, in addition to interlayer shear and breathing modes [168, 169]. Of the three peaks, the 2D peak undergoes the greatest changes when shifting from monolayer to bulk samples [168]; Fig. 3.21b, c. Therefore, the peak position and shape can be used to determine the number of layers [168]; Fig. 3.21c. The relative intensity of the D and G peaks allows a measure of the defect density in the graphene flakes [168].

When working with solution-exfoliated graphene, two key differences emerge in the measured Raman spectrum. The first is that the relatively small size of flakes produced via techniques such as LPE means that the measured spectrum inevitably includes contributions from multiple flakes with different numbers of layers. The measured spectra are therefore less well defined than those from mechanically

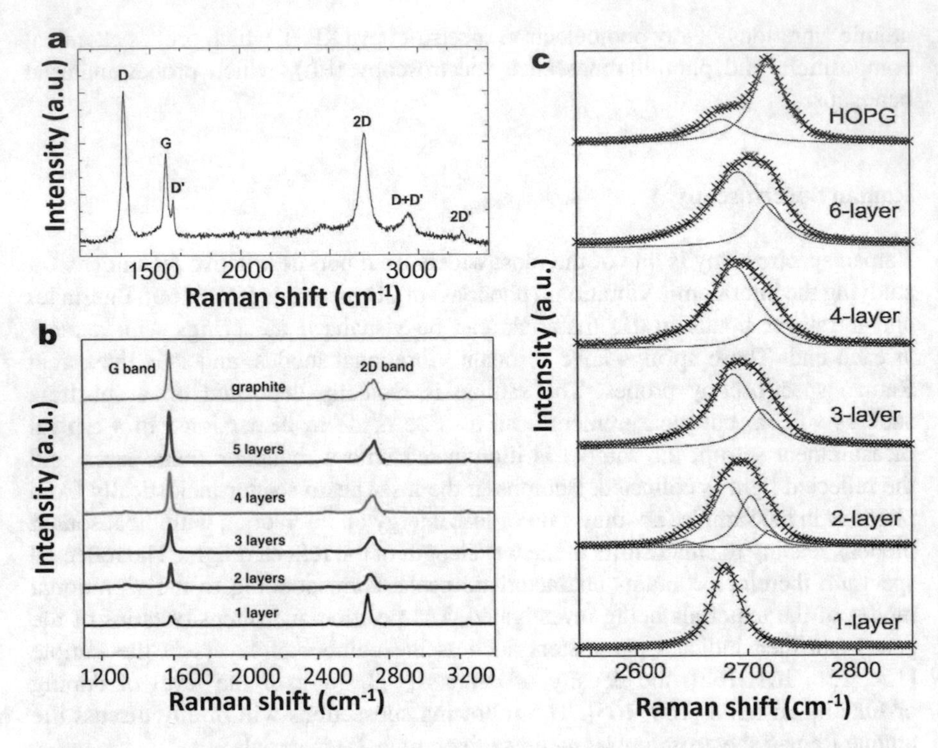

Fig. 3.21 Raman characterisation of graphene. (**a**) Typical Raman spectrum of graphene. Evolution of (**b**) Raman spectrum and (**c**) 2D peak for graphene with regard to the number of graphene layers. (**b**) Reproduced with permission from Ref. [166]. Copyright (2013) Springer publishing. (**c**) Adapted with permission from Ref. [167]. Copyright (2007) American Chemical Society. Excitation wavelength: 532 nm in all cases

cleaved graphene, and give a qualitative rather than quantitative analysis of the distribution of flake dimensions [1, 89, 119, 120]. Secondly, the small dimensions of the flakes can also lead to a higher than expected D peak even for well-ordered samples due to the influence of the flake edges [170, 171]. This can be differentiated from disorder or defects from the absence of broadening in the G peak, and by the linear increase in the D-peak intensity relative to that of the G peak with decreasing average lateral flake dimensions [170, 172].

Raman Spectroscopy of Transition Metal Dichalcogenides

Among the TMDs, semiconducting TMDs (s-TMDs) are the most widely studied, including in terms of their Raman spectra. s-TMDs such as MoS_2 and WS_2 have two major peaks in the spectrum of both few-layer and bulk material, corresponding to the in-plane (E^1_{2g}) and out-of-plane (A_{1g}) vibration modes [156, 160, 161, 173, 174]. This can be seen by the corresponding peaks in Fig. 3.22. The positions of these

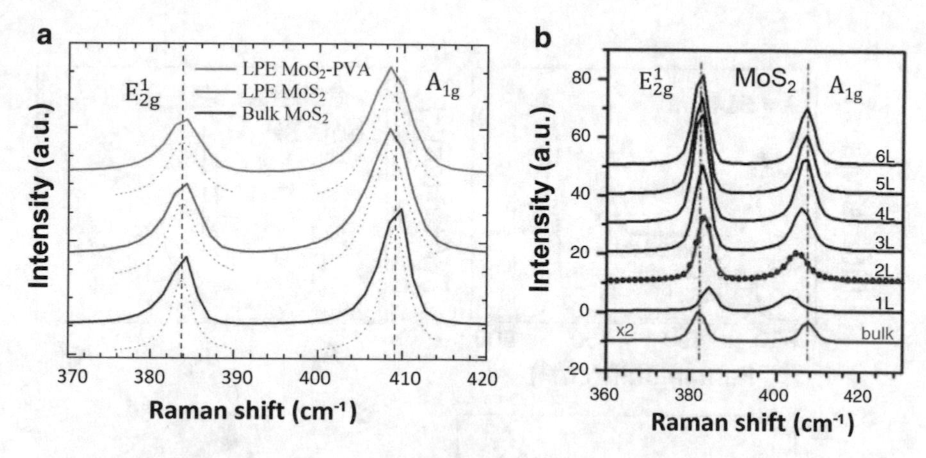

Fig. 3.22 Raman characterisation of TMDs. (**a**) Raman spectra of bulk MoS_2, LPE MoS_2 and MoS_2 composited in polyvinyl alcohol. Reproduced with permission from Ref. [175]. Copyright (2014) Tsinghua University Press and Springer-Verlag Berlin Heidelberg. (**b**) Raman spectrum and change in E^1_{2g} and A_{1g} peaks from bulk to single layer MoS_2. Adapted with permission from Ref. [161]. Copyright (2010) American Chemical Society. Excitation wavelength: 514 nm in both cases

two peaks vary with the number of layers [156, 160, 161, 173, 174]; Fig. 3.22a. However, the small shifts in the peak positions can often be challenging to determine (Fig. 3.22b). In general, A_{1g} blue-shifts [160, 161, 174] with increasing layer count as the interlayer van der Waals forces stiffen the vibration mode in thicker samples [161, 169]. Meanwhile, E^1_{2g} red-shifts [160, 161, 174, 176], due to increased dielectric screening of the interlayer Coulomb interactions, leading to softening of the mode [176]. The transition leads to a progressive increase in the difference between the peak positions (δw) with increasing layer count [160, 160, 161, 173]. Therefore, δw may be used to assess the number of layers in certain s-TMDs [160, 160, 161, 173].

As with graphene, Raman spectroscopy of solution-exfoliated 2D materials shows peaks that are typically broadened relative to those seen in mechanically cleaved samples. Since the peak shifts are small between bulk and monolayer, it is important to collect a large number of spectra to avoid ambiguity in the peak positions [175].

Raman Spectroscopy of Black Phosphorus

The Raman spectrum for BP has three major peaks close to 400 cm^{-1}, corresponding to one out-of-plane (A^1_g) and two in-plane (B_{2g} and A^2_g) vibrational modes [177, 178]; Fig. 3.23a. The additional in-plane mode compared to TMDs arises from the anisotropy of BP that leads to directionality of the vibration [177, 178]. All three of the peaks red-shift with increasing number of BP layers, by 2–6 cm^{-1} from

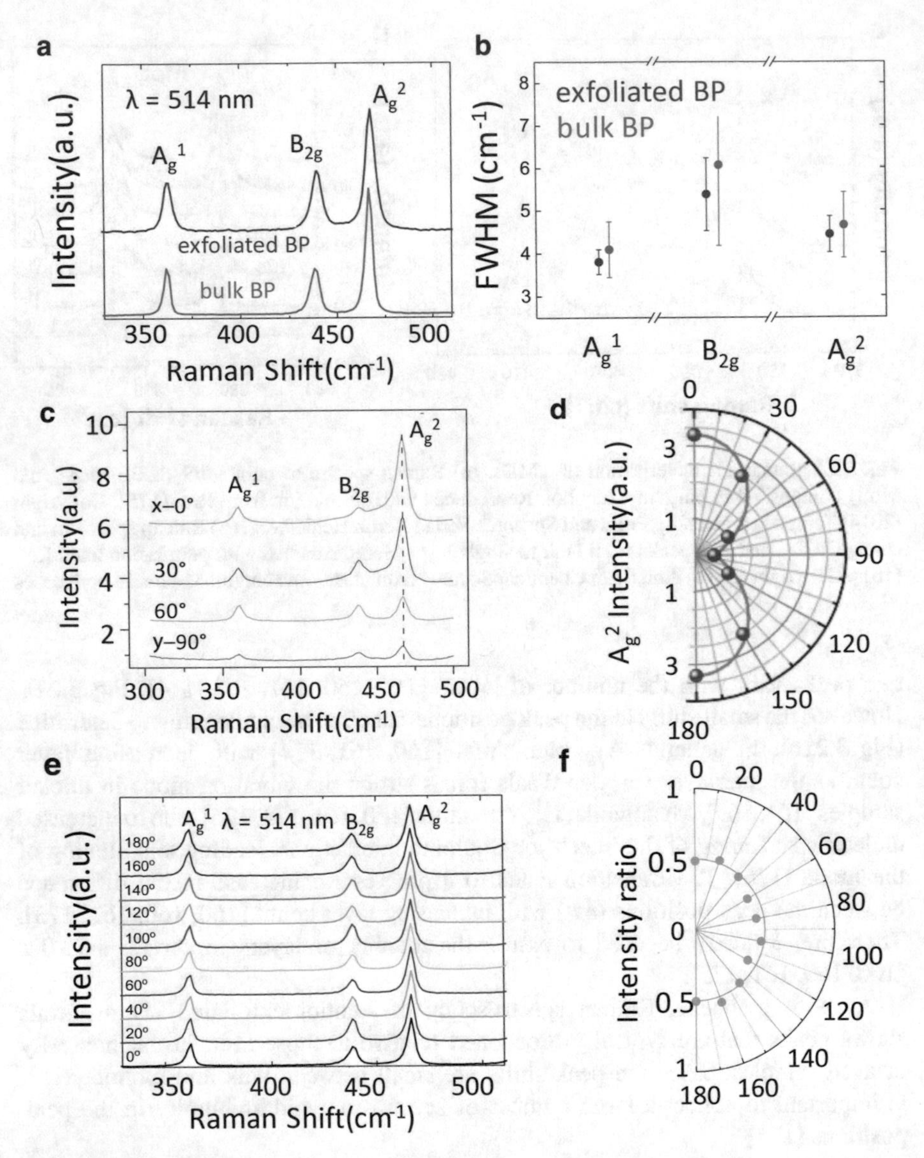

Fig. 3.23 Raman characterisation of BP. (**a**) Raman spectra of exfoliated (LPE in NMP) and bulk BP, and (**b**) the corresponding statistical FWHM. Adapted with permission from Ref. [140]. Copyright (2017), Nature Publishing Group. (**c**) Polarisation-resolved Raman spectra of mechanically-exfoliated monolayer BP, and (**d**) the corresponding intensity of A_g^2 peak as a function of the excitation laser polarisation angle. Adapted with permission from Ref. [150]. Copyright (2015), Nature Publishing Group. (**e**) Polarisation-resolved Raman spectra of LPE BP, and (**f**) the corresponding peak intensity ratio of A_g^1 to A_g^2 as a function of the excitation laser polarisation angle. Adapted with permission from Ref. [179]. Copyright (2017), Nature Publishing Group. Excitation wavelength: 514 nm in (**a**), (**b**), (**e**), (**f**), 532 nm in (**c**), (**d**)

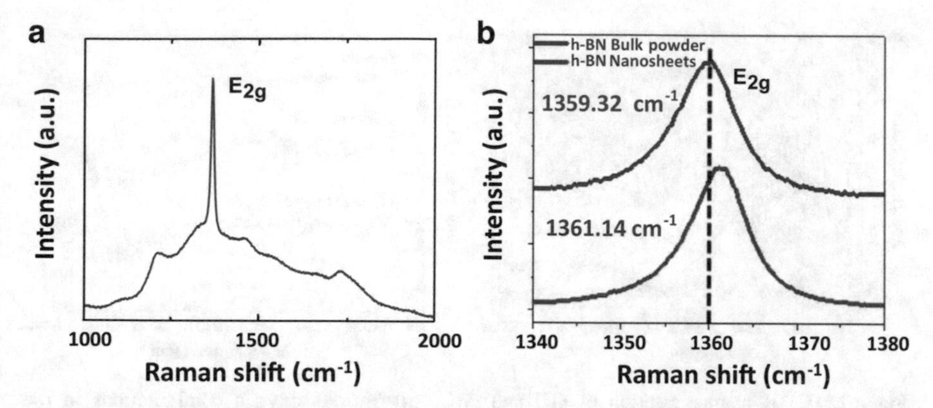

Fig. 3.24 Raman characterisation of *h*-BN. (**a**) E_{2g} peak of *h*-BN in bulk form. (**b**) Shift in E_{2g} peak after exfoliation. Excitation wavelength: 514 nm in both cases

monolayer to bulk [178]. The FWHM of the three peaks also undergoes variations with the BP thickness. For example, Fig. 3.23b shows that BP exfoliated by LPE has narrower peaks compared to the bulk [140].

A feature of BP is the angle of polarisation-dependent change in the peak intensity due to its anisotropic crystal structure [140, 179]. This is also observed in other anisotropic 2D materials, such as ReS_2 [180]. Figure 3.23c, d present such angle dependent physical properties observed in mechanically cleaved isolated BP flakes [179]. However, LPE BP flakes may not exhibit such behaviour; Fig. 3.23e, f [140]. This is because the laser spot during typical confocal Raman measurements (a few microns) includes many randomly oriented flakes, the ensemble thereby cancels the anisotrophic nature [140].

Raman Spectroscopy of Hexagonal Boron Nitride

Bulk *h*-BN exhibits a characteristic Raman peak, the E_{2g} phonon mode, analogous to the G peak in graphene; Fig. 3.24a [181]. This is arising from the similar hexagonal arrangement of boron and nitrogen atoms within its molecular structure [181]. Figure 3.24b shows the difference between bulk *h*-BN and monolayer *h*-BN. As presented, exfoliation from bulk *h*-BN into nanosheets usually shows a shift of up to 4 cm^{-1} [181].

Fourier Transform Infrared Spectroscopy

FTIR probes the atomic bonds in a material by recording the infrared energy absorbed when light passes through the sample. It is based on the principle that different atomic bonds absorb at different wavelengths in the infrared, causing them

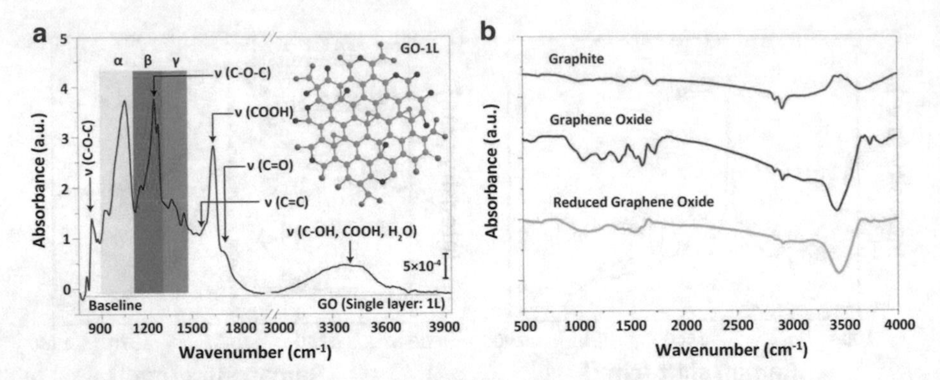

Fig. 3.25 FTIR characterisation of GO and rGO. (**a**) Various oxygen configurations in the structure of graphene oxide. Inset, Schematic representation of functional groups including epoxide (green), C=C (aqua), C–O (red), C–OH (blue), COOH (brown) and C=O (grey). (**b**) Difference in FTIR spectrum between graphite, GO and rGO. Reproduced with permission from Ref. [182]. Copyright (2010) Nature Publishing group

to bend and stretch. This means that the absorption peaks can be correlated to recognise the presence or absence of such bonds in the material [182–184]. This is particularly useful in the detection of impurities or structural deviations, as these will be visible as unexpected absorption peaks in the FTIR spectrum.

FTIR is commonly used to study the changes in defects from GO and rGO [182–184]. Figure 3.25a presents an FTIR spectrum that reflects the presence of various oxygen configurations in the structure of graphene oxide. These include the vibration modes of epoxide (C–O–C) (1,230–1,320 cm^{-1}, asymmetric stretching; ~850 cm^{-1}, bending motion), sp^2-hybridised C=C (1,500–1,600 cm^{-1}, in-plane vibrations), carboxyl (COOH) (1,650–1,750 cm^{-1} including C–OH vibrations at 3,530 cm^{-1} and 1,080 cm^{-1}), ketonic species (C=O) (1,600–1,650 cm^{-1}, 1,750–1,850 cm^{-1}) and hydroxyl (namely phenol, C–OH) (3,050–3,800 cm^{-1} and 1,070 cm^{-1}) with all C–OH vibrations from COOH and H_2O. Regions of spectral overlap involving mostly C–O and C=O contributions (850–1,500 cm^{-1}) are broken down into three regions: the α-region (900–1,100 cm^{-1}), β-region (1,100–1,280 cm^{-1}) and γ-region (1,280–1,500 cm^{-1}) [182].

FTIR can also be used to determine the difference and effectiveness of reduction of GO to rGO by studying the presence of functional groups within the structure of rGO. Figure 3.25b presents the difference in FTIR spectra between pristine graphite, GO and rGO [183]. The peaks of the rGO spectra are noticeably less intense as compared to the peaks of GO. This corresponds to the incomplete removal of functional groups during the reduction step of GO.

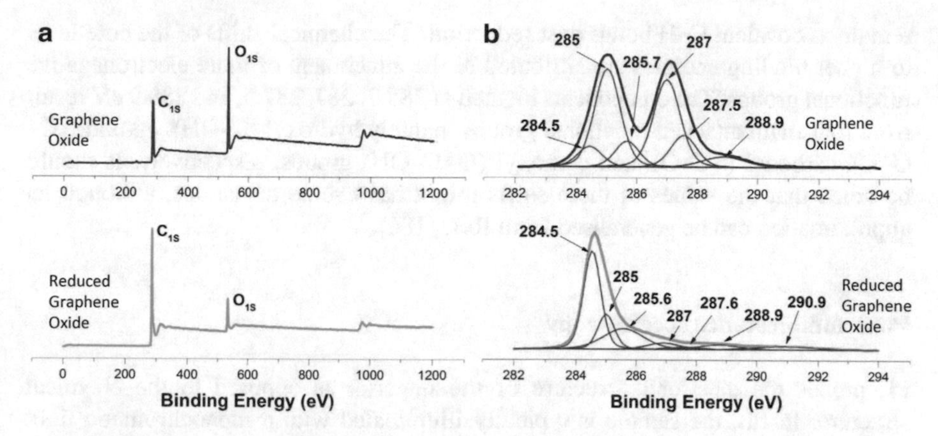

Fig. 3.26 XPS characterisation of GO and rGO. (**a**) XPS survey spectra of GO and rGO. (**b**) Deconvoluted C_{1S} signals of GO and rGO. Reproduced with permission from Ref. [183]. Copyright (2015) Nature Publishing group

X-Ray Photoelectron Spectroscopy

XPS relies on the ability of high-energy (X-ray) photons to generate a flow of electrons by 'knocking' them out of their orbits around nuclei. The intensity and energy of the emitted electrons reveals the binding energy (eV) of the materials present (i.e. the energy required to free electrons). Peaks in the electron intensity at certain energies can then be correlated to the different elements and bond types present in the sample. XPS is often used in conjunction with FTIR to study the hybridisation and functional groups attached to a 2D material.

Figure 3.26 presents an XPS spectrum conducted on GO and rGO powder. The oxygen content of rGO is mostly expressed as the carbon to oxygen ratio (C/O) in the evaluation of the reduction process. According to the XPS survey spectra in Fig. 3.26a, the C/O ratio has increased from 2.4 in GO to 8 in rGO, confirming a successful reduction process. In order to construct a background sensitive to the change in data, an algorithm called a Shirley background correction and Gaussian-Lorentzian functions can be applied to get a higher resolution readout of specific peaks [183, 185]. In the example given in Fig. 3.26, the high resolution C_{1S} peaks were deconvoluted using these methods; Fig. 3.26b. The assignments of the deconvoluted components were based on theoretical predictions of core level shifts and on reported spectra containing the particular oxygen functional groups [185].

In Fig. 3.26b, the main peaks located at 284.5 and 285 eV are assigned to the sp^2 and sp^3 hybridisations of carbon atoms, respectively. The majority of carbon atoms in GO have sp^3 hybridisation due to the formation of covalent bonds during the oxidation stage while the sp^2 component originates from unoxidised carbon atoms. Although the reduction process recovers most of the sp^2 bonds, some carbon atoms remain in sp^3 hybridisation. This is largely because a fraction of sp^3 bonds remain as covalent C–H bonds post reduction. The chemical shifts of the core level to higher binding energies are attributed to the attachment of more electronegative functional groups. The components located at 285.7, 287, 287.5, and 288.9 eV result from four main attached functional groups, namely hydroxyl (C–OH), epoxide (C–O–C), carbonyl (>C=O), and carboxyl (O=C–OH) groups, respectively. It should be noted that the values of these shifts might have some deviations, although an approximation can be generalised from Ref. [186].

Photoluminescence spectroscopy

PL probes the electronic structure of the material as opposed to the chemical structure. In PL, the sample is typically illuminated with a monochromatic light source with a wavelength sufficiently shorter than the smallest direct bandgap of the material. Electrons are then photoexcited into the conduction band, but not freed from their atoms entirely. The electrons then relax back to a lower energy state, and the additional energy is released as light. Studying the wavelength and intensity of this emitted light can therefore allow the material bandgap to be determined. This is of particular importance for materials such as s-TMDs like MoS_2, which show shifts in their PL wavelength and intensity as the bandgap increases and becomes direct with decreasing layer count. In the characterisation of 2D materials, PL is of particular significance to s-TMDs [187], BP [179] and GO [188].

For example, Fig. 3.27a, b presents the photoluminescence of a single area of an exfoliated MoS_2 film [187]. This area contains MoS_2 films having a variety of thicknesses, as shown in the optical photograph inset of Fig. 3.27b. This area was divided into two regions to collect PL spectra as indicated by the dotted white lines. The subsequent PL spectra is plotted in order to show the difference in the PL spectra between 1L, 2L, 3L, 4L and few layer (FL) MoS_2. The strongest PL intensity is produced by the direct bandgap of the monolayer MoS_2 [187]. PL intensity decreases substantially with increasing atomic thickness due to the direct-to-indirect bandgap transition; Fig. 3.27b [187].

PL can also be used to characterise the oxidation of graphene and the reduction of GO to rGO. This feature was first reported by Gokus et al. who induced strong photoluminescence in single-layer graphene by using oxygen plasma treatment [189]. In doing so, Gokus et al. successfully introduced peroxide (–O) groups on the top layers in a mechanically cleaved graphene/few layer graphene sample [189]. This is depicted in Fig. 3.27c–e which shows that strong photoluminescence is induced in monolayer graphene but is subsequently quenched in increasing layers of graphene [189]. This phenomenon was further explored when GO, with its attached functional groups, was found to have blue photoluminescence upon excitation by Eda et al.[188]. It is likely that this is down to the radiative recombination of electron-hole pairs within the localised states consisting of sp^2 and sp^3 clusters [188]. Upon reduction of GO into rGO, less radiative recombination occurs due to the reduction of functional groups, therefore reducing the intensity of blue photoluminescence. This makes PL a useful tool in determining successful reduction of GO into rGO.

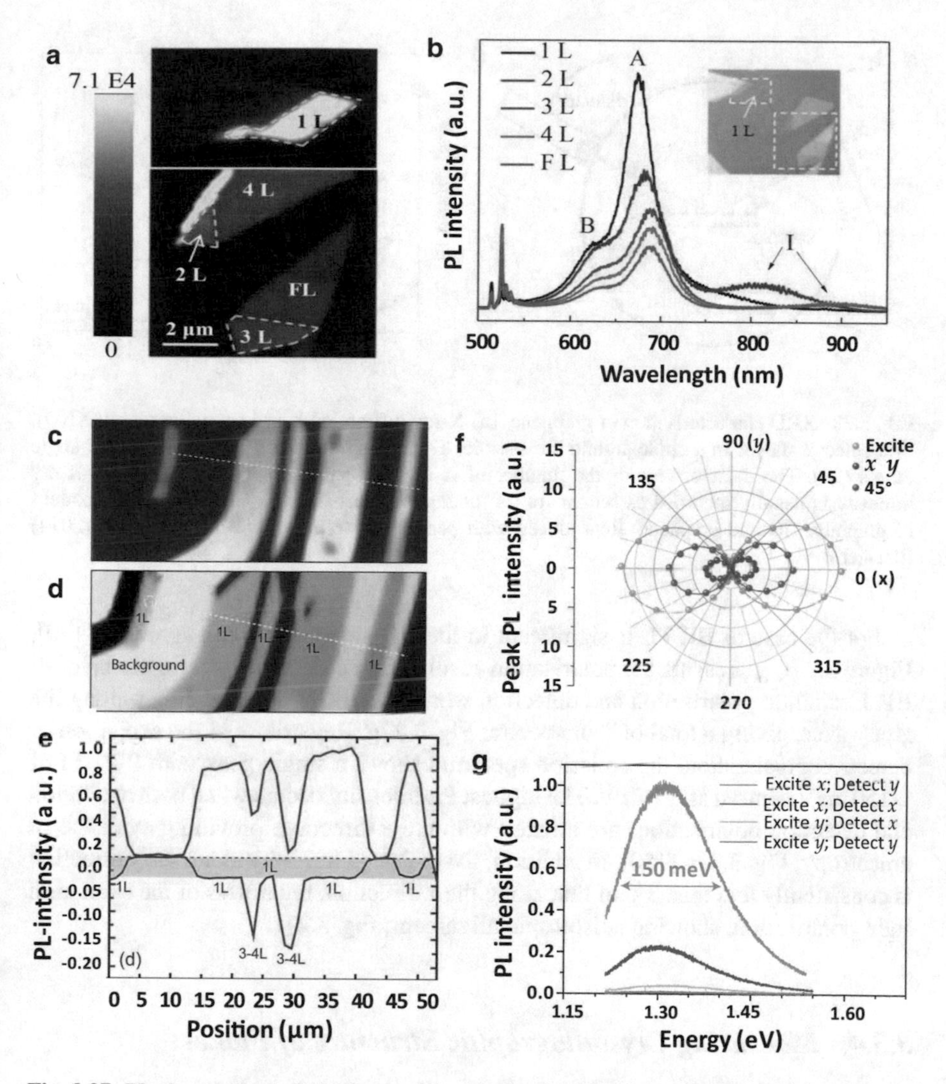

Fig. 3.27 PL characterisation of 2D materials. (**a**) PL mapping image of MoS_2 film showing the (**b**) integrated intensities of the A and B exciton peaks and their corresponding change in PL spectra from monolayer to few layer MoS_2. Reproduced with permission from Ref. [187]. Copyright (2014) Royal Society of Chemistry. Correlation between PL and layer thickness: (**c**) PL image; (**d**) elastic scattering image of the same sample area. (**e**) Corresponding cross sections taken along the dashed lines in (**c**, **d**). PL is only observed from treated mono-layer graphene, marked 1L. Reproduced with permission from Ref. [189]. Copyright (2009) Nature Publishing Group. (**f**) BPL peak intensity as a function of polarisation detection angle for excitation laser polarised along the *x* (grey), 45° (magenta) and *y* (blue) directions. (**g**) Polarisation-resolved PL spectra, revealing the excitonic nature of emission from the monolayer BP. The excitation laser is linearly polarised along either the *x* (grey curves) or *y* (blue curves) direction. On the detection side, a half-wave plate and linear polariser selects *x*- or *y*-polarised components of the emitted light, leading to a total of four different combinations as shown. Reproduced with permission from Ref. [179]. Copyright (2015) Nature Publishing Group

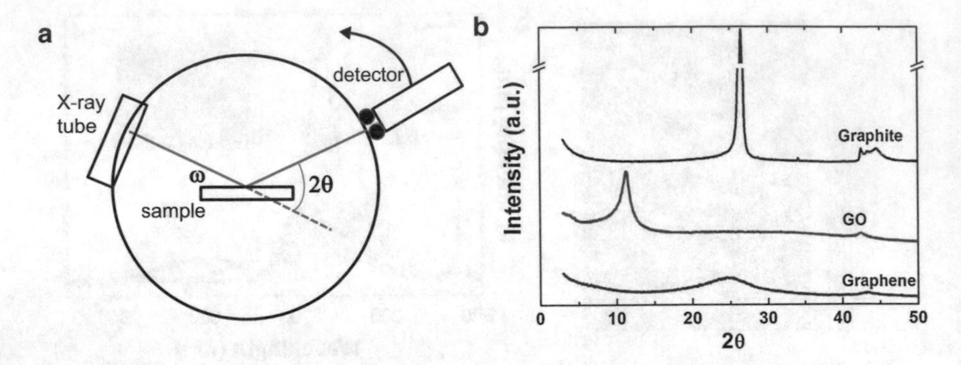

Fig. 3.28 XRD characterisation of graphene. (**a**) X-ray diffraction-based crystallography (XRD). The detector moves in a circle around the sample. The detector position is recorded as the angle 2theta (2θ). The detector records the number of X-rays observed at each angle 2θ. The X-ray intensity is usually recorded as 'count' or as 'counts per second', (**b**) XRD diffraction pattern of graphite, GO and graphene. Reproduced with permission from Ref. [191]. Copyright (2014) Elsevier B.V.

For the case of BP, PL is significant in identifying anisotropic alignment [179]. Figure 3.27f, g presents the polarisation-resolved spectra of a typical monolayer of BP. Excitation polarisation and detection were selectively oriented either along the x or y axes, giving a total of four spectra; Fig. 3.27g. Regardless of the excitation or detection polarisation, the emission spectrum shows a single peak with FWHM of ~150 meV centred at ~1.3 eV. The highest PL intensity occurs when both excitation and detection polarisations are aligned with the x direction, providing evidence of anisotropy; Fig. 3.27g [150]. In addition, the emission intensity along the y direction is consistently less than 3% of that along the x direction, regardless of the excitation light polarisation, showing anisotropic alignment; Fig. 3.27f.

3.3.4 Measuring Crystallographic Structure of Flakes

In addition to spectroscopic measurements of flake structure, it may also be important to measure the crystal structure of 2D material samples. This allows any effect of the exfoliation process on the material structure to be determined. Repeating structures such as the uniformly spaced atoms in a 2D material stack can cause constructive/destructive interference in reflected radiation at points where the repetition distance, d is equal to an integer multiple of the incident wavelength λ [190]. Constructive interference, observed as peaks in the reflected spectrum, occurs according to Bragg's law: $n\lambda = 2dsin(\theta)$, where n is an integer, θ is the incident angle, and the remaining terms are as described above [190]. An example is shown in Fig. 3.28, demonstrating the structural differences between graphene, GO, and graphite.

Two of the most commonly used sources of radiation for this purpose are X-rays (XRD) or electrons. XRD patterns are generally used to determine what crystalline phases there are in a sample. Peak broadening may indicate either a smaller crystallite size in nanocrystalline materials, more stacking faults, microstrain, and other defects in the crystal structure. The X-ray diffraction patterns therefore form a 'fingerprint' allowing observation of changes within a crystallographic sample through the broadening and narrowing of peaks.

3.4 Conclusion

In this chapter, we have described the key production methods and processes for 2D materials, with a particular focus on solution-based methods. We have also outlined a number of standard characterisation techniques for these materials and briefly discussed the useful data that can be gathered and interpreted. The production methods we have described are intended to be mostly low-cost, scalable and ideally do not require extensive modification of existing machinery. Barriers do still remain in establishing the parameters needed for the large-scale production of fit-for-purpose 2D materials. There are currently no industry recognised, standardised test procedures in place to adequately compare different sources of materials. This will be required before any manufacturing method becomes particularly dominant. The methods described in this chapter are of particular interest for producing and characterising 2D materials to function as precursors to printing inks which we will explore in the following two chapters.

References

1. F. Bonaccorso, A. Lombardo, T. Hasan, Z. Sun, L. Colombo, A.C. Ferrari, Production and processing of graphene and 2D crystals. Mater. Today **15**(12), 564–589 (2012)
2. K.S. Novoselov, A.K. Geim, S.V. Morozov, D. Jiang, Y. Zhang, S.V. Dubonos, I.V. Grigorieva, A.A. Firsov, Electric field effect in atomically thin carbon films. Science **306**(5696), 666–669 (2004)
3. K.S. Novoselov, D. Jiang, F. Schedin, T.J. Booth, V.V. Khotkevich, S.V. Morozov, A.K. Geim, Two-dimensional atomic crystals. Proc. Natl. Acad. Sci. **102**(30), 10451–10453 (2005)
4. Y. Zhang, L. Zhang, C. Zhou, Review of chemical vapor deposition of graphene and related applications. Acc. Chem. Res. **46**(10), 2329–2339 (2013)
5. J. Yu, J. Li, W. Zhang, H. Chang, Synthesis of high quality two-dimensional materials via chemical vapor deposition. Chem. Sci. **6**, 6705–6716 (2015)
6. Z.A. Munir, U. Anselmi-Tamburini, M. Ohyanagi, The effect of electric field and pressure on the synthesis and consolidation of materials: a review of the spark plasma sintering method. J. Mater. Sci. **41**(3), 763–777 (2006)
7. G. Cunningham, M. Lotya, C.S. Cucinotta, S. Sanvito, S.D. Bergin, R. Menzel, M.S.P. Shaffer, J.N. Coleman, Solvent exfoliation of transition metal dichalcogenides: dispersibility of exfoliated nanosheets varies only weakly between compounds. ACS Nano **6**(4), 3468–3480 (2012)

8. H. Li, J. Wu, X. Huang, G. Lu, J. Yang, X. Lu, Q. Xiong, H. Zhang, Rapid and reliable thickness identification of two-dimensional nanosheets using optical microscopy. ACS Nano **7**(11), 10344–10353 (2013)
9. R.F. Frindt, A.D. Yoffe, Physical properties of layer structures: optical properties and photoconductivity of thin crystals of molybdenum disulphide. Proc. R. Soc. A **273**(1352), 69–83 (1963)
10. R.F. Frindt, Optical absorption of a few unit-cell layers of MoS_2. Phys. Rev. **140**(2A), A536–A539 (1965)
11. R.F. Frindt, Single crystals of MoS_2 several molecular layers thick. J. Appl. Phys. **37**(4), 1928–1929 (1966)
12. X. Lu, M. Yu, H. Huang, R.S. Ruoff, Tailoring graphite with the goal of achieving single sheets. Nanotechnology **10**(3), 269–272 (1999)
13. A.K. Geim, K.S. Novoselov, The rise of graphene. Nat. Mater. **6**(3), 183–191 (2007)
14. S. Bae, H. Kim, Y. Lee, X. Xu, J.-S. Park, Y. Zheng, J. Balakrishnan, T. Lei, H.R. Kim, Y.I. Song, Y.-J. Kim, K.S. Kim, B. Ozyilmaz, J.-H. Ahn, B.H. Hong, S. Iijima, Roll-to-roll production of 30-inch graphene films for transparent electrodes. Nat. Nanotechnol. **5**(8), 574–578 (2010)
15. X. Chen, B. Wu, Y. Liu, Direct preparation of high quality graphene on dielectric substrates. Chem. Soc. Rev. **45**(8), 2057–2074 (2016)
16. Y. Chen, X.L. Gong, J.-G. Gai, Progress and challenges in transfer of large-area graphene films. Adv. Sci. **3**(8), 1500343 (2016)
17. N. Petrone, C.R. Dean, I. Meric, A.M. van der Zande, P.Y. Huang, L. Wang, D. Muller, K.L. Shepard, J. Hone, Chemical vapor deposition-derived graphene with electrical performance of exfoliated graphene. Nano Lett. **12**(6), 2751–2756 (2012)
18. G. Ruan, Z. Sun, Z. Peng, J.M. Tour, Growth of graphene from food, insects, and waste. ACS Nano **5**(9), 7601–7607 (2011)
19. Y.-H. Lee, X.-Q. Zhang, W. Zhang, M.-T. Chang, C.-T. Lin, K.-D. Chang, Y.-C. Yu, J.T.-W. Wang, C.-S. Chang, L.-J. Li, T.-W. Lin, Synthesis of large-area MoS_2 atomic layers with chemical vapor deposition. Adv. Mater. **24**(17), 2320–2325 (2012)
20. K.-K. Liu, W. Zhang, Y.-H. Lee, Y.-C. Lin, M.-T. Chang, C.-Y. Su, C.-S. Chang, H. Li, Y. Shi, H. Zhang, C.-S. Lai, L.-J. Li, Growth of large-area and highly crystalline MoS_2 thin layers on insulating substrates. Nano Lett. **12**(3), 1538–1544 (2012)
21. Y.-H. Lee, L. Yu, H. Wang, W. Fang, X. Ling, Y. Shi, C.-T. Lin, J.-K. Huang, M.-T. Chang, C.-S. Chang, M. Dresselhaus, T. Palacios, L.-J. Li, J. Kong, Synthesis and transfer of single-layer transition metal disulfides on diverse surfaces. Nano Lett. **13**(4), 1852–1857 (2013)
22. S. Najmaei, Z. Liu, W. Zhou, X. Zou, G. Shi, S. Lei, B.I. Yakobson, J.-C. Idrobo, P.M. Ajayan, J. Lou, Vapour phase growth and grain boundary structure of molybdenum disulphide atomic layers. Nat. Mater. **12**(8), 754–759 (2013)
23. Z.J. Qi, S.J. Hong, J.A. Rodríguez-Manzo, N.J. Kybert, R. Gudibande, M. Drndić, Y.W. Park, A.T. Charlie Johnson, Electronic transport in heterostructures of chemical vapor deposited graphene and hexagonal boron nitride. Small **11**(12), 1402–1408 (2015)
24. W.J. Zhang, C.Y. Chan, K.M. Chan, I. Bello, Y. Lifshitz, S.T. Lee, Deposition of large-area, high-quality cubic boron nitride films by ECR-enhanced microwave-plasma CVD. Appl. Phys. Mater. Sci. Process. **76**(6), 953–955 (2003)
25. S.J. Cartamil-Bueno, M. Cavalieri, R. Wang, S. Houri, S. Hofmann, H.S.J. van der Zant, Mechanical characterization and cleaning of CVD single-layer *h*-BN resonators. NPJ 2D Mater. Appl. **1**(1), 16 (2017)
26. A. Gurarslan, Y. Yu, L. Su, Y. Yu, F. Suarez, S. Yao, Y. Zhu, M. Ozturk, Y. Zhang, L. Cao, Surface-energy-assisted perfect transfer of centimeter-scale monolayer and few-layer MoS_2 films onto arbitrary substrates. ACS Nano **8**(11), 11522–11528 (2014)
27. G.H. Han, J.A. Rodríguez-Manzo, C.W. Lee, N.J. Kybert, M.B. Lerner, Z.J. Qi, E.N. Dattoli, A.M. Rappe, M. Drndic, A.T.C. Johnson, Continuous growth of hexagonal graphene and boron nitride in-plane heterostructures by atmospheric pressure chemical vapor deposition. ACS Nano **7**(11), 10129–10138 (2013)

28. M. Wang, S.K. Jang, W.J. Jang, M. Kim, S.Y. Park, S.W. Kim, S.J. Kahng, J.Y. Choi, R.S. Ruoff, Y.J. Song, S. Lee, A platform for large-scale graphene electronics - CVD growth of single-layer graphene on CVD-grown hexagonal boron nitride. Adv. Mater. **25**(19), 2746–2752 (2013)
29. K.H. Lee, H.J. Shin, J. Lee, I.Y. Lee, G.H. Kim, J.Y. Choi, S.W. Kim, Large-scale synthesis of high-quality hexagonal boron nitride nanosheets for large-area graphene electronics. Nano Lett. **12**(2), 714–718 (2012)
30. K.K. Kim, A. Hsu, X. Jia, S.M. Kim, Y. Shi, M. Dresselhaust, T. Palacios, J. Kong, Synthesis and characterization of hexagonal boron nitride film as a dielectric layer for graphene devices. ACS Nano **6**(10), 8583–8590 (2012)
31. B.C. Bayer, S. Caneva, T.J. Pennycook, J. Kotakoski, C. Mangler, S. Hofmann, J.C. Meyer, Introducing overlapping grain boundaries in chemical vapor deposited hexagonal boron nitride monolayer films. ACS Nano **11**(5), 4521–4527 (2017)
32. T. Kobayashi, M. Bando, N. Kimura, K. Shimizu, K. Kadono, N. Umezu, K. Miyahara, S. Hayazaki, S. Nagai, Y. Mizuguchi, Y. Murakami, D. Hobara, Production of a 100-m-long high-quality graphene transparent conductive film by roll-to-roll chemical vapor deposition and transfer process. Appl. Phys. Lett. **102**(2), 023112 (2013)
33. J. Chan, A. Venugopal, A. Pirkle, S. McDonnell, D. Hinojos, C.W. Magnuson, R.S. Ruoff, L. Colombo, R.M. Wallace, E.M. Vogel, Reducing extrinsic performance-limiting factors in graphene grown by chemical vapor deposition. ACS Nano **6**(4), 3224–3229 (2012)
34. L. Banszerus, M. Schmitz, S. Engels, J. Dauber, M. Oellers, F. Haupt, K. Watanabe, T. Taniguchi, B. Beschoten, C. Stampfer, Ultrahigh-mobility graphene devices from chemical vapor deposition on reusable copper. Sci. Adv. **1**(6), e1500222 (2015)
35. Y. Wang, Y. Zheng, X. Xu, E. Dubuisson, Q. Bao, J. Lu, K.P. Loh, Electrochemical delamination of CVD-grown graphene film: toward the recyclable use of copper catalyst. ACS Nano **5**(12), 9927–9933 (2011)
36. H. Zhang, T. Cao, Y. Cheng, Preparation of few-layer graphene nanosheets by radio-frequency induction thermal plasma. Carbon **86**, 38–45 (2015)
37. M. Tian, S. Batty, C. Shang, Synthesis of nanostructured carbons by the microwave plasma cracking of methane. Carbon **51**(1), 243–248 (2013)
38. K.S. Kim, S.H. Hong, K.-S. Lee, W.T. Ju, Continuous synthesis of nanostructured sheetlike carbons by thermal plasma decomposition of methane. IEEE Trans. Plasma Sci. **35**(2), 434–443 (2007)
39. R. Pristavita, J.L. Meunier, D. Berk, Carbon nano-flakes produced by an inductively coupled thermal plasma system for catalyst applications. Plasma Chem. Plasma Process. **31**(2), 393–403 (2011)
40. A. Dato, V. Radmilovic, Z. Lee, J. Phillips, M. Frenklach, Substrate-free gas-phase synthesis of graphene sheets. Nano Lett. **8**(7), 2012–2016 (2008)
41. V. Nicolosi, M. Chhowalla, M.G. Kanatzidis, M.S. Strano, J.N. Coleman, Liquid exfoliation of layered materials. Science **340**(6139), 1226419 (2013)
42. M.S. Dresselhaus, G. Dresselhaus, Intercalation compounds of graphite. Adv. Phys. **30**(2), 139–326 (1981)
43. P. Joensen, R.F. Frindt, S.R. Morrison, Single-layer MoS_2. Mater. Res. Bull. **21**(4), 457–461 (1986)
44. X. Li, G. Zhang, X. Bai, X. Sun, X. Wang, E. Wang, H. Dai, Highly conducting graphene sheets and Langmuir-Blodgett films. Nat. Nanotechnol. **3**(9), 538–542 (2008)
45. C. Valles, C. Drummond, H. Saadaoui, C.A. Furtado, M. He, O. Roubeau, L. Ortolani, M. Monthioux, Alain Pénicaud, Solutions of negatively charged graphene sheets and ribbons. J. Am. Chem. Soc. **130**(47), 15802–15804 (2008)
46. Z. Zeng, T. Sun, J. Zhu, X. Huang, Z. Yin, G. Lu, Z. Fan, Q. Yan, H.H. Hng, H. Zhang, An effective method for the fabrication of few-layer-thick inorganic nanosheets. Angew. Chem. Int. Ed. **51**(36), 9052–9056 (2012)

47. J. Zheng, H. Zhang, S. Dong, Y. Liu, C. Tai Nai, H. Suk Shin, H. Young Jeong, B. Liu, K. Ping Loh, High yield exfoliation of two-dimensional chalcogenides using sodium naphthalenide. Nat. Commun. **5**, 2995 (2014)
48. G. Eda, H. Yamaguchi, D. Voiry, T. Fujita, M. Chen, M. Chhowalla, Photoluminescence from chemically exfoliated MoS_2. Nano Lett. **11**(12), 5111–5116 (2011)
49. Y. Jung, Y. Zhou, J.J. Cha, Intercalation in two-dimensional transition metal chalcogenides. Inorg. Chem. Front. **3**(4), 452–463 (2016)
50. J. Liu, H. Yang, S.G. Zhen, C.K. Poh, A. Chaurasia, J. Luo, X. Wu, E.K.L. Yeow, N.G. Sahoo, J. Lin, Z. Shen, A green approach to the synthesis of high-quality graphene oxide flakes via electrochemical exfoliation of pencil core. RSC Adv. **3**(29), 11745–11750 (2013)
51. Z.Y. Xia, S. Pezzini, E. Treossi, G. Giambastiani, F. Corticelli, V. Morandi, A. Zanelli, V. Bellani, V. Palermo, The exfoliation of graphene in liquids by electrochemical, chemical, and sonication-assisted techniques: a nanoscale study. Adv. Funct. Mater. **23**(37), 4684–4693 (2013)
52. K. Parvez, R. Li, S.R. Puniredd, Y. Hernandez, F. Hinkel, S. Wang, X. Feng, K. Müllen, Electrochemically exfoliated graphene as solution-processable, highly conductive electrodes for organic electronics. ACS Nano **7**(4), 3598–3606 (2013)
53. K. Parvez, Z.-S. Wu, R. Li, X. Liu, R. Graf, X. Feng, K. Müllen, Exfoliation of graphite into graphene in aqueous solutions of inorganic salts. J. Am. Chem. Soc. **136**(16), 6083–6091 (2014)
54. Z. Zeng, Z. Yin, X. Huang, H. Li, Q. He, G. Lu, F. Boey, H. Zhang, Single-layer semiconducting nanosheets: high-yield preparation and device fabrication. Angew. Chem. **50**(47), 11093–11097 (2011)
55. C.-Y. Su, A.-Y. Lu, Y. Xu, F.-R. Chen, A.N. Khlobystov, L.-J. Li, High-quality thin graphene films from fast electrochemical exfoliation. ACS Nano **5**(3), 2332–2339 (2011)
56. N. Liu, F. Luo, H. Wu, Y. Liu, C. Zhang, J. Chen, One-step ionic-liquid-assisted electrochemical synthesis of ionic-liquid-functionalized graphene sheets directly from graphite. Adv. Funct. Mater. **18**, 1518–1525 (2008)
57. J.H. Lee, D.W. Shin, V.G. Makotchenko, A.S. Nazarov, V.E. Fedorov, Y.H. Kim, J.-Y. Choi, J.M. Kim, J.-B. Yoo, One-step exfoliation synthesis of easily soluble graphite and transparent conducting graphene sheets. Adv. Mater. **21**(43), 4383–4387 (2009)
58. S. Stankovich, D.A. Dikin, R.D. Piner, K.A. Kohlhaas, A. Kleinhammes, Y. Jia, Y. Wu, S.T. Nguyen, R.S. Ruoff, Synthesis of graphene-based nanosheets via chemical reduction of exfoliated graphite oxide. Carbon **45**(7), 1558–1565 (2007)
59. G. Eda, M. Chhowalla, Chemically derived graphene oxide: towards large-area thin-film electronics and optoelectronics. Adv. Mater. **22**(22), 2392–2415 (2010)
60. C. Mattevi, G. Eda, S. Agnoli, S. Miller, K.A. Mkhoyan, O. Celik, D. Mastrogiovanni, G. Granozzi, E. Garfunkel, M. Chhowalla, Evolution of electrical, chemical, and structural properties of transparent and conducting chemically derived graphene thin films. Adv. Funct. Mater. **19**(16), 2577–2583 (2009)
61. G. Eda, G. Fanchini, M. Chhowalla, Large-area ultrathin films of reduced graphene oxide as a transparent and flexible electronic material. Nat. Nanotechnol. **3**(5), 270–274 (2008)
62. S. Park, R.S. Ruoff, Chemical methods for the production of graphenes. Nat. Nanotechnol. **4**(4), 217–224 (2009)
63. K.P. Loh, Q. Bao, P.K. Ang, J. Yang, The chemistry of graphene. J. Mater. Chem. **20**(12), 2277 (2010)
64. D.R. Dreyer, S. Park, C.W. Bielawski, R.S. Ruoff, The chemistry of graphene oxide. Chem. Soc. Rev. **39**(1), 228–240 (2010)
65. W.S. Hummers Jr., R.E. Offeman, Preparation of graphitic oxide. J. Am. Chem. Soc. **80**(6), 1339 (1958)
66. A. Buchsteiner, A. Lerf, J. Pieper, Water dynamics in graphite oxide investigated with neutron scattering. J. Phys. Chem. B **110**(45), 22328–22338 (2006)
67. Y. Si, E.T. Samulski, Synthesis of water soluble graphene. Nano Lett. **8**(6), 1679–1682 (2008)

68. S. Wang, P.K. Ang, Z. Wang, A.L.L. Tang, J.T.L. Thong, K.P. Loh, High mobility, printable, and solution-processed graphene electronics. Nano Lett. **10**(1), 92–98 (2010)
69. L.T. Le, M.H. Ervin, H. Qiu, B.E. Fuchs, W.Y. Lee, Graphene supercapacitor electrodes fabricated by inkjet printing and thermal reduction of graphene oxide. Electrochem. Commun. **13**(4), 355–358 (2011)
70. J.D. Fowler, M.J. Allen, V.C. Tung, Y. Yang, R.B. Kaner, B.H. Weiller, Practical chemical sensors from chemically derived graphene. ACS Nano **3**(2), 301–306 (2009)
71. J. Wang, M. Liang, Y. Fang, T. Qiu, J. Zhang, L. Zhi, Rod-coating: towards large-area fabrication of uniform reduced graphene oxide films for flexible touch screens. Adv. Mater. **24**(21), 2874–2878 (2012)
72. Y. Yang, Z. Liu, Z. Yin, Z. Du, L. Xie, M. Yi, J. Liu, W. Huang, Rod-coating all-solution fabrication of double functional graphene oxide films for flexible alternating current (AC)-driven light-emitting diodes. RSC Adv. **4**(98), 55671–55676 (2014)
73. S. Stankovich, D.A. Dikin, G.H.B. Dommett, K.M. Kohlhaas, E.J. Zimney, E.A. Stach, R.D. Piner, S.T. Nguyen, R.S. Ruoff, Graphene-based composite materials. Nature **442**(7100), 282–286 (2006)
74. J. Xu, J. Liu, S. Wu, Q.-H. Yang, P. Wang, Graphene oxide mode-locked femtosecond erbium-doped fiber lasers. Opt. Express **20**(14), 15474–15480 (2012)
75. Y.J. Noh, H.-I. Joh, J. Yu, S.H. Hwang, S. Lee, C.H. Lee, S.Y. Kim, J.R. Youn, Ultra-high dispersion of graphene in polymer composite via solvent free fabrication and functionalization. Sci. Rep. **5**, 9141 (2015)
76. R. Sengupta, M. Bhattacharya, S. Bandyopadhyay, A.K. Bhowmick, A review on the mechanical and electrical properties of graphite and modified graphite reinforced polymer composites. Prog. Polym. Sci. **36**(5), 638–670 (2011)
77. S. Park, J. An, J.R. Potts, A. Velamakanni, S. Murali, R.S. Ruoff, Hydrazine-reduction of graphite- and graphene oxide. Carbon **49**(9), 3019–3023 (2011)
78. S. Pei, H.M. Cheng, The reduction of graphene oxide. Carbon **50**(9), 3210–3228 (2012)
79. Y. Zhou, Q. Bao, L.A.L. Tang, Y. Zhong, K.P. Loh, Hydrothermal dehydration for the "green" reduction of exfoliated graphene oxide to graphene and demonstration of tunable optical limiting properties. Chem. Mater. **21**(13), 2950–2956 (2009)
80. A. Bagri, C. Mattevi, M. Acik, Y.J. Chabal, M. Chhowalla, V.B. Shenoy, Structural evolution during the reduction of chemically derived graphene oxide. Nat. Chem. **2**(7), 581–587 (2010)
81. Y.B. Tan, J.-M. Lee, Graphene for supercapacitor applications. J. Mater. Chem. A **1**(47), 14814 (2013)
82. N. Yang, J. Zhai, D. Wang, Y. Chen, L. Jiang, Two-dimensional graphene bridges enhanced photoinduced charge transport in dye-sensitized solar cells. ACS Nano **4**(2), 887–894 (2010)
83. K.P. Loh, Q. Bao, G. Eda, M. Chhowalla, Graphene oxide as a chemically tunable platform for optical applications. Nat. Chem. **2**(12), 1015–1024 (2010)
84. K.K.H. De Silva, H.-H. Huang, R.K. Joshi, M. Yoshimura, Chemical reduction of graphene oxide using green reductants. Carbon **119**, 190–199 (2017)
85. L.G. Guex, B. Sacchi, K.F. Peuvot, R.L. Andersson, A.M. Pourrahimi, V. Ström, S. Farris, R.T. Olsson, Experimental review: chemical reduction of graphene oxide (GO) to reduced graphene oxide (rGO) by aqueous chemistry. Nanoscale **9**(27), 9562–9571 (2017)
86. F.J. Tölle, M. Fabritius, R. Mülhaupt, Emulsifier-free graphene dispersions with high graphene content for printed electronics and freestanding graphene films. Adv. Funct. Mater. **22**, 1136–1144 (2012)
87. T. Hasan, V. Scardaci, P. Tan, A.G. Rozhin, W.I. Milne, A.C. Ferrari, Stabilization and "debundling" of single-wall carbon nanotube dispersions in N-Methyl-2-pyrrolidone (NMP) by polyvinylpyrrolidone (PVP). J. Phys. Chem. C **111**(34), 12594–12602 (2007)
88. T. Hasan, Z. Sun, F. Wang, F. Bonaccorso, P.H. Tan, A.G. Rozhin, A.C. Ferrari, Nanotube-polymer composites for ultrafast photonics. Adv. Mater. **21**(38), 3874–3899 (2009)

89. Y. Hernandez, V. Nicolosi, M. Lotya, F.M. Blighe, Z. Sun, S. De, I.T. McGovern, B. Holland, M. Byrne, Y.K. Gun'Ko, J.J. Boland, P. Niraj, G. Duesberg, S. Krishnamurthy, R. Goodhue, J. Hutchison, V. Scardaci, A.C. Ferrari, J.N. Coleman, High-yield production of graphene by liquid-phase exfoliation of graphite. Nat. Nanotechnol. **3**(9), 563–568 (2008)
90. J.N. Coleman, Liquid-phase exfoliation of nanotubes and graphene. Adv. Funct. Mater. **19**(23), 3680–3695 (2009)
91. F. Bonaccorso, A. Bartolotta, J.N. Coleman, C. Backes, 2D-crystal-based functional inks. Adv. Mater. **28**(29), 6136–6166 (2016)
92. J.N. Coleman, Liquid exfoliation of defect-free graphene. Acc. Chem. Res. **46**(1), 14–22 (2013)
93. J.N. Coleman, M. Lotya, A. O'Neill, S.D. Bergin, P.J. King, U. Khan, K. Young, A. Gaucher, S. De, R.J. Smith, I.V. Shvets, S.K. Arora, G. Stanton, H.-Y. Kim, K. Lee, G.T. Kim, G.S. Duesberg, T. Hallam, J.J. Boland, J.J. Wang, J.F. Donegan, J.C. Grunlan, G. Moriarty, A. Shmeliov, R.J. Nicholls, J.M. Perkins, E.M. Grieveson, K. Theuwissen, D.W. McComb, P.D. Nellist, V. Nicolosi, Two-dimensional nanosheets produced by liquid exfoliation of layered materials. Science **331**(6017), 568–571 (2011)
94. D. Hanlon, C. Backes, E. Doherty, C.S. Cucinotta, N.C. Berner, C. Boland, K. Lee, A. Harvey, P. Lynch, Z. Gholamvand, S. Zhang, K. Wang, G. Moynihan, A. Pokle, Q.M. Ramasse, N. McEvoy, W.J. Blau, J. Wang, G. Abellan, F. Hauke, A. Hirsch, S. Sanvito, D.D. O'Regan, G.S. Duesberg, V. Nicolosi, J.N. Coleman, Liquid exfoliation of solvent-stabilized few-layer black phosphorus for applications beyond electronics. Nat. Commun. **6**, 8563 (2015)
95. T. Hasan, F. Torrisi, Z. Sun, D. Popa, V. Nicolosi, G. Privitera, F. Bonaccorso, A.C. Ferrari, Solution-phase exfoliation of graphite for ultrafast photonics. Phys. Status Solidi B **247**(11–12), 2953–2957 (2010)
96. S. Santra, G. Hu, R.C.T. Howe, A. De Luca, S.Z. Ali, F. Udrea, J.W. Gardner, S.K. Ray, P.K. Guha, T. Hasan, CMOS integration of inkjet-printed graphene for humidity sensing. Sci. Rep. **5**, 17374 (2015)
97. R.C.T. Howe, G. Hu, Z. Yang, T. Hasan, Functional inks of graphene, metal dichalcogenides and black phosphorus for photonics and (opto)electronics. Proc. SPIE **9553**, 95530R (2015)
98. Z. Sun, T. Hasan, F. Torrisi, D. Popa, G. Privitera, F. Wang, F. Bonaccorso, D.M. Basko, A.C. Ferrari, Graphene mode-locked ultrafast laser. ACS Nano **4**(2), 803–810 (2010)
99. A. Harvey, J.B. Boland, I. Godwin, A.G. Kelly, B.M. Szydłowska, G. Murtaza, A. Thomas, D.J. Lewis, P. O'Brien, J.N. Coleman, Exploring the versatility of liquid phase exfoliation: producing 2D nanosheets from talcum powder, cat litter and beach sand. 2D Mater. **4**(2), 25054 (2017)
100. J.M. Hughes, D. Aherne, J.N. Coleman, Generalizing solubility parameter theory to apply to one- and two-dimensional solutes and to incorporate dipolar interactions. J. Appl. Polym. Sci. **127**(6), 4483–4491 (2013)
101. K.R. Paton, E. Varrla, C. Backes, R.J. Smith, U. Khan, A. O'Neill, C. Boland, M. Lotya, O.M. Istrate, P. King, T. Higgins, S. Barwich, P. May, P. Puczkarski, I. Ahmed, M. Moebius, H. Pettersson, E. Long, J. Coelho, S.E. O'Brien, E.K. McGuire, B.M. Sanchez, G.S. Duesberg, N. McEvoy, T.J. Pennycook, C. Downing, A. Crossley, V. Nicolosi, J.N. Coleman, Scalable production of large quantities of defect-free few-layer graphene by shear exfoliation in liquids. Nat. Mater. **13**(6), 624–630 (2014)
102. E. Varrla, C. Backes, K.R. Paton, A. Harvey, Z. Gholamvand, J. McCauley, J.N. Coleman, Large-scale production of size-controlled MoS_2 nanosheets by shear exfoliation. Chem. Mater. **27**(3), 1129–1139 (2015)
103. E. Varrla, K.R. Paton, C. Backes, A. Harvey, R.J. Smith, J. McCauley, J.N. Coleman, Turbulence-assisted shear exfoliation of graphene using household detergent and a kitchen blender. Nanoscale **6**(20), 11810–11819 (2014)
104. J. Shang, F. Xue, E. Ding, The facile fabrication of few-layer graphene and graphite nanosheets by high pressure homogenization. Chem. Commun. **51**(87), 15811–15814 (2015)
105. F. Xue, E. Ding, J. Shang, Efficient exfoliation of molybdenum disulphide nanosheets by a high-pressure homogeniser. Micro Nano Lett. **10**(10), 589–591 (2015)

106. T.J. Nacken, C. Damm, J. Walter, A. Rüger, W. Peukert, Delamination of graphite in a high pressure homogenizer. RSC Adv. **5**(71), 57328–57338 (2015)
107. P.G. Karagiannidis, S.A. Hodge, L. Lombardi, F. Tomarchio, N. Decorde, S. Milana, I. Goykhman, Y. Su, S.V. Mesite, D.N. Johnstone, R.K. Leary, P.A. Midgley, N.M. Pugno, F. Torrisi, A.C. Ferrari, Microfluidization of graphite and formulation of graphene-based conductive inks. ACS Nano **11**(3), 2742–2755 (2017)
108. M.A. Ibrahem, T.-W. Lan, J.K. Huang, Y.-Y. Chen, K.-H. Wei, L.-J. Li, C.W. Chu. High quantity and quality few-layers transition metal disulfide nanosheets from wet-milling exfoliation. RSC Adv. **3**(32), 13193 (2013)
109. W. Zhao, M. Fang, F. Wu, H. Wu, L. Wang, G. Chen, Preparation of graphene by exfoliation of graphite using wet ball milling. J. Mater. Chem. **20**(28), 5817 (2010)
110. Y. Yao, Z. Lin, Z. Li, X. Song, K.-S. Moon, C.-P. Wong, Large-scale production of two-dimensional nanosheets. J. Mater. Chem. **22**(27), 13494 (2012)
111. L.H. Li, Y. Chen, G. Behan, H. Zhang, M. Petravic, A.M. Glushenkov, Large-scale mechanical peeling of boron nitride nanosheets by low-energy ball milling. J. Mater. Chem. **21**(32), 11862 (2011)
112. T.J. Mason, J.P. Lorimer, *Applied Sonochemistry* (Wiley-VCH, Weinheim, 2002)
113. S.Y. Tang, P. Shridharan, M. Sivakumar, Impact of process parameters in the generation of novel aspirin nanoemulsions - comparative studies between ultrasound cavitation and microfluidizer. Ultrason. Sonochem. **20**(1), 485–497 (2013)
114. A. Posch, *2D PAGE: Sample Preparation and Fractionation* (Humana Press, Clifton, 2008)
115. T. Panagiotou, S.V. Mesite, J.M. Bernard, K.J. Chomistek, R.J. Fisher, Production of polymer nanosuspensions using microfluidizer processor based technologies, in *NSTI-Nanotech 2008*, vol. 1 (2008), pp. 688–691
116. T. Lajunen, K. Hisazumi, T. Kanazawa, H. Okada, Y. Seta, M. Yliperttula, A. Urtti, Y. Takashima, Topical drug delivery to retinal pigment epithelium with microfluidizer produced small liposomes. Eur. J. Pharm. Sci. **62**, 23–32 (2014)
117. S.M. Jafari, Y. He, B. Bhandari, Production of sub-micron emulsions by ultrasound and microfluidization techniques. J. Food Eng. **82**(4), 478–488 (2007)
118. D. Lee, B. Lee, K.H. Park, H.J. Ryu, S. Jeon, S.H. Hong, Scalable exfoliation process for highly soluble boron nitride nanoplatelets by hydroxide-assisted ball milling. Nano Lett. **15**(2), 1238–1244 (2015)
119. T. Hasan, F. Torrisi, Z. Sun, D. Popa, V. Nicolosi, G. Privitera, F. Bonaccorso, A.C. Ferrari, Solution-phase exfoliation of graphite for ultrafast photonics. Phys. Status Solidi **247**(11), 2953–2957 (2010)
120. Y. Hernandez, M. Lotya, D. Rickard, S.D. Bergin, J.N. Coleman, Measurement of multi-component solubility parameters for graphene facilitates solvent discovery. Langmuir **26**(5), 3208–3213 (2010)
121. C.M. Hansen, *Hansen Solubility Parameters: A User's Handbook* (CRC Press, West Palm Beach, 2007)
122. A. Ciesielski, P. Samorì, Graphene via sonication assisted liquid-phase exfoliation. Chem. Soc. Rev. **43**(1), 381–398 (2014)
123. C.L. Yaws, *The Yaws Handbook of Physical Properties for Hydrocarbons and Chemicals*, 2nd edn. (Elsevier Science, New York , 2015)
124. M. Yi, Z. Shen, S. Ma, X. Zhang, A mixed-solvent strategy for facile and green preparation of graphene by liquid-phase exfoliation of graphite. J. Nanoparticle Res. **14**(8), 1003 (2012)
125. K.-G. Zhou, N.-N. Mao, H.-X. Wang, Y. Peng, H.-L. Zhang, A mixed-solvent strategy for efficient exfoliation of inorganic graphene analogues. Angew. Chem. **50**(46), 10839–10842 (2011)
126. R.C.T. Howe, F. Torrisi, F. Tomarchio, S. Mignuzzi, A.C. Ferrari, T. Hasan, Large-scale exfoliation of molybdenum disulphide in solvent mixtures, in *ImagineNano* (2013)

127. E.B. Secor, P.L. Prabhumirashi, K. Puntambekar, M.L. Geier, M.C. Hersam, Inkjet printing of high conductivity, flexible graphene patterns. J. Phys. Chem. Lett. **4**(8), 1347–1351 (2013)
128. E.B. Secor, B.Y. Ahn, T.Z. Gao, J.A. Lewis, M.C. Hersam, Rapid and versatile photonic annealing of graphene inks for flexible printed electronics. Adv. Mater. **27**(42), 6683–6688 (2015)
129. D. Dodoo-Arhin, R.C.T. Howe, G. Hu, Y. Zhang, P. Hiralal, A. Bello, G. Amaratunga, T. Hasan, Inkjet-printed graphene electrodes for dye-sensitized solar cells. Carbon **105**, 33–41 (2016)
130. F. Bonaccorso, Z. Sun, Solution processing of graphene, topological insulators and other 2D crystals for ultrafast photonics. Opt. Mater. Express **4**(1), 63–78 (2014)
131. H.-J. Butt, K. Graff, M. Kappl, *Physics and Chemistry of Interfaces*, 3rd edn. (Wiley-VCH, Weinheim, 2013)
132. M.J. Rosen, J.T. Kunjappu, *Surfactants and Interfacial Phenomena*, 4th edn. (Wiley, Hoboken, 2012)
133. R.J. Smith, P.J. King, M. Lotya, C. Wirtz, U. Khan, S. De, A. O'Neill, G.S. Duesberg, J.C. Grunlan, G. Moriarty, J. Chen, J. Wang, A.I. Minett, V. Nicolosi, J.N. Coleman, Large-scale exfoliation of inorganic layered compounds in aqueous surfactant solutions. Adv. Mater. **23**(34), 3944–3948 (2011)
134. R.C.T. Howe, R.I. Woodward, G. Hu, Z. Yang, E.J.R. Kelleher, T. Hasan, Surfactant-aided exfoliation of molybdenum disulfide for ultrafast pulse generation through edge-state saturable absorption. Phys. Status Solidi **253**(5), 911–917 (2016)
135. P. Ramalingam, S.T. Pusuluri, S. Periasamy, R. Veerabahu, J. Kulandaivel, Role of deoxy group on the high concentration of graphene in surfactant/water media. RSC Adv. **3**, 2369 (2013)
136. M. Lotya, Y. Hernandez, P.J. King, R.J. Smith, V. Nicolosi, L.S. Karlsson, F.M. Blighe, S. De, Z. Wang, I.T. McGovern, G.S. Duesberg, J.N. Coleman, Liquid phase production of graphene by exfoliation of graphite in surfactant/water solutions. J. Am. Chem. Soc. **131**(10), 3611–3620 (2009)
137. M.S. Kang, K.T. Kim, J.U. Lee, W.H. Jo, Direct exfoliation of graphite using a non-ionic polymer surfactant for fabrication of transparent and conductive graphene films. J. Mater. Chem. C **1**(9), 1870 (2013)
138. A.A. Green, M.C. Hersam, Solution phase production of graphene with controlled thickness via density differentiation. Nano Lett. **9**(12), 4031–4036 (2009)
139. M.S. Arnold, S.I. Stupp, M.C. Hersam, Enrichment of single-walled carbon nanotubes by diameter in density gradients. Nano Lett. **5**(4), 713–718 (2005)
140. G. Hu, T. Albrow-Owen, X. Jin, A. Ali, G. Hu, C.T. Richard, Z. Yang, X. Zhu, R. Woodward, T.-C. Wu, H. Jussila, P. Tan, Z. Sun, E. Kelleher, Y. Xu, M. Zhang, Black phosphorus ink formulation for inkjet printing of optoelectronics and photonics. Nat. Commun. **8**, 278 (2017)
141. U. Khan, H. Porwal, A. O'Neill, K. Nawaz, P. May, J.N. Coleman, Solvent-exfoliated graphene at extremely high concentration. Langmuir **27**(15), 9077–9082 (2011)
142. J.N. Coleman, Liquid exfoliation of defect-free graphene. Acc. Chem. Res. **46**(1), 14–22 (2013)
143. A. Ciesielski, P. Samor, Graphene via sonication assisted liquid-phase exfoliation. Chem. Soc. Rev. **43**(1), 381–398 (2014)
144. J.N. Coleman, Liquid-phase exfoliation of nanotubes and graphene. Adv. Funct. Mater. **19**(23), 3680–3695 (2009)
145. R. Marchesini, A. Bertoni, S. Andreola, E. Melloni, A.E. Sichirollo, Extinction and absorption coefficients and scattering phase functions of human tissues in vitro. Appl. Opt. **28**(12), 2318 (1989)
146. L. Yang, *Materials Characterization: Introduction to Microscopic and Spectroscopic Methods*, 2nd edn. (Wiley, Hoboken, 2009)

147. C. Backes, R.J. Smith, N. McEvoy, N.C. Berner, D. McCloskey, H.C. Nerl, A. O'Neill, P.J. King, T. Higgins, D. Hanlon, N. Scheuschner, J. Maultzsch, L. Houben, G.S. Duesberg, J.F. Donegan, V. Nicolosi, J.N. Coleman, Edge and confinement effects allow in situ measurement of size and thickness of liquid-exfoliated nanosheets. Nat. Commun. **5**, 4576 (2014)
148. K.S. Aneja, S. Bohm, A.S. Khanna, H.L. Mallika Bohm, Graphene based anticorrosive coatings for Cr(VI) replacement. Nanoscale **7**(42), 17879–17888 (2015)
149. R. Erni, M.D. Rossell, C. Kisielowski, Ulrich Dahmen, Atomic-resolution imaging with a Sub-50-pm electron probe. Phys. Rev. Lett. **102**(9), 096101 (2009)
150. N. Wang, Q. Xu, S. Xu, Y. Qi, M. Chen, H. Li, High-efficiency exfoliation of layered materials into 2D nanosheets in switchable CO_2/Surfactant/H_2O system. Sci. Rep. **5**, 16764 (2015)
151. F. Torrisi, T. Hasan, W. Wu, Z. Sun, A. Lombardo, T.S. Kulmala, G.-W. Hsieh, S. Jung, F. Bonaccorso, P.J. Paul, D. Chu, A.C. Ferrari, Inkjet-printed graphene electronics. ACS Nano **6**(4), 2992–3006 (2012)***
152. J.C. Meyer, A.K. Geim, M.I. Katsnelson, K.S. Novoselov, D. Obergfell, S. Roth, C. Girit, A. Zettl, On the roughness of single- and bi-layer graphene membranes. Solid State Commun. **143**(1–2), 101–109 (2007)
153. J.C. Meyer, A.K. Geim, M.I. Katsnelson, K.S. Novoselov, T.J. Booth, S. Roth, The structure of suspended graphene sheets. Nature **446**(7131), 60–63 (2007)
154. A.C. Ferrari, J.C. Meyer, V. Scardaci, C. Casiraghi, M. Lazzeri, F. Mauri, S. Piscanec, D. Jiang, K.S. Novoselov, S. Roth, A.K. Geim, Raman spectrum of graphene and graphene layers. Phys. Rev. Lett. **97**(18), 187401 (2006)
155. S. Zhang, J. Yang, R. Xu, F. Wang, W. Li, M. Ghufran, Y.-W. Zhang, Z. Yu, G. Zhang, Q. Qin, Y. Lu, Extraordinary photoluminescence and strong temperature/angle-dependent Raman responses in few-layer phosphorene. ACS Nano **8**(9), 9590–9596 (2014)
156. H. Li, Q. Zhang, C.C.R. Yap, B.K. Tay, T.H.T. Edwin, A. Olivier, D. Baillargeat, From bulk to monolayer MoS_2: evolution of Raman scattering. Adv. Funct. Mater. **22**(7), 1385–1390 (2012)
157. C. Backes, K.R. Paton, D. Hanlon, S. Yuan, M.I. Katsnelson, J. Houston, R.J. Smith, D. McCloskey, J.F. Donegan, J.N. Coleman, Spectroscopic metrics allow in situ measurement of mean size and thickness of liquid-exfoliated few-layer graphene nanosheets. Nanoscale **8**(7), 4311–4323 (2016)
158. A.C Ferrari, D.M. Basko, Raman spectroscopy as a versatile tool for studying the properties of graphene. Nat. Nanotechnol. **8**(4), 235–246 (2013)
159. K.N. Kudin, B. Ozbas, H.C. Schniepp, R.K. Prud'homme, I.A. Aksay, R. Car, Raman spectra of graphite oxide and functionalized graphene sheets. Nano Lett. **8**(1), 36–41 (2008)
160. X. Zhang, X.-F. Qiao, W. Shi, J.-B. Wu, D.-S. Jiang, P.-H. Tan, Phonon and Raman scattering of two-dimensional transition metal dichalcogenides from monolayer, multilayer to bulk material. Chem. Soc. Rev. **44**(9), 2757–2785 (2015)
161. C. Lee, H. Yan, L.E. Brus, T.F. Heinz, J. Hone, S. Ryu, Anomalous lattice vibrations of single- and few-layer MoS_2. ACS Nano **4**(5), 2695–2700 (2010)
162. Z. Guo, H. Zhang, S. Lu, Z. Wang, S. Tang, J. Shao, Z. Sun, H. Xie, H. Wang, X.-F. Yu, P.K. Chu, From black phosphorus to phosphorene: basic solvent exfoliation, evolution of Raman scattering, and applications to ultrafast photonics. Adv. Funct. Mater. **25**(45), 6996–7002 (2015)
163. L.G. Cançado, A. Jorio, E.H.M. Ferreira, F. Stavale, C.A. Achete, R.B. Capaz, M.V.O. Moutinho, A. Lombardo, T.S. Kulmala, A.C. Ferrari, Quantifying defects in graphene via Raman spectroscopy at different excitation energies. Nano Lett. **11**(8), 3190–3196 (2011)
164. A.C. Ferrari, Raman spectroscopy of graphene and graphite: disorder, electron-phonon coupling, doping and nonadiabatic effects. Solid State Commun. **143**(1–2), 47–57 (2007)
165. B. Chakraborty, A. Bera, D.V.S. Muthu, S. Bhowmick, U.V. Waghmare, A.K. Sood, Symmetry-dependent phonon renormalization in monolayer MoS_2 transistor. Phys. Rev. B: Condens. Matter Mater. Phys. **85**(16), 161403(r) (2012)

166. Y. Liu, Z. Liu, W.S. Lew, Q.J. Wang, Temperature dependence of the electrical transport properties in few-layer graphene interconnects. Nanoscale Res. Lett. **8**(1), 335 (2013)
167. D. Graf, F. Molitor, K. Ensslin, C. Stampfer, A. Jungen, C. Hierold, L. Wirtz, Spatially resolved Raman spectroscopy of single- and few-layer graphene. Nano Lett. **7**(2), 238–242 (2007)
168. A.C. Ferrari, J.C. Meyer, V. Scardaci, C. Casiraghi, M. Lazzeri, F. Mauri, S. Piscanec, D. Jiang, K.S. Novoselov, S. Roth, A.K. Geim, Raman spectrum of graphene and graphene layers. Phys. Rev. Lett. **97**(18), 187401 (2006)
169. X.-L. Li, W.-P. Han, J.-B. Wu, X.-F. Qiao, J. Zhang, P.-H. Tan, Layer-number dependent optical properties of 2D materials and their application for thickness determination. Adv. Funct. Mater. **27**(19), 1604468 (2017)
170. F. Torrisi, T. Hasan, W.P. Wu, Z.P. Sun, A. Lombardo, T.S. Kulmala, G.W. Hsieh, S.J. Jung, F. Bonaccorso, P.J. Paul, D.P. Chu, A.C. Ferrari, Inkjet-printed graphene electronics. ACS Nano **6**(4), 2992–3006 (2012)
171. C. Casiraghi, Raman spectroscopy of graphene edges. Nano Lett. **9**(4), 1433–1441 (2009)
172. U. Khan, A. O'Neill, H. Porwal, P. May, K. Nawaz, J.N. Coleman, Size selection of dispersed, exfoliated graphene flakes by controlled centrifugation. Carbon **50**(2), 470–475 (2012)
173. W. Zhao, Z. Ghorannevis, K.K. Amara, J.R. Pang, Lattice dynamics in mono-and few-layer sheets of WS_2 and WSe_2. Nanoscale **5**(20), 9677–9683 (2013)
174. X. Zhang, W.P. Han, J.B. Wu, S. Milana, Y. Lu, Q.Q. Li, A.C. Ferrari, P.H. Tan, Raman spectroscopy of shear and layer breathing modes in multilayer MoS_2. Phys. Rev. B: Condens. Matter Mater. Phys. **87**(11), 115413 (2013)
175. M. Zhang, R.C.T. Howe, R.I. Woodward, E.J.R. Kelleher, F. Torrisi, G. Hu, S.V. Popov, J.R. Taylor, T. Hasan, Solution processed MoS_2-PVA composite for sub-bandgap mode-locking of a wideband tunable ultrafast Er:fiber laser. Nano Res. **8**(5), 1522–1534 (2015)
176. L. Liang, V. Meunier, First-principles Raman spectra of MoS_2, WS_2 and their heterostructures. Nanoscale **6**(10), 5394–5401 (2014)
177. D. Li, H. Jussila, L. Karvonen, G. Ye, H. Lipsanen, X. Chen, Z. Sun, Polarization and thickness dependent absorption properties of black phosphorus: new saturable absorber for ultrafast pulse generation. Sci. Rep. **5**, 15899 (2015)
178. A. Castellanos-Gomez, L. Vicarelli, E. Prada, J.O. Island, K.L. Narasimha-Acharya, S.I. Blanter, D.J. Groenendijk, M. Buscema, G.A. Steele, J.V. Alvarez, H.W. Zandbergen, J.J. Palacios, H.S.J. van der Zant, Isolation and characterization of few-layer black phosphorus. 2D Mater. **1**(2), 025001 (2014)
179. X. Wang, A.M. Jones, K.L. Seyler, V. Tran, Y. Jia, H. Zhao, H. Wang, L. Yang, X. Xu, F. Xia, Highly anisotropic and robust excitons in monolayer black phosphorus. Nat. Nanotechnol. **10**(6), 517–521 (2015)
180. H. Yang, H. Jussila, A. Autere, H.-P. Komsa, G. Ye, X. Chen, T. Hasan, Z. Sun, Optical waveplates based on birefringence of anisotropic two-dimensional layered materials. ACS Photon. **4**(12), 3023–3030 (2017)
181. R.V. Gorbachev, I. Riaz, R.R. Nair, R. Jalil, L. Britnell, B.D. Belle, E.W. Hill, K.S. Novoselov, K. Watanabe, T. Taniguchi, A.K. Geim, P. Blake, Hunting for monolayer boron nitride: optical and Raman signatures. Small **7**(4), 465–468 (2011)
182. M. Acik, G. Lee, C. Mattevi, M. Chhowalla, K. Cho, Y.J. Chabal, Unusual infrared-absorption mechanism in thermally reduced graphene oxide. Nat. Mater. **9**(10), 840–845 (2010)
183. S. Abdolhosseinzadeh, H. Asgharzadeh, H. Seop Kim, Fast and fully-scalable synthesis of reduced graphene oxide. Sci. Rep. **5**, 10160 (2015)
184. S.N. Alam, N. Sharma, L. Kumar, Synthesis of graphene oxide (GO) by modified Hummers method and its thermal reduction to obtain reduced graphene oxide (rGO). J. Graphene **6**(1), 73348 (2017)
185. A. Ganguly, S. Sharma, P. Papakonstantinou, J. Hamilton, Probing the thermal deoxygenation of graphene oxide using high-resolution in situ X-ray-based spectroscopies. J. Phys. Chem. C **115**(34), 17009–17019 (2011)

186. S. Yumitori, Correlation of C_{1s} chemical state intensities with the O_{1s} intensity in the XPS analysis of anodically oxidized glass-like carbon samples. J. Mater. Sci. **35**(1), 139–146 (2000)
187. K.P. Dhakal, D.L. Duong, J. Lee, H. Nam, M. Kim, M. Kan, Y.H. Lee, J. Kim, Confocal absorption spectral imaging of MoS_2: optical transitions depending on the atomic thickness of intrinsic and chemically doped MoS_2. Nanoscale **6**(21), 13028–13035 (2014)
188. G. Eda, Y.Y. Lin, C. Mattevi, H. Yamaguchi, H.A. Chen, I.S. Chen, C.W. Chen, M. Chhowalla, Blue photoluminescence from chemically derived graphene oxide. Adv. Mater. **22**(4), 505–509 (2010)
189. T. Gokus, R.R. Nair, A. Bonetti, M. Böhmler, A. Lombardo, K.S. Novoselov, A.K. Geim, A.C. Ferrari, A. Hartschuh, Making graphene luminescent by oxygen plasma treatment. ACS Nano **3**(12), 3963–3968 (2009)
190. C. Kittel, *Introduction to Solid State Physics*, 8th edn. (Wiley, New York, 2004)
191. F.T. Johra, J.W. Lee, W.G. Jung, Facile and safe graphene preparation on solution based platform. J. Ind. Eng. Chem. **20**(5), 2883–2887 (2014)

Chapter 4
2D Ink Design

Abstract The use of two-dimensional (2D) materials in low-cost, high-throughput printing as a scalable production method relies heavily on its successful incorporation into printable ink systems. However, industrial scale production of 2D material printing inks has proven to be a significant challenge. This chapter begins with a brief history of inks and progresses to the basic ink formulations and the techniques used in traditional printing technologies. This chapter also discusses ink properties, processing techniques and optimisation of formulation parameters using methods such as Design of Experiments (DOE), and useful characterisation techniques when formulating 2D material inks. Finally, the chapter covers ink–substrate interaction and optimisation strategies for ink formulation.

4.1 A Brief History of Inks

The earliest evidence of recognisable printing comes from third-century China in the form of hand-carved wooden blocks for letter reproduction [1]. The inks used at that time were made by combining lamp-black (essentially soot from inefficient burning of lamps, acting as a dark pigment) and gum (tree-resin, acting as a binder) dissolved in water (the simplest of all solvents) [1, 2]. This simple process is the foundation of letterpress relief printing and its usage spread around the world over many centuries [1].

Letterpress printing proved to be very effective in (re)producing standard-sized manuscripts. However, it was a cumbersome process which required hours of skilled carving of wooden blocks which had only a single effective use. Inks printed from wooden blocks could often appear washed out and brown due to the absorbent nature of the wood [1]. In the mid-fifteenth century, German goldsmith Johannes Gutenberg addressed these challenges by inventing a method for mass producing metallic movable type and adjustable moulds to hold the type firmly in place. The printing press he developed was based on agricultural screw presses which was able to utilise these epochal innovations. The finely featured, non-absorbent metal blocks could be rearranged in different sequences and re-used, producing high quality and uniform prints [1]. The process is perhaps best known for the printing of the

L. W. T. Ng et al., *Printing of Graphene and Related 2D Materials*,
https://doi.org/10.1007/978-3-319-91572-2_4

Gutenberg Bible in 1456. As the surface of the type was made of metal, it was unable to use conventional water-based inks [3]. This necessitated the development of oil-based inks, initially formulated from linseed oil, walnut oil, turpentine, rosin and pitch while still using lamp-black as a pigment [1, 2].

Evolutions of these oil-based inks would go on to dominate the printing market for the next several centuries. These high-viscosity paste inks had binders made from drying oils and resins. Some of these formulations are still in use today as 'green' oil-based inks often made from soya oil or linseed oil [1].

The unique properties of oil-based inks also enabled the lithographic printing process, invented in 1790 by the German actor and playwright, Johann Alois Senefelder. Lithography utilises oil-based inks and a planar printing plate with hydrophobic image areas and hydrophilic non-image regions. This allowed the printing of more complex images, in particular pictorial images considered too challenging to achieve with a letterpress. Within the lithographic process the plate is wetted prior to ink application, the oil-based ink repelled by the wetted hydrophilic area and transfers only on to the hydrophobic image portions. This inked image then transfers to the paper substrate. The lithographic process greatly increased the demand for coloured inks and spurred new development of synthetic oils as the ink binders [1].

More recent developments have seen the field of printing processes diversify greatly as different technologies and applications enter the market. The printing industry today comprises of a number of widely used printing processes, including offset lithographic, flexographic gravure, letterpress, screen and inkjet printing. A steady evolution of ink formulation and the development of photocurable binders have allowed the printing on non-porous substrates such as metal substrates and plastics [1, 3]. The increasing demands on the printing industry for ever higher-quality prints have been supported by continuous research into the chemistry and performance of inks. Meanwhile, the reliable and well-developed printing processes are beginning to find new applications in the drive towards low-cost flexible electronics, some of which we summarise in this book.

4.2 Basic Ink Composition and Formulations

A simple ink formulation is typically composed of pigments, a binder system, solvents and various additives [1, 3, 4]. The ink composition is usually adjusted to allow suitable rheological properties for each specific printing technology. Table 4.1 is a basic list of ink composition and viscosity designed for commonly used printing technologies.

In conventional graphics printing, the pigment works as the coloured component to the ink. Typical historical pigments included soot, coloured earth and plant or mineral extracts. More recent inks use a wide variety of coloured inorganic and organic pigments such as carbon black (black), and organic lithol (red), phthalocyanine (green) and indanthrene (blue). The modern industry also uses

Table 4.1 Typical composition and viscosity of inks for graphics printing technologies

Printing method	Ink composition (wt%)				Viscosity (mPa s)
	Pigment	Resin	Solvent	Additive	
Inkjet	5–10	5–20	65–95	1-5	4–30
Flexographic	12–17	40–45	25–45	1-5	1,000–2,000
Gravure	12–17	20–35	60–65	1-2	100–1,000
Screen	12–20	40–65	20–30	1-5	1,000–10,000

The data is collected from Refs. [1, 3–13]

materials designed to impart optical effects such as pearlescence and metallic effects. For example, extenders like calcium carbonate ($CaCO_3$) are transparent pigments that make the colour of other pigments less intense, while opacifiers such as titanium dioxide (TiO_2) are white pigments to improve ink opacity. There are also other pigments that can be used to provide gloss, abrasiveness, or even resistance to certain ambient conditions (e.g. light, heat, solvents, chemicals) [1, 3].

Over the past 20 years, there has been a growing interest in incorporating functional materials, such as conducting, semiconducting and dielectric materials, into the inks as active pigments. A wide range of materials have been studied for this purpose, including metallic [14] and dielectric nanoparticles, organic semiconductors [15, 16], carbon materials (e.g. carbon black, carbon nanotubes (CNTs)) [17–19], and more recently solution-processed 2D materials [20–40]. These materials can lend their functionalities to the prints, allowing the development of diverse applications [21, 41–43].

The binder portion of the ink formulation is essentially polymeric resins to function as film formers to 'bind' the pigments together and provide adhesion of the pigments to the substrate. Commonly used resins include acrylics, alkyds, cellulose and its derivatives, and rubber resins [1, 3]. The binder also contributes to the properties of the final prints, affecting gloss, resistance to weathering, and chemical attack or abrasion. As examples, water insoluble polymers such as ethyl cellulose and epoxies can provide resistance to water damage whereas high glass transition temperature polymers such as polyimides allow resistance to high temperature.

The solvent within the ink formulation is the volatile active component of the ink. Together with the binder, they form the liquid carrier of the other ink components (i.e. pigments, resins and additives) and are sometimes referred to as the ink varnish. The primary function of the solvent is to keep the ink in a liquid form during printing until it is deposited onto the substrate. The type and amount of solvent in the ink usually significantly influences its viscosity and rheology (i.e. the rate of flow under shear). Depending on the binder system, solvents can range from water to a variety of organic solvents [1, 3]. The selection of the solvent and binder depends on the specific printing technologies, substrates and end applications. For instance, printing processes such as solvent-based flexographic printing and gravure printing that run at high speeds require rapid ink drying. Hence, the suitable solvents typically have low boiling points, for instance ethyl acetate (boiling point 77 °C) and isopropanol

(IPA, boiling point 82.6 °C). Other printing processes, such as screen printing, need solvents with moderately low evaporation rates, such as cyclohexanone (boiling point 156 °C) and butoxyethanol (boiling point 171 °C).

Additives are used to alter the properties of the inks and the printed films. Examples of additives include some types of surfactants that can be used to improve the wetting of either the pigment or the substrate. Alkalis can also be introduced into water-based inks to develop a mildly basic pH to dissolve polymer resins (e.g. cellulose), such that the printed films can resist moisture after they solidify. Additives can also include materials such as wax that can form a protective layer to improve the rub or scuff resistance of the printed film, and defoamers (typically silicone based) that can prevent bubble formation during printing [1]. They are added only in small proportions, but can have a significant influence on the printability of the ink, and the final characteristics of the dried film.

The inks can be broadly categorised into two general groups based on their viscosities:

- Paste inks: High viscosity inks used for offset, screen and letterpress printing;
- Liquid inks: Low viscosity inks used for gravure, inkjet and flexographic printing;

Paste inks are high viscosity colloidal suspensions that are non-Newtonian in behaviour, meaning that shear force must be applied before printing to make the ink 'flow'. The high viscosity of paste inks typically prevents pigment aggregation or agglomeration, provided that a homogeneous dispersion can be achieved during ink formulation [1, 3]. Paste inks may also require the addition of a dryer, which is a catalyst that increases the rate at which the printed film solidifies. In contrast, liquid inks are naturally far lower in viscosity than paste inks and therefore pigments are more prone to agglomeration. In order to prevent formation of aggregations of pigment particles, separating agents may be introduced as part of the additive package to ensure that agglomeration does not take place [1, 4]. Typical formulations for each ink type are shown in Fig. 4.1.

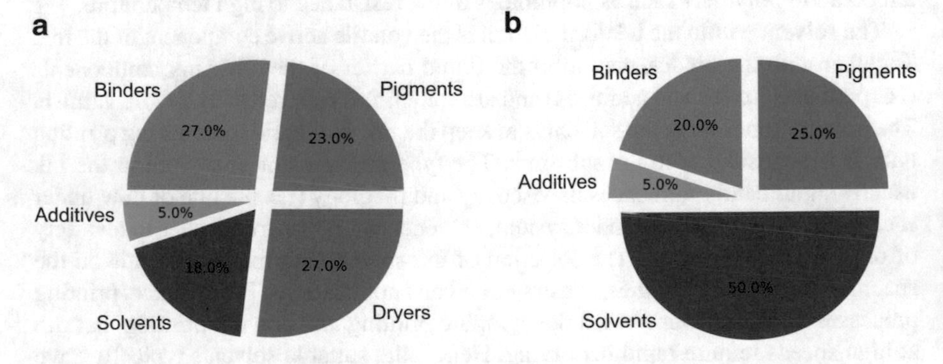

Fig. 4.1 Typical formulation compositions for graphics (**a**) paste inks and (**b**) liquid inks

4.2.1 2D Material Ink Systems

The development of 2D printing ink systems in general has followed a similar three-component package to the above traditional graphics ink systems by simply replacing the pigment component [43]. This allows the inks to behave similarly to the graphics ink systems in printing. 2D materials that have already been introduced to ink systems include graphene [36, 44, 45], reduced graphene oxide (rGO) [46], molybdenum disulphide (MoS_2) [24, 47], hexagonal boron nitride (*h*-BN) [38, 48] and black phosphorus (BP) [38]. As 2D ink systems are designed to have functional rather than graphical properties, characteristics such as gloss and colour strength are usually not vital considerations in the formulation of the ink.

The nature of 2D materials means that their methods of inclusion into ink formulations may sometimes differ from traditional pigmented systems. For example, many 2D materials are not supplied in powdered form unlike traditional forms of pigment. Instead they are supplied either pre-dispersed in a liquid form, or require exfoliation from their bulk. The materials pre-dispersed in liquid may be simply mixed with the ink varnish to produce an ink, although alterations may be required for the ink formulation to maintain the correct solvent-binder balance. 2D materials that come in bulk crystal form can be exfoliated in situ, using the techniques such as liquid phase exfoliation (LPE) [38, 40] described in the previous chapter; with the ink varnish in place of the pure solvents more widely used. Low viscosity liquid inks, such as for inkjet printing, are examples of this. Addition of surfactants or dispersing agents may be necessary to prevent aggregation of the 2D material flakes within the ink system [22, 30, 49–52].

4.3 Viscosity and Rheology of Ink Systems

Having established what constitutes an ink, we will now move on to describe two key parameters of ink formulation that determine the printability of an ink in a print system. The complexity of modern day printing systems generally means that multiple parameters specific to printing processes such as drying time, water pickup and dot gain are used and characterised in varying ways. Although these other parameters play an important role, only ink viscosity and rheology are the two factors generally taken into account during initial ink design.

4.3.1 Ink Viscosity

The viscosity of a system is the interaction of shear, shear stress and shear strain, and may be determined by the rate of flow under the influence of an applied shear stress [1, 53–56]. To help understand the concept of viscosity, it is useful to imagine

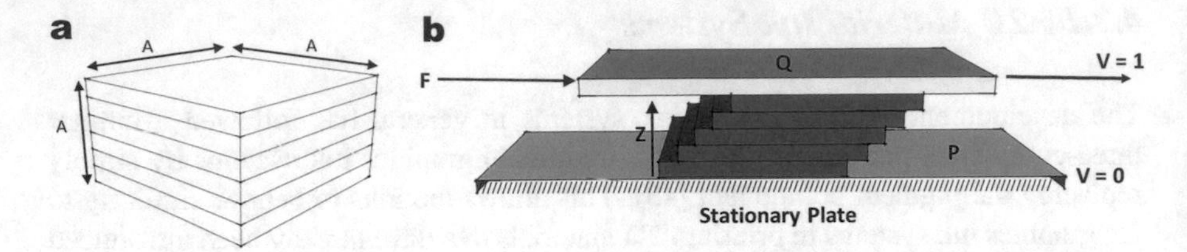

Fig. 4.2 Laminar flow of liquid under shear. (**a**) Multi-layered cube of fluid. (**b**) Flow under constant shear

Table 4.2 Approximate viscosities of common materials at room temperature

Material	Viscosity (mPa s)
Water	1
Milk	3
SAE 10 motor oil	85–140
SAE 20 motor oil	140–420
SAE 30 motor oil	420–650
SAE 40 motor oil	650–900
Castrol oil	1,000
Honey	10,000
Chocolate	25,000
Ketchup	50,000
Mustard	70,000
Sour cream	100,000
Peanut butter	250,000

Table 4.3 Conversion between different unit systems used for viscosity

	Pa s	mPa s	P	cP	S	cS
1 Pa s	–	1,000	10	1,000	10ρ	$1{,}000\rho$
1 mPa s	0.001	–	0.01	1	0.01ρ	ρ
1 P	0.1	100	–	100	ρ	100ρ
1 cP	0.001	1	0.01	–	0.01ρ	ρ
1 S	$\frac{0.1}{\rho}$	$\frac{100}{\rho}$	$\frac{1}{\rho}$	$\frac{100}{\rho}$	–	100
1 cS	$\frac{1{,}000}{\rho}$	$\frac{1}{\rho}$	$\frac{0.01}{\rho}$	$\frac{1}{\rho}$	0.01	–

liquids as a cube of fluid; Fig. 4.2a. A flowing liquid will show a velocity gradient perpendicular to its rate of flow; Fig. 4.2b. This gradient in flow velocity has units $\frac{\mathrm{m\,s^{-1}}}{\mathrm{m}} = \mathrm{s^{-1}}$ and is termed as the shear rate or shear strain. The shear stress is defined as the force/unit area of liquid, and has units N m^{-2} (= Pa). The viscosity is then defined as $\eta = \frac{\text{Shear stress}}{\text{Shear rate}}$, with units Pa s [1, 54, 55, 57–59]. Table 4.2 gives the approximate viscosity of common liquids for reference.

We note that many alternative unit systems have been used to describe viscosity, including 'milli Pascal seconds' (mPa s), poise (P) and centipoise (cP), and Stokes (S) and centiStokes (cS). The Stokes system is related to the other two via the fluid density ρ (g cm^{-3}), and is used by some viscometers as described later in the chapter. Conversion between the different unit systems is shown in Table 4.3.

Characterisation of Ink Viscosity

Measurement of ink viscosity is usually done using simple viscometers that take an immediate viscosity measurement at a defined temperature. Equipment for viscometry is usually low-cost, simple and mechanical, requiring no digital readouts. Hence, these methods are popular for use on production lines for quick and simple measurements. The choice of viscometer depends on the nature or the starting viscosity of the ink. We will discuss three methods of measuring viscosity, namely using flow cups, the capillary viscometer and the falling sphere viscometer.

Flow Cups

One of the most common viscometer types is the flow cup, a schematic and photograph of which are shown in Fig. 4.3a, b [54]. Common flow cups include the Ford, Shell, Zahn and ISO flow cup systems. Flow cups consist of a reservoir of known volume above an aperture of known area. The time taken for the liquid to completely pass through the cavity, typically observed by recording the time at which a constant stream of ink breaks into small droplets [54] is recorded, and converted to a viscosity via conversion tables supplied by the manufacturer. Different sized flow cup apertures may be used depending on the approximate viscosity of the measured liquid. Flow cups are widely used to characterise liquid (e.g. flexographic and gravure) inks, as they provide a rapid and simple measurement technique that can be carried out immediately alongside the printing/ink formulation process; Fig. 4.3b.

Falling Sphere Viscometers

A falling sphere viscometer is another flow-based system, but one that uses an object moving through the fluid, rather than the movement of the fluid itself. The working principle is based on Stokes' law for frictional force which describes the maximum velocity of an object falling under gravity, moving through the resistive force presented by a viscous fluid. In this method, a sphere of known dimension and mass falls between the start and end points of the viscometer; Fig. 4.3c, d, and the time taken is used to measure the viscosity of the liquid [3, 54, 56].

Capillary Viscometers

Capillary viscometers (also known as Ostwalk or U-tube viscometers) consist of a U-shaped tube of known cross-sectional area; Fig. 4.3e, f. They can be used to measure liquids under low-shear conditions. A fixed amount of liquid is drawn up from the reservoir past two marks on the capillary—the start and stop marks. Once the suction is removed, the time taken for the fluid to flow back through the capillary (or the time between the meniscus passing the start and stop marks) is measured,

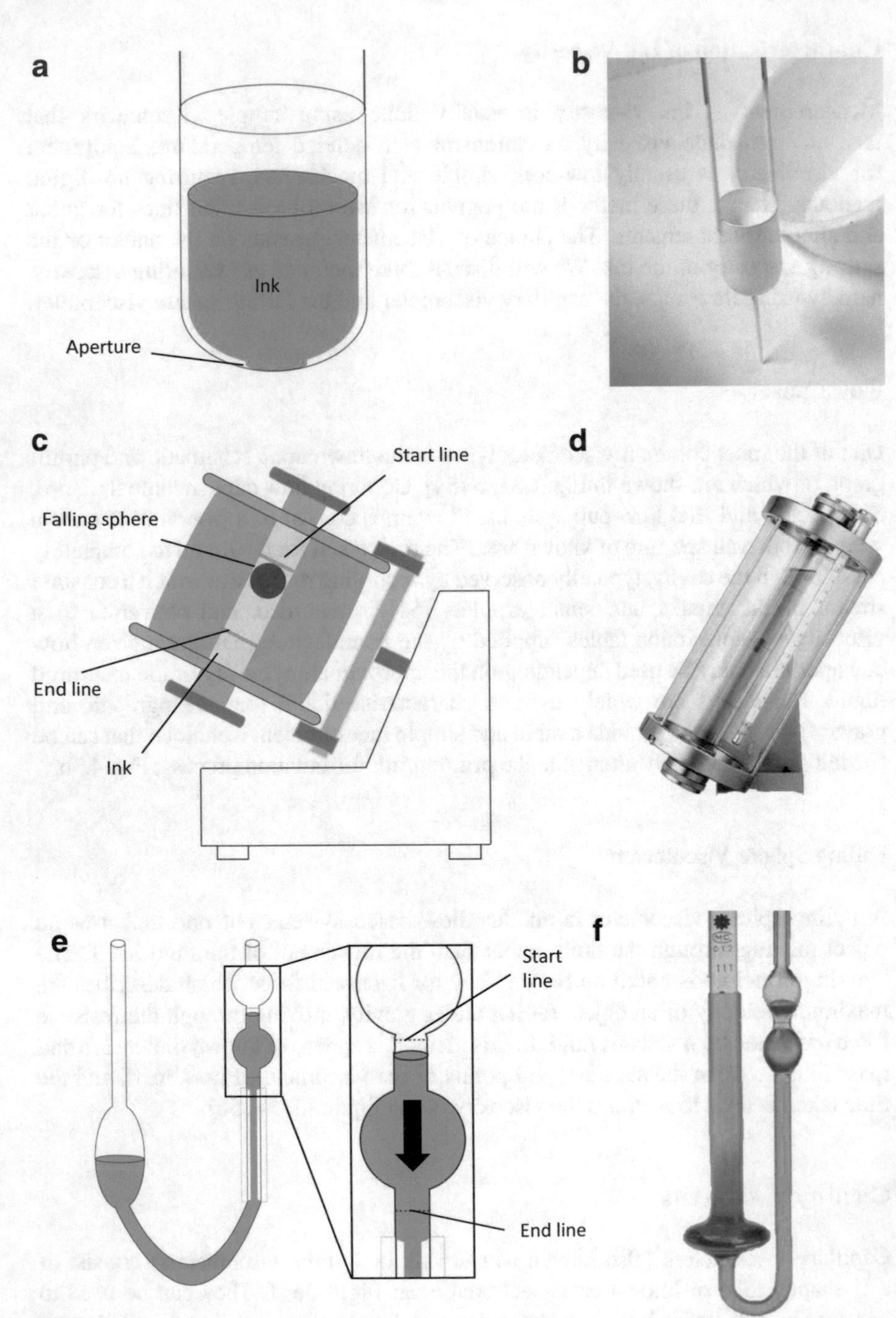

Fig. 4.3 Common viscometers. Schematic of (**a**) a flow cup, (**c**) a falling sphere viscometer and (**e**) a capillary viscometer, with (**b**), (**d**), (**f**) corresponding photographs

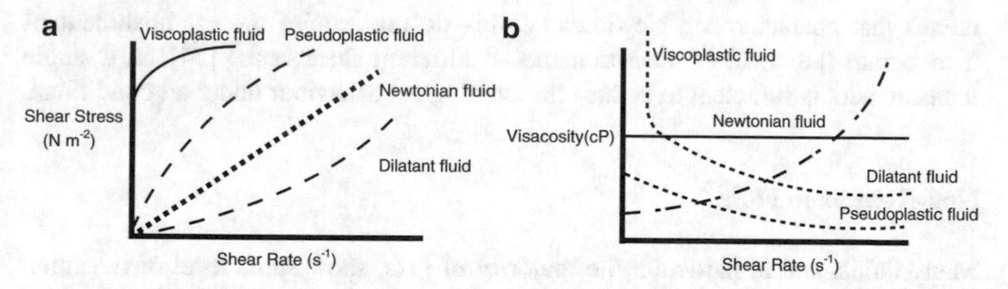

Fig. 4.4 (**a**) Shear stress and shear strain flow curves for different fluids and (**b**) viscosity profiles for different fluid systems

allowing the viscosity to be interpreted. The shear force is dependent on the density of the fluid, and the viscosity is therefore measured in Stokes/centistokes [3, 54, 56, 60].

4.3.2 *Rheological Flow in Different Fluid Systems*

The rheology of an ink can be simply described as how an ink 'flows', or indeed doesn't, at different stages of the printing process. The term 'rheology' was first used by Professor Bingham of Lafayette College, Easton, PA. Bingham's definition of rheology as 'the science of deformation and flow of matter' was largely accepted upon the formation of the American Society of Rheology in 1929 [61]. The rheological profile directly affects the performance of inks on the printing press. It is important to distinguish between the often confused terms of 'viscosity' and 'rheology'.

Rheology is the study of a fluid over time and shear strain and stress whereas viscosity describes the resistance to flow of a liquid at a specific, given point of time [54, 62, 63]. The viscosity of an ink at any given time depends on its prior exposure to external forces, whereas its rheology governs how these external forces affect the ink, and is itself invariant.

The flow properties of a fluid may be represented graphically to depict a rheological relationship between shear stress (σ) and shear strain ($\dot{\gamma}$); Fig. 4.4 [1, 53–57, 59, 64–68].

Newtonian Flow

An ideal fluid (also known as a Newtonian fluid) has a linear relationship between shear stress and shear rate, meaning that the viscosity remains constant for all rates of shear; Fig. 4.4a, b. While no fluid is entirely Newtonian, liquids such as water, solvents and mineral oils can be considered as such for practical purposes [1, 55, 64]. The linear relationship between shear stress and shear strain of Newtonian fluids means that characterising Newtonian fluids do not require the establishment of a rheogram (i.e. a set of measurements at different shear rates) [54], as a single measurement is sufficient to deduce the rheological behaviour under all conditions.

Non-Newtonian Flow

Many fluids, and in particular the majority of inks, show some level of deviation from Newtonian behaviour. Non-Newtonian fluids may be pseudoplastic (shear thinning), dilatant (shear-thickening), viscoplastic (Bingham fluids), thixotropic or rheopectic [53, 56, 58, 59, 65–68]. Deviations from Newtonian behaviour are due to the interactions between the components of a fluid system and the mechanical phenomena (i.e. the emergence or loss of structure within the liquid under the influence of shear).

Pseudoplastic Fluids

Pseudoplastic fluids show reduced shear stress under increasing shear rate (i.e. require less force to maintain flow at higher shear rates), observed as a drop in viscosity; Fig. 4.4b. Such liquids are termed shear-thinning, and include many emulsions, suspensions and dispersions. Shear-thinning may be a desired property in an ink system, as it allows flow through the printing elements (high shear), but prevents movement or smudging of the ink once deposited onto the substrate (low shear). At rest, shear-thinning liquids typically have a viscous, pasty structure due to the entanglement of powders and small particles within the system. Under shear, these components can orientate, stretch, or disentangle, allowing them to slip past one another more easily, leading to the observed reduction in viscosity [3, 53, 57–59, 68, 69].

Dilatant Fluids

Dilatant (shear-thickening) fluids show increased viscosity under high-shear conditions; Fig. 4.4b. Systems exhibiting dilatant behaviour are typically highly concentrated suspensions in colloidal form [3, 53, 56–59, 66–69], with examples including cornstarch in water and silly putty. At rest, the liquid component fully wets the solid content of the fluid, leading to low viscosity. Under increasing shear, the particles can interact readily, forming a solid structure within the fluid to increase the viscosity.

Viscoplastic Flow

Viscoplastic fluids are those that will not flow until a certain shear threshold is reached. Examples of such systems include toothpaste in a tube, gels, greases and tomato ketchup; Fig. 4.4b. This is usually caused by inter-particle/inter-molecular binding forces that require a certain amount of shear stress to overcome before achieving liquid flow.

Time-Dependent Behaviour

In many cases, a liquid will take a finite time to respond to increasing shear, or to ‘relax’ to its rest state after the shear is removed, leading to time-dependent behaviour. Thixotropy is the time-dependent equivalent of pseudoplasticity (i.e. fluids show shear-thinning over time for a constant shear rate, and gradual recovery of viscosity once the shear is removed). Such behaviour is desirable in applications such as paints, since it allows limited flow to remove brush marks, but not enough to allow drips to form. The reverse effect (termed rheopexy) is observed as an increase in viscosity under constant shear rate, and gradual reduction on cessation of movement [54, 57–59].

Characterisation of Ink Rheology

This section will discuss the most widely used methods for rheology measurement. There are a multitude of methods to characterise ink rheology. Plate-to-plate and cone-to plate systems are predominantly the most expensive but offer the highest quality information on rheological performance of non-Newtonian fluids [54]. Shear or rotation-based rheology measurement methods are designed to deliver and measure a controlled shear rate or shear stress to the fluid in order to examine the fluid’s rheological flow over time [1, 3, 54, 56, 70].

Cone-to-Plate and Plate-to-Plate Rheometers

Cone-to-plate (CTP) and plate-to-plate (PTP) rheometers are the most sophisticated tools available in characterising the rheology of inks. The principle behind CTP and PTP viscometers is shown in Fig. 4.5a and b, respectively. These are essentially identical in configuration, except that the CTP set-up has a conical top plate, meaning that the shear rate is constant across the whole plate (noting that the shear rate is the velocity gradient across the gap between the plates), while the PTP set-up uses a flat top plate. The sample fluid is loaded into the gap between the two plates, and the torque required to maintain a set shear rate is measured.

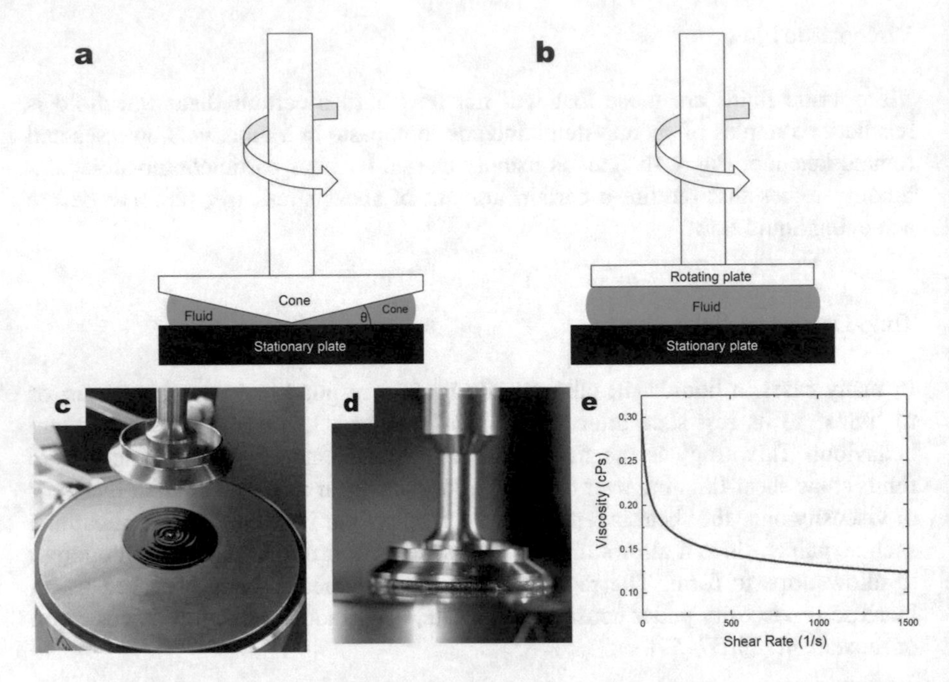

Fig. 4.5 Schematics (not to scale) of (**a**) CTP and (**b**) PTP viscometers. Photograph of a PTP set-up preparing for measurement with (**c**) plate retracted and ink loaded and (**d**) plate down on an ink sample. (**e**) Rheogram of a screen-printable graphene ink

By varying the size of the cone and speed of torque, a variety of measurements can be taken. PTP systems are best suited for inks with larger particle size. CTP set-ups require more precise machining of the cone, and are less well-suited to inks with large particles due to the small gap at the centre, but provide more accurate measurement of fluids with a highly shear-rate dependent viscosity. PTP set-ups offer the advantage of simpler manufacture, and can operate with larger plate gaps, but may suffer from inaccuracy due to the varying shear rates across the plate face [1, 3, 54].

CTP and PTP rheometers are highly effective tools in aiding ink formulation and understanding the rheological behaviour of inks. For instance, Fig. 4.5e depicts a rheogram of a graphene screen printing ink that displays shear thinning behaviour. The flow curve was measured by increasing the shear stress from 1 to 1,000 s^{-1} at a gap of 0.5 mm. This is equivalent to the shear stress applied during screen printing [51]. This indicates thixotropic behaviour due to the decrease in viscosity with an increasing shear rate [71]. Such shear-thinning behaviour is caused by the disentanglement of polymer chains and increased orientation in the direction of flow [54].

Coaxial Rotational Rheometers

A slightly more dated method of characterising rheology is using coaxial rotational viscometers. The most common variant of these is the cup and bob viscometer. This usually consists of a 'bob' suspended from a torsion spring and inserted into the ink sample. The fluid in the ink reservoir is sheared relative to a cylindrical 'bob', either by rotating the reservoir or the 'bob' [1, 3, 54]. The extension or contraction of this spring indicates the drag force and hence the measured viscosity of the fluid [1, 3, 54]. A rheogram can then be plotted over time from the drag experienced by the torsion spring [1, 3, 54].

4.4 Ink Production

Printing ink production and processing generally follows a 3-step scheme:

1. Creation of an optimal dispersion varnish
2. Dispersion by mechanical action
3. Dilution or letdown with ink varnishes

This approach is typical for most kinds of printing inks. This section will discuss these three main processes.

4.4.1 Creation of an Optimal Millbase

A varnish can be described as a higher viscosity, non-pigmented ink. It is generally made up of resins, solvents and additives without the pigment and is specifically designed for pigment dispersion. In the ink industry the combination of varnish and pigment, it can be termed as 'millbase' [1, 3, 72]. Creating a high concentration millbase is vital to creating optimal conditions for shear stress during dispersion. The millbase is significantly higher in viscosity than the final ink, to offer optimum transfer of energy from the milling process through the liquid medium in order to break up any agglomerates of pigments. In addition, the high shear promotes consistent dispersion of surface modifier additives such as liquid silicone and surfactants [1, 3, 71–74].

Creating a stable ink formulation requires that as much as possible pigment powder is converted from agglomerates and aggregates into primary particles in the shortest time and at the lowest cost. An optimised millbase allows complete adsorption of the liquid component on the primary particle. In general, the higher the viscosity of the millbase, the greater the amount of shear force that can be transmitted from the dispersion machinery to the pigment particles. An accepted method for finding the ideal pigment concentration for an optimised millbase is the 'flow-point' method. In this method, the varnish resin is titrated into a known

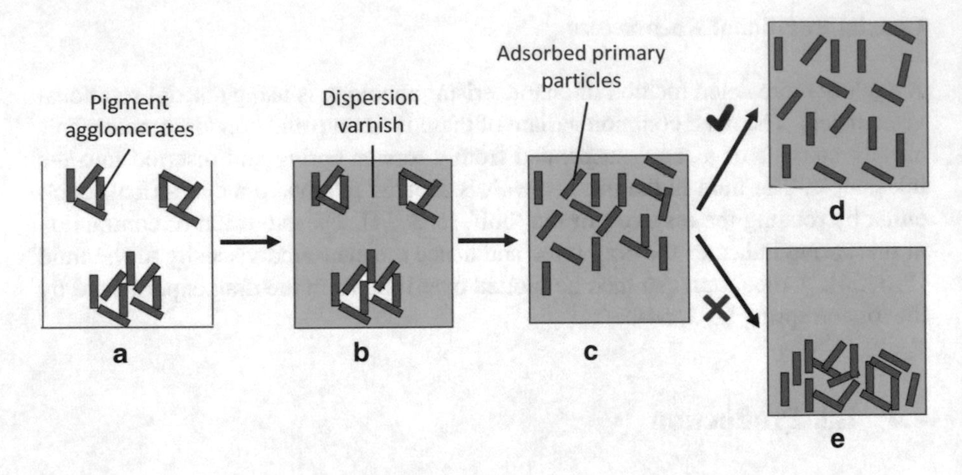

Fig. 4.6 The dispersion process. (**a**) Agglomerates are (**b**) wetted from outside. (**c**) During dispersion, resin solution penetrates the cavities, partially separating the particles, which results in pigment wetting. This can lead to either (**d**) successful processing forming a homogeneous dispersion of primary particles, or (**e**) unsuccessful dispersion due to improper choice of chemicals and procedures, leading to poor stabilisation and therefore sedimentation

quantity of pigment, until the 'flow-point' (i.e. the point at which the millbase starts to flow) is reached [1, 3, 71, 73]. This then represents the starting point for ink dispersion.

4.4.2 Ink Dispersion

Figure 4.6 depicts a typical dispersion process. This process describes the breaking down of aggregates and agglomerates of pigments into primary particles. At a macroscopic level, agglomerates and aggregates contain air pockets in their structure due to the arrangement of primary particles. Complete dispersion can then be seen as having all adsorbed gases on the primary pigment particles replaced by the liquid vehicle in the millbase. Complete adsorption of the liquid phase by the primary particles is a phenomenon called pigment wetting. Figure 4.6 depicts the typical dispersion process.

The efficiency of the dispersion process is based on the dispersing machinery and the associated forces; Fig. 4.7. It is therefore important to ensure that the millbase is formulated for each individual milling machine to optimise dispersion. Impact and shear forces act as the main dispersion forces in the millbase. Shear forces occur when two surfaces, particles, or a combination move past each other in similar or opposite directions, while impact forces occur when they collide. Both physical machinery and fluids may be able to generate shear forces within

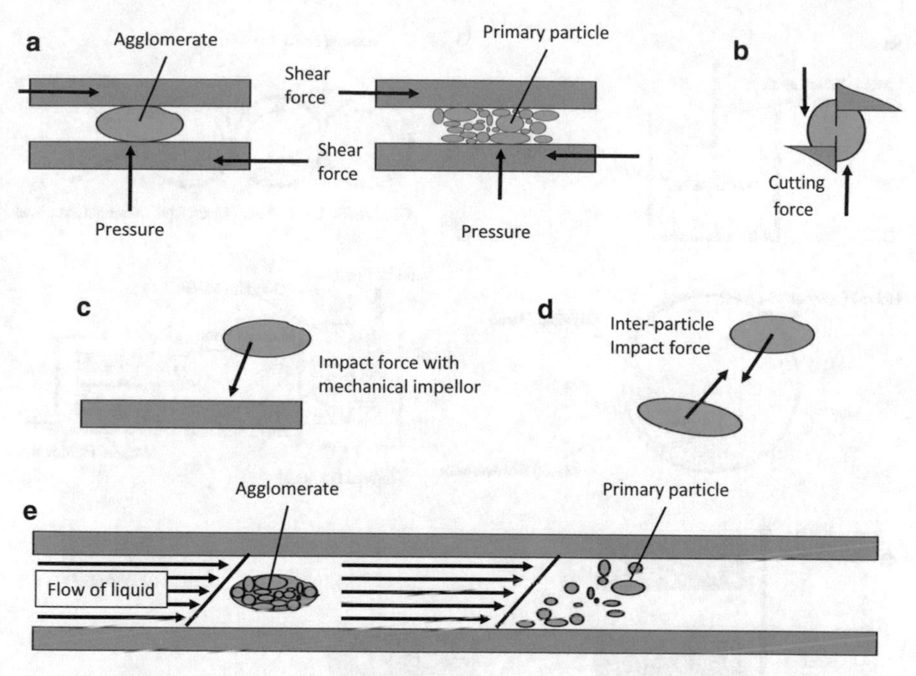

Fig. 4.7 Summary of milling and dispersion forces from mechanical dispersion. (**a**) Shear forces from mechanical impeller blade acting on agglomerates to break them down into primary particles. (**b**) Mechanical cutting forces acting on agglomerates. (**c**) Impact forces with mechanical impeller blades. (**d**) Inter-particulate impact forces. (**e**) Shear forces from flow of dispersion varnish

the millbase, whereas impact forces require solid surfaces (e.g. particles, machine parts)[1, 3, 56, 72, 73, 75, 76].

An appropriate selection of the procedures, chemicals and efficient milling operations will likely lead to well-dispersed primary particles suspended in a liquid vehicle. However, an inappropriate selection of chemicals and procedures or insufficient energy transferred to the primary particles during the milling process will likely lead to poor dispersion and therefore an ink with printability issues.

Dispersion Machinery

Many different systems exist for generating the required forces for pigment dispersion. Selection of the dispersing machines depends on the initial viscosity of the millbase and the desired viscosity of the end ink. While the design of different systems varies, they share the same operating principles—transmitting shear force through laminar flow gradients and impact forces using tensile-compressive loads. Figure 4.8 shows schematics of the most common mill types, which include high speed dissolvers, triple-roll mills, ball mills and attritor mills.

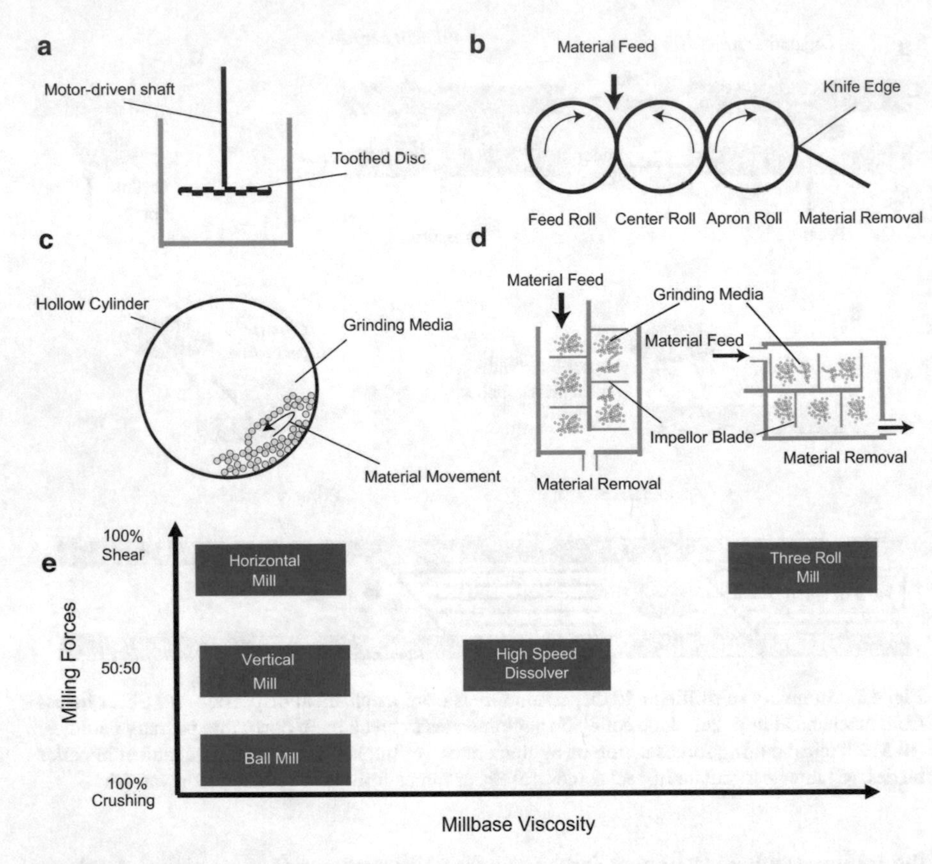

Fig. 4.8 Simplified schematics of (**a**) high speed dissolver, (**b**) three-roll mill, (**c**) ball mill and (**d**) horizontal (attritor) and vertical mills. (**e**) Comparison between the millbase viscosity and dispersing forces for each mill

High Speed Dissolvers

High speed dissolvers use a toothed disc which is rotated at high speeds in a smooth-walled container; Fig. 4.8a. The rotating disc drives particles outwards, breaking up agglomerates through impact forces with the container wall [1, 3, 72]. A modification of this kind of disperser is the rotor-stator mill, sometimes referred to as a 'Silverson' or 'batch ranger', which has a meshed screen surrounding the toothed disc. This is the same rotor-stator mill mentioned in the previous chapter used in the liquid phase exfoliation of 2D materials. The rotor-stator mill generates additional shear forces as the millbase passes through the gaps in the mesh.

Three-Roll Mills

Three-roll mills consist of a set of cylindrical rollers (feed roll, centre roll, and apron roll) separated by precisely controlled gaps; Fig. 4.8b. The gaps between rolls can be mechanically or hydraulically adjusted and maintained depending on the viscosity of the millbase [3, 72]. Typically, the gap distance is far greater than the desired particle size. In some operations, the gap distance starts large and is gradually decreased to achieve the desired level of dispersion. The rollers rotate in opposing directions at progressively faster speeds. The millbase is fed into the wedge-shaped area formed by the feed and the central rolls. From there, it passes through the two roll gaps to the apron roll, where it can then be removed by a doctor blade for collection. This milling cycle can be repeated several times to maximise dispersion.

Ball Mills

Ball mills consist of a hollow cylinder or jar, which rotates about a horizontal axis when in operation; Fig. 4.8c [1, 3, 72]. They are filled with a grinding media, usually zirconium or steel balls. As the jar rotates, the movement of the grinding media promotes dispersion via impact and shear forces.

Agitator Mills or Attritors

Agitator mills or attritors consist of a container and an impeller blade, and may be vertically or horizontally orientated; Fig. 4.8d [1, 3, 72]. As with ball mills, grinding media are used to aid dispersion, although in this case, they tend to be smaller.

4.4.3 Letdown with Ink Varnish

Letdown ink varnishes are in general of lower viscosity than the dispersion varnishes but are of similar composition. Some additives, such as surfactants and amines, may have a larger proportion in the letdown varnish in order to maintain stable ink dispersion [1, 4]. The letdown process is generally simple without any need for sophisticated machinery. Often a high speed dissolver with a toothed disc is used for low viscosity inks while a three-roll mill may be used for higher viscosity inks [1, 3]. At this stage, the ink takes on its final characteristics. Hence, it is important to constantly monitor viscosities required for the final ink [1, 3, 4].

4.5 Ink–Substrate Interactions

Substrates for functional printing can often be classified into absorbent (e.g. paper, cardboard, textiles) and non-absorbent (e.g. plastic, glass, metals) types. The substrate will have a considerable effect on ink characteristics such as ink spread, drying, and adhesion, and must therefore be taken into account when selecting/designing an ink for a specific application. A good understanding of the ink–substrate interactions will therefore assist in the successful design of a printing ink. This section will discuss both substrate wetting of inks and theories of adhesion; two components that govern the ink–substrate interactions.

4.5.1 Wetting and Spreading of Inks on Surfaces

For any specific printing and substrate combination, the rate of wetting depends on a number of factors, including the viscosity of the ink and the roughness, porosity and surface energy of the substrate. The higher the viscosity, or the rougher the substrate, the less readily the ink will flow after printing. Meanwhile, the spreading of the ink and the ability of adjacent droplets to coalesce are governed by wetting.

In general, the wetting phenomenon between a liquid and a substrate is governed by their interfacial tensions and can be expressed by Young's equation [77–79]

$$\gamma_{s-v} = \gamma_{s-l} + \gamma_{l-v} \cos\theta$$

where γ_{s-v}, γ_{s-l} and γ_{l-v} are the interfacial tensions between the solid surface (s), the ink/liquid (l) and the vapour (v). The critical parameter that is often taken as a good indication of wetting is the θ of the liquid, or contact angle; Fig. 4.9. $\theta > 90°$ indicates poor wetting and the likelihood of printing defects such as adhesion failure and non-uniform deposition and a $\theta < 90°$ will indicate a more successful print with a uniform deposition. To attain a 100% wetting of a substrate is to achieve a θ of 0°, a phenomenon known as *ideal spreading* [80]. Thus, when $\theta = 0°$, or when $\cos\theta = 1$.

$$\gamma_{l-v} = \gamma_{s-v} - \gamma_{s-l}$$

or

$$\gamma_{l-v} \leq \gamma_{s-v}$$

This provides guidance for printing.

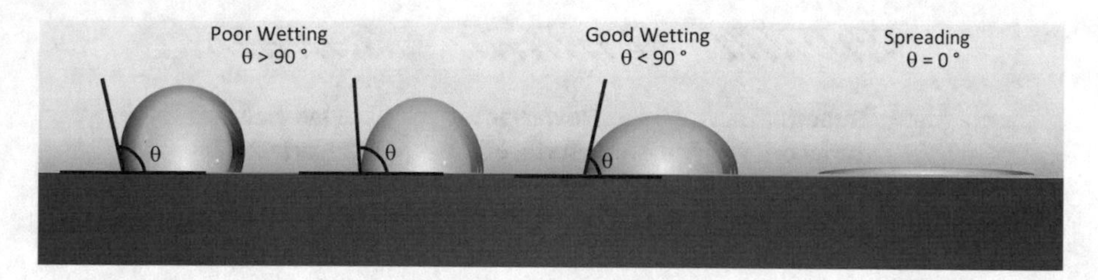

Fig. 4.9 Examples of different contact angles on substrates

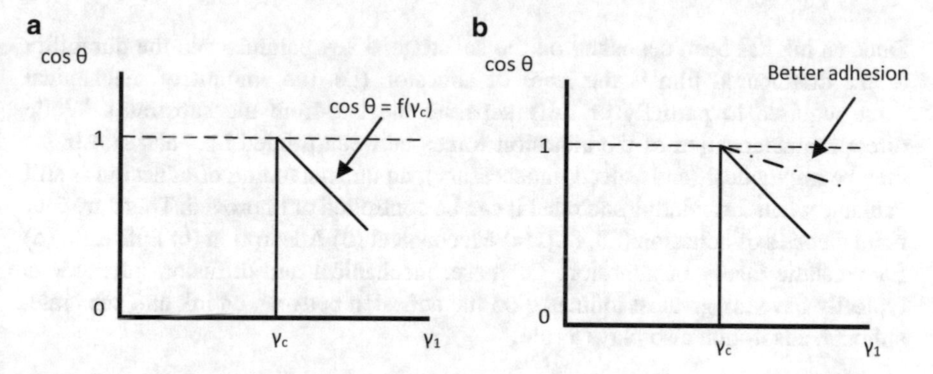

Fig. 4.10 (**a**) Zisman plot for determination of CST, with the flatter slope in (**b**) indicating a higher probability of adhesion

The wetting properties of a substrate are typically predicted by deriving the critical surface tension (CST) of a substrate, γ_C. The CST of a surface is defined as the equivalent surface tension of the liquid which on contact with the substrate will give a zero contact angle. The CST of a solid surface is an indication of its relative hydrophobic or hydrophilic character. A low CST means a surface with a low energy per unit area, hence indicating a less wettable surface. In general, a partial spreading of an ink ($0 < \theta < 90°$) should require a lower surface tension of the ink than the CST of its corresponding substrate [80]. Indeed, a guideline given in a packaging manual by Robertson states that the ink surface tension should be at least 7–10 m Nm^{-1} lower than that of the critical surface tension of the substrate to ensure good, printable wetting [81].

The CST is measured by studying the wetting using liquids with known surface tensions. A small droplet of each test liquid is placed onto the surface, and the contact angle is then measured and plotted against its surface tension in a Zisman plot; Fig. 4.10 [82–98]. Both the CST and the slope of the Zisman graph can indicate likely wettability, with a flatter slope on the Zisman graph generally suggesting a stronger ink adhesion.

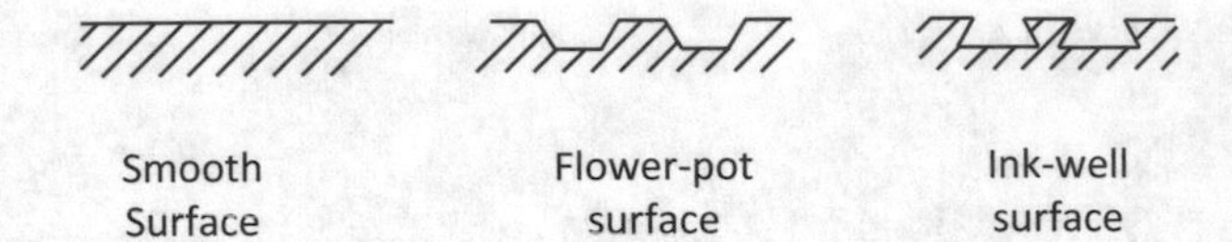

Fig. 4.11 Smooth, flower-pot and ink-well profiles of typical surfaces

4.5.2 Theories of Adhesion

Once an ink has been deposited on the substrate, a key parameter in the durability of the dried/cured film is the level of adhesion (i.e. the amount of mechanical force required to partially or fully separate the ink from the substrate). While direct characterisation of the adhesion forces between printed inks and substrates may be impractical (and indeed unnecessary), an understanding of adhesion is still valuable when formulating such that it can be controlled or improved. There are four main theories of adhesion [83, 88]: (a) Mechanical (b) Adsorption (c) Diffusion (d) Electrostatic theory of adhesion. Of these, mechanical and diffusion interactions typically have the greatest influence on the adhesion between an ink and substrate, although adsorption also plays a role.

Mechanical Theory of Adhesion

The mechanical theory of adhesion refers to the interlocking of the fluid with the texture on the surface. At a micro-level, the surface of all substrates can be seen as either having a smooth profile, a 'flower-pot' profile or an 'ink well' profile. The success of adhesion of an ink to the surface depends on how well it is adsorbed onto the cavities of the surface. In general, fibrous and porous substrates facilitate micro-interlocking of the ink to the surface, while smoother surfaces require some assistance in facilitating adhesion (Fig. 4.11).

In practice, it is often impractical to alter the surface texture to improve adhesion. Therefore, the ink must instead be formulated to maximise the interaction with any surface features. This might include minimising the contact angle by adjusting the surface energy of both the ink and the surface, or controlling the rheological properties of the ink. Inks are often designed to be shear-thinning, allowing them to be of lower viscosity at time of deposition in order to fill the substrate cavities before forming a film. Lower molecular mass resins in general have a lower viscosity and hence make an ideal choice of binder if maximising adhesion is the objective [83, 88]. In addition, applying higher temperatures in the form of either a heated roller or printing bed can result in lower viscosity of the ink during deposition.

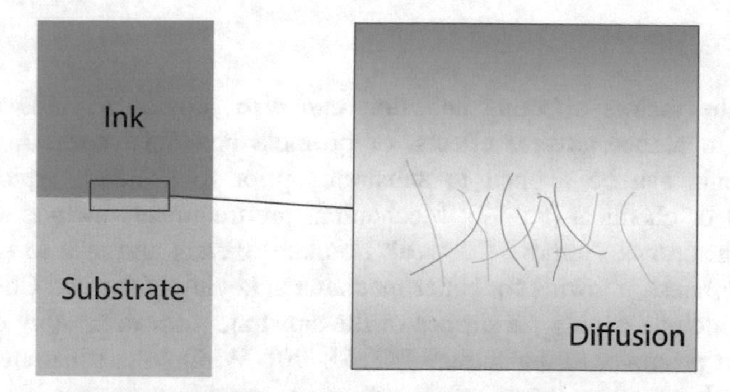

Fig. 4.12 Demonstration of the diffusion theory of adhesion, showing the interaction of ink and substrate molecules at the interface

Diffusion Theory of Adhesion

The diffusion theory of adhesion describes an adhesion between ink and substrate which develops over time. This is due to the diffusion of long molecular chains present in both the inks and the plastic substrates. Over time, two polymers placed close enough to each other will see the polymer chains merging into one. This is also useful in understanding multi-layer ink adhesion whereby a layer of ink printed on another can result in the diffusion into a single layer of ink (Fig. 4.12).

Adsorption Theory of Adhesion

The adsorption theory of adhesion can be explained as adhesion due to short range (0.7–10 Å) intermolecular forces between the ink and the substrate. Effective wetting and spreading of the ink on the substrate surface is essential for such forces to occur. Inks with a low contact angle on the substrate are therefore best suited to utilise this form of adhesion.

Factors Affecting Adhesion

When considering adhesion, it is important to understand the topography of the substrate. Absorbent substrates, as well as treated polymers, typically have more porous topographical profile when compared to non-absorbent substrates such as glass, allowing improved adhesion from mechanical influences. Meanwhile, adhesion may be adversely affected by weak boundary layers between the ink and the substrates [83, 84]. Examples of boundary layers include dust, grease and weak metal oxides such as rust. It is also important to consider the surface chemistry of the ink and the substrate for good adhesion to occur, the printed inks must be of lower surface energy than that of the substrate.

Improving Adhesion

Knowing the factors affecting adhesion can also provide a guide to suitable treatments to reduce adverse effects, or promote beneficial ones. A number of pre-treatments can be applied to substrates prior to printing, typically either mechanical or chemical [85–87]. Mechanical pre-treatments include sanding the substrates in order to remove the weak boundary layers and also to increase the surface roughness, allowing for better mechanical keying of the ink. Chemical pre-treatments include wiping the surface of the substrate prior to printing or applying an adhesion promoter to the surface [82, 88–90]. Wiping the substrates serves to remove weak boundary layers, while adhesion promoters are capable of altering the surface chemistry of the substrate by increasing its surface tension and also by increasing surface roughness.

Plastics, widely used in the printing industry, very often have low surface energy and hence are notoriously difficult to print on. The most common method for improving ink adhesion onto plastics on press is a corona discharge unit whereby plastic substrates are passed under high voltage (e.g. 20 kV) at high speed (e.g. 1–2 ms^{-1}); Fig. 4.13a–c [91–95]. A continuous discharge is created on the substrate surface, producing a plasma in which gas molecules are broken into individual atoms, free radicals, ions and electrons. These are highly reactive and disrupt the carbon–hydrogen and carbon–carbon bonds at the plastic surface to a depth of about 5 to 50 nm (for corona discharge). This results in a more polar surface, hence enhancing adhesion.

Other commonly employed surface treatments which operate in a similar manner to corona treatment include oxygen plasma treatment that makes a surface more hydrophilic. This is demonstrated in Fig. 4.13d, e, where oxygen plasma treatment is demonstrated to allow better wetting of a polydimethylsiloxane (PDMS) substrate [99].

Metal surfaces have a more limited range of adhesion pre-treatments and are generally restricted to only removal of contaminants by chemical means. However, metal surfaces are often rough by nature and hence easier to print on than plastics[96–98]. Post-treatments can also be useful in enhancing adhesion of inks. Increasing impression pressure during printing or adding an extra calendaring step (passing the printed substrate through two high pressure rollers) can enhance the adhesion of inks. Increasing the curing temperature of the ink can also cause the viscosity of the ink to temporarily lower, allowing the ink to better fill out the topographic defects of the substrate for improved adhesion. Finally, increasing curing time allows increased adhesion by diffusion.

4.6 Ink Formulation Optimisation

A critical component of ink formulation is optimising the final formulation to deliver the desired characteristics (e.g. rheology, pigment concentration, print uniformity). The final characteristics depend on a multitude of factors unique to

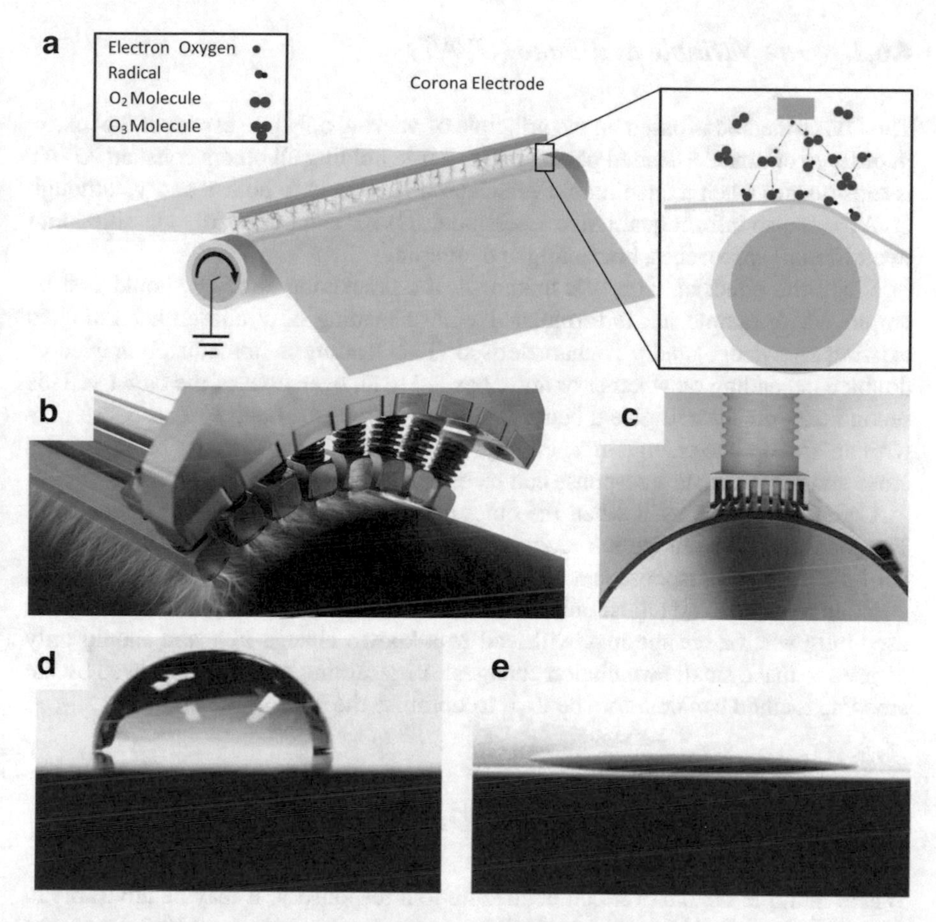

Fig. 4.13 (**a**) Schematic showing corona treatment of plastic surface. Inset: substrate interactions with corona electrode and surface modification using O_2 and O_3 molecules. (**b**, **c**) Photographic close-up of corona treatment. Copyright Vetaphone Corona and Plasma (http://www.vetaphone.com). (**d**) Contact angle of a water droplet on PDMS surface before oxygen plasma treatment and (**e**) after oxygen plasma treatment. Reproduced with permission from Ref. [99]. Copyright (2012) The Royal Society of Chemistry

each individual printer and press set up. Even environmental conditions, such as ambient temperature and humidity can have an effect on the performance of the ink system on press. There are a number of different strategies employed to optimise an ink formulation. However, in general, they can all be summarised as achieving an outcome (or outcomes) y by adjusting a variable (or set of variables) x. Depending on the ink formulation, adjusting one variable may affect multiple outcomes, and so any changes must be balanced against each other. The choice of any of these strategies is dependent on the starting formula of every ink recipe. In general, formulation optimisation strategies can be categorised into one variable at a time (OVAT) strategy [71] and design of experiments (DOE) [100].

4.6.1 One Variable at a Time (OVAT)

The OVAT method is based on the principle of varying only one aspect of the ink, x, in order to optimise a desired parameter, y, while holding all others constant. OVAT is most useful when a quantifiable value can be assigned to both x and y, although OVAT can also inform qualitative assessment. OVAT can be broadly classified into two different approaches, bracketing and creeping.

When the effect of x on y is unknown, the bracketing strategy should first be employed. A sample ink is formulated with a loading of x chosen according to existing knowledge, and y is characterised. This loading is immediately halved or doubled depending on whether or not x has led to an overshoot of the target y. This second experiment will give a better understanding of the degree of response of y with an appropriate change in x. Finally, the required change in x to bring about a close and approximate y response can be estimated.

Creeping may be used when the outcome of varying x is broadly understood. The proportion of x in the ink recipe is increased or decreased by a small amount until the desired y response is achieved. The creeping method is very useful if it is possible to isolate and adjust only one variable at a time. This approach should be used only when a change in x will lead to a known change in y and should only be used to make small formulation changes. A bracketing approach followed by the creeping method can therefore be used to optimise the ink formulation.

4.6.2 Design of Experiments (DOE)

Where multiple factors of x can contribute to a response y, it may be advisable to adopt the design of experiments (DOE) approach (also known as Yates analysis) [100]. This is a statistical approach to experimental formulation, and is typically less time-consuming than OVAT or 'trial-and-error' methods when attempting to optimise multiple factors [100, 101]. DOE looks at interdependent variables and can take into account the weight of individual x factors in eliciting a y response. A series of planned experiments are carried out, allowing a mathematical model to be developed that predicts how changes in input or other controlling variables (such as duration, time, temperature) affect the outcome. This sequence of experiments is called a design [100]. DOE has long been used in many industries including diesel, paint and food manufacturing and more applicably, the printing industry [100, 101].

A factor is a variable condition in the experiment (e.g. amount of solvent in a system) and a level refers to either a high or low loading of said variable). The range of factors encompassed by a design can be visualised graphically, as in the example in Fig. 4.14, which shows a 2-level, 3-factorial design. When the design involves more than three variables, it produces what can be described as a 'hypercube' [100] with more than three dimensions which, although un-depictable in graphical form, still adheres to the basic principles of DOE. An experiment (e.g.

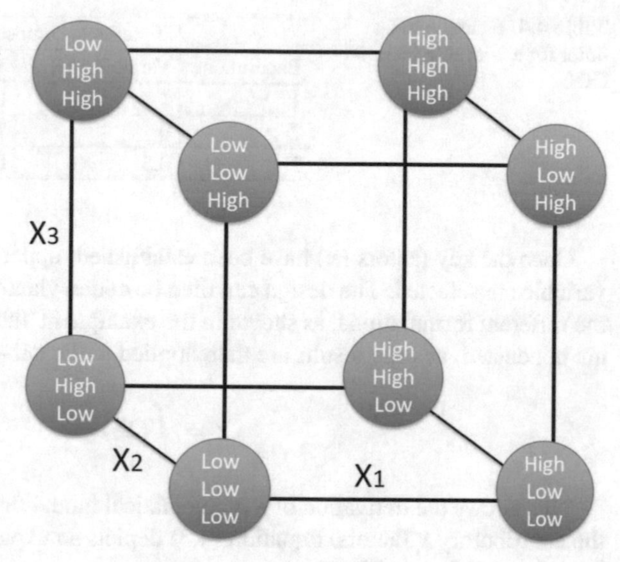

Fig. 4.14 DOE cube depicting a 2-level, 3-factorial experiment

an ink formulation) can be located anywhere in the cube, with its coordinates defined by the three variables (e.g. pigment, solvent, and binder concentration). The value of y (the desired output) associated with this point can be estimated statistically, or determined experimentally as time and resources allow. Many variations of design exist, with different numbers of factors (the axes on the cube), the extreme values (end points of the axes), and the number of settings for each factor (divisions along the axes). The choice of design affects the minimum number of experimental runs required to establish a predictive mathematical model reflecting the contribution of all the factors in the DOE [100].

The most applicable form of DOE for ink formulation is 2-level, 3-factorial design of experiments. A 2-level design means that only a high and low loading of each factor x is needed when constructing the predictive DOE model [100]. DOE is most effective when the main contributors affecting the outcome y are already known, as the number of experiments required increases rapidly with increased numbers of factors. The minimum number of experiments that need to be conducted in a DOE is:

$$z = 2^n - 1 \tag{4.1}$$

where z is the number of experiments required in order to form a complete mathematical model and n is the number of factors, x, in the design. Hence, a 2-level, 3-factorial design would minimally require seven independent trials in order to form a mathematical model. A 2-level, 5-factorial design would require 31 independent trials. For more complex formulations containing up to 19 ingredients, 524,287 independent trials would be required. It is therefore important to identify the two or three main contributory factors in order to use DOE effectively.

Table 4.4 Example Yates order for a 2-level, 2-factorial DOE

Experiment	Fractional composition of mixture		Response
	Variable 1 (X_1)	Variable 2 (X_1)	(Y)
1	1	0	Y_1
2	0	1	Y_2
3	0.5	0.5	Y_3

Once the key factors (x) have been established, upper and lower limits for each variable are selected. The design can then be coded via a Yates order which sets out the different formulations, as shown in the example in Table 4.4. These experiments are conducted, and the results are then applied to the universal formula of mixtures:

$$Y = \int (X_i) \tag{4.2}$$

This allows the derivation of a mathematical model that reflects the weight of all the contributory x factors. Equation (4.3) depicts an expanded formula of a 2-level, 2-component factorial DOE:

$$Y = a_1X_1 + a_2X_2 + a_{12}X_1X_2 \tag{4.3}$$

where a_1 is an indication of the level of influence of X_1, a_2 is a reflection of the level of influence of X_2 and a_{12} is a coefficient reflecting the level of influence of a mixture of X_1 and X_2.

The results from the experiments are then substituted into the equation relevant to the mixture in order to derive the necessary coefficients. A larger number of factors to be considered will also mean a more complex model, as indicated in Eqs. (4.4) and (4.5), which show the models for a 2-level, 3-component, and 2-level, 4-component DOE, respectively.

$$Y = a_1X_1 + a_2X_2 + a_3X_3 + a_{12}X_1X_2 + a_{13}X_1X_3 + a_{23}X_2X_3 + a_{123}X_1X_2X_3 \tag{4.4}$$

$$\begin{aligned} Y = {} & a_1X_1 + a_2X_2 + a_3X_3 + a_4X_4 + a_{12}X_1X_2 + a_{13}X_1X_3 + a_{14}X_1X_4 \\ & + a_{23}X_2X_3 + a_{24}X_2X_4 + a_{34}X_3X_4 + a_{123}X_1X_2X_3 + a_{124}X_1X_2X_4 \\ & + a_{134}X_1X_3X_4 + a_{234}X_2X_3X_4 + a_{1234}X_1X_2X_3X_4 \end{aligned} \tag{4.5}$$

It is therefore key to minimise the number of factors to avoid an overly complex final model. In fact, DOE can also be used here as well, screening variables to decide which are significant ones. In screening designs, it is assumed that real-world processes are driven mostly by only a few factors, the others may be contributory but are relatively unimportant to the desired outcome. When dealing with potentially

19 or 20 different factors, screening designs are able to easily identify the factors which most significantly affect the desired outcome as compared to the classical OVAT strategy [102].

It should be noted that fractional and full factorial designs are not the only DOE models available. Some of the more popular models for DOE include [100–104]:

- **Plackett-Burman:** This DOE has a resolution higher than simple factorial designs and is a good technique for a large numbers of variables.
- **Box-Behnken:** An efficient technique for modelling quantitative 3-level factors.
- **Box-Wilson:** A design mostly used when significant interaction levels are suspected.

However, a more in-depth discussion of DOE and of the merits of the different models falls beyond the scope of this book. If desired, further reading may be found in works such as Ref. [100].

4.7 Conclusion

This chapter has given a brief overview on the traditional ink formulation process and how it can be used to design ink systems for 2D materials.

Inks are generally made up of the same components across all printing methods-pigments, binders, solvents and additives. The two generic categories of inks, paste and liquid inks, generally have varying quantities of solvents depending on the desired viscosity and final print characteristics. The largest difference in the design of ink systems for specific printing processes lies in its viscosity and rheology. In general, shear-thinning behaviour is desired for inks as this facilitates the transfer of ink to the substrate.

The large-scale manufacturing of inks generally follows a three-step schema which starts with the creation of a dispersion millbase, followed by the subsequent dispersion using dispersion machinery, and finally the letdown or dilution with lower viscosity varnishes in order to achieve the desired final viscosity parameters. In ink formulation, it is also important to take into account ink–substrate interactions, requiring understanding of the wetting of surfaces, and how inks adhere onto substrates. Statistical methods, such as DOE, may also be used to optimise certain parameters in ink formulations.

The parameters and methods described in this chapter are generally applicable to all forms of ink for all printing processes. The next chapter will describe ink formulations and considerations specific to each printing process, and will further discuss the use of printing as a fabrication method for 2D material devices.

References

1. R.H. Leach, R.J. Pierce, *The Printing Ink Manual* (Springer, Dordrecht, 1993)
2. D.N. Carvalho, *Forty Centuries of Ink* (BiblioLife, Charleston, 2008)
3. A. Goldschmidt, H.-J. Streitburger, *BASF Handbook on Basics of Coating Technology* (William Andrew, Norwich, 2003)
4. E.W. Flick, *Printing Ink and Overprint Varnish Formulations*, 2nd edn. (William Andrew, Norwich, 1999)
5. A. Tracton (ed.), *Coatings Technology Handbook*, 3rd edn. (CRC Press, West Palm Beach, 2005)
6. I.M. Hutchings, G.D. Martin (eds.), *Inkjet Technology for Digital Fabrication* (Wiley, Chichester, 2012)
7. H. Lievens, Wide web coating of complex materials. Surf. Coat. Technol. **76–77**, 744–753 (1995)
8. M. Lahti, S. Leppävuori, V. Lantto, Gravure-offset-printing technique for the fabrication of solid films. Appl. Surf. Sci. **142**(1–4), 367–370 (1999)
9. H.A.D. Nguyen, C. Lee, K.-H. Shin, D. Lee, An investigation of the ink-transfer mechanism during the printing phase of high-resolution roll-to-roll gravure printing. IEEE Trans. Compon. Packag. Manuf. Technol. **5**(10), 1516–1524 (2015)
10. T. Smith, Flexographic inks. Pigm. Resin Technol. **15**(3), 11–12 (1986)
11. F.C. Krebs, Fabrication and processing of polymer solar cells: a review of printing and coating techniques. Sol. Energy Mater. Sol. Cells **93**(4), 394–412 (2009)
12. H. Kipphan (ed.), *Handbook of Print Media* (Springer, Berlin, 2001)
13. K. Suganuma, *Printing Technology* (Springer, New York, 2014)
14. H.-H. Lee, K.-S. Chou, K.-C. Huang, Inkjet printing of nanosized silver colloids. Nanotechnology **16**(10), 2436–2441 (2005)
15. H. Sirringhaus, T. Kawase, R.H. Friend, T. Shimoda, M. Inbasekaran, W. Wu, E.P. Woo, High-resolution inkjet printing of all-polymer transistor circuits. Science **290**(5499), 2123–2126 (2000)
16. B.-J. de Gans, P.C. Duineveld, U.S. Schubert, Inkjet printing of polymers: state of the art and future developments. Adv. Mater. **16**(3), 203–213 (2004)
17. P. Beecher, P. Servati, A. Rozhin, A. Colli, V. Scardaci, S. Pisana, T. Hasan, A.J. Flewitt, J. Robertson, G.W. Hsieh, F.M. Li, A. Nathan, A.C. Ferrari, W.I. Milne, Ink-jet printing of carbon nanotube thin film transistors. J. Appl. Phys. **102**(4), 043710 (2007)
18. Z. Liu, Z.-B. Zhang, Q. Chen, L.-R. Zheng, S.-L. Zhang, Solution-processable nanotube/polymer composite for high-performance TFTs. IEEE Electron. Device Lett. **32**(9), 1299–1301 (2011)
19. Z. Liu, H. Li, Z. Qiu, S.-L. Zhang, Z.-B. Zhang, Small-hysteresis thin-film transistors achieved by facile dip-coating of nanotube/polymer composite. Adv. Mater. **24**(27), 3633–3638 (2012)
20. D. McManus, S. Vranic, F. Withers, V. Sanchez-Romaguera, M. Macucci, H. Yang, R. Sorrentino, K. Parvez, S.-K. Son, G. Iannaccone, K. Kostarelos, G. Fiori, C. Casiraghi, Water-based and biocompatible 2D crystal inks for all-inkjet-printed heterostructures. Nat. Nanotechnol. **12**(4), 343–350 (2017)
21. F. Bonaccorso, A. Bartolotta, J.N. Coleman, C. Backes, 2D-crystal-based functional inks. Adv. Mater. **28**(29), 6136–6166 (2016)
22. R.C.T. Howe, G. Hu, Z. Yang, T. Hasan, Functional inks of graphene, metal dichalcogenides and black phosphorus for photonics and (opto)electronics. Proc. SPIE 9553, 95530R (2015)
23. A.G. Kelly, T. Hallam, C. Backes, A. Harvey, A.S. Esmaeily, I. Godwin, J. Coelho, V. Nicolosi, J. Lauth, A. Kulkarni, S. Kinge, L.D.A. Siebbeles, G.S. Duesberg, J.N. Coleman, All-printed thin-film transistors from networks of liquid-exfoliated nanosheets. Science **356**(6333), 69–73 (2017)

24. D.J. Finn, M. Lotya, G. Cunningham, R.J. Smith, D. McCloskey, J.F. Donegan, J.N. Coleman, Inkjet deposition of liquid-exfoliated graphene and MoS_2 nanosheets for printed device applications. J. Mater. Chem. C **2**(5), 925–932 (2014)
25. V. Bianchi, T. Carey, L. Viti, L. Li, E.H. Linfield, A.G. Davies, A. Tredicucci, D. Yoon, P.G. Karagiannidis, L. Lombardi, F. Tomarchio, A.C. Ferrari, F. Torrisi, M.S. Vitiello, Terahertz saturable absorbers from liquid phase exfoliation of graphite. Nat. Commun. **8**, 15763 (2017)
26. F. Torrisi, J.N. Coleman, Electrifying inks with 2D materials. Nat. Nanotechnol. **9**(10), 738–739 (2014)
27. F. Torrisi, T. Hasan, W. Wu, Z. Sun, A. Lombardo, T.S. Kulmala, G.-W. Hsieh, S. Jung, F. Bonaccorso, P.J. Paul, D. Chu, A.C. Ferrari, Inkjet-printed graphene electronics. ACS Nano **6**(4), 2992–3006 (2012)
28. J. Li, F. Ye, S. Vaziri, M. Muhammed, M.C. Lemme, M. Östling, Efficient inkjet printing of graphene. Adv. Mater. **25**(29), 3985–3992 (2013)
29. E.B. Secor, P.L. Prabhumirashi, K. Puntambekar, M.L. Geier, M.C. Hersam, Inkjet printing of high conductivity, flexible graphene patterns. J. Phys. Chem. Lett. **4**(8), 1347–1351 (2013)
30. E.B. Secor, S. Lim, H. Zhang, C.D. Frisbie, L.F. Francis, M.C. Hersam, Gravure printing of graphene for large-area flexible electronics. Adv. Mater. **26**(26), 4533–4538 (2014)
31. E.B. Secor, T.Z. Gao, A.E. Islam, R. Rao, S.G. Wallace, J. Zhu, K.W. Putz, B. Maruyama, M.C. Hersam, Enhanced conductivity, adhesion, and environmental stability of printed graphene inks with nitrocellulose. Chem. Mater. **29**(5), 2332–2340 (2017)
32. L. Huang, Y. Huang, J. Liang, X. Wan, Y. Chen, Graphene-based conducting inks for direct inkjet printing of flexible conductive patterns and their applications in electric circuits and chemical sensors. Nano Res. **4**(7), 675–684 (2011)
33. C. Sriprachuabwong, C. Karuwan, A. Wisitsorrat, D. Phokharatkul, T. Lomas, P. Sritongkham, A. Tuantranont, Inkjet-printed graphene-PEDOT:PSS modified screen printed carbon electrode for biochemical sensing. J. Mater. Chem. **22**(12), 5478 (2012)
34. T. Vuorinen, J. Niittynen, T. Kankkunen, T.M. Kraft, M. Mäntysalo, Inkjet-printed graphene/PEDOT:PSS temperature sensors on a skin-conformable polyurethane substrate. Sci. Rep. **6**(1), 35289 (2016)
35. T.-L. Chang, Z.-C. Chen, S.-F. Tseng, Laser micromachining of screen-printed graphene for forming electrode structures. Appl. Surf. Sci. **374**, 305–311 (2016)
36. J. Baker, D. Deganello, D.T. Gethin, T.M. Watson, Flexographic printing of graphene nanoplatelet ink to replace platinum as counter electrode catalyst in flexible dye sensitised solar cell. Mater. Res. Innov. **18**(2), 86–90 (2014)
37. W.J. Hyun, E.B. Secor, M.C. Hersam, C.D. Frisbie, L.F. Francis, High-resolution patterning of graphene by screen printing with a silicon stencil for highly flexible printed electronics. Adv. Mater. **27**(1), 109–115 (2015)
38. G. Hu, T. Albrow-Owen, X. Jin, A. Ali, Y. Hu, R.C.T. Howe, K. Shehzad, Z. Yang, X. Zhu, R.I. Woodward, T.-C. Wu, H. Jussila, J.-B. Wu, P. Peng, P.-H. Tan, Z. Sun, E.J.R. Kelleher, M. Zhang, Y. Xu, T. Hasan, Black phosphorus ink formulation for inkjet printing of optoelectronics and photonics. Nat. Commun. **8**(1), 278 (2017)
39. S. Santra, G. Hu, R.C.T. Howe, A. De Luca, S.Z. Ali, F. Udrea, J.W. Gardner, S.K. Ray, P.K. Guha, T. Hasan, CMOS integration of inkjet-printed graphene for humidity sensing. Sci. Rep. **5**(1), 17374 (2015)
40. D. Dodoo-Arhin, R.C.T. Howe, G. Hu, Y. Zhang, P. Hiralal, A. Bello, G. Amaratunga, T. Hasan, Inkjet-printed graphene electrodes for dye-sensitized solar cells. Carbon **105**, 33–41 (2016)
41. M. Singh, H.M. Haverinen, P. Dhagat, G.E. Jabbour, Inkjet printing-process and its applications. Adv. Mater. **22**(6), 673–685 (2010)
42. Z. Cui, *Printed Electronics* (Wiley, Singapore, 2016)
43. G. Hu, J. Kang, L.W.T. Ng, X. Zhu, R.C.T. Howe, C. Jones, M.C. Hersam, T. Hasan, Functional inks and printing of two-dimensional materials. Chem. Soc. Rev. **47**(9), 3265–3300 (2018)

44. Y. Gao, W. Shi, W. Wang, Y. Leng, Y. Zhao, Inkjet printing patterns of highly conductive pristine graphene on flexible substrates. Ind. Eng. Chem. Res. **53**(43), 16777–16784 (2014)
45. K. Arapov, E. Rubingh, R. Abbel, J. Laven, G. de With, H. Friedrich, Conductive screen printing inks by gelation of graphene dispersions. Adv. Funct. Mater. **26**(4), 586–593 (2016)
46. V. Dua, S.P. Surwade, S. Ammu, S.R. Agnihotra, S. Jain, K.E. Roberts, S. Park, R.S. Ruoff, S.K. Manohar, All-organic vapor sensor using inkjet-printed reduced graphene oxide. Angew. Chem. Int. Ed. **49**(12), 2154–2157 (2010)
47. J. Li, M.M. Naiini, S. Vaziri, M.C. Lemme, M. Östling, Inkjet printing of MoS_2. Adv. Funct. Mater. **24**(41), 6524–6531 (2014)
48. A.G. Kelly, D. Finn, A. Harvey, T. Hallam, J.N. Coleman, All-printed capacitors from graphene-BN-graphene nanosheet heterostructures. Appl. Phys. Lett. **109**(2), 023107 (2016)
49. J.-L. Capelo-Martnez (ed.), *Ultrasound in Chemistry* (Wiley-VCH Verlag, Weinheim, 2008)
50. A. Pattammattel, C.V. Kumar, Kitchen chemistry 101: multigram production of high quality biographene in a blender with edible proteins. Adv. Funct. Mater. **25**(45), 7088–7098 (2015)
51. P.G. Karagiannidis, S.A. Hodge, L. Lombardi, F. Tomarchio, N. Decorde, S. Milana, I. Goykhman, Y. Su, S.V. Mesite, D.N. Johnstone, R.K. Leary, P.A. Midgley, N.M. Pugno, F. Torrisi, A.C. Ferrari, Microfluidization of graphite and formulation of graphene-based conductive inks. ACS Nano **11**(3), 2742–2755 (2017)
52. R.J. Smith, P.J. King, M. Lotya, C. Wirtz, U. Khan, S. De, A. O'Neill, G.S. Duesberg, J.C. Grunlan, G. Moriarty, J. Chen, J. Wang, A.I. Minett, V. Nicolosi, J.N. Coleman, Large-scale exfoliation of inorganic layered compounds in aqueous surfactant solutions. Adv. Mater. **23**(34), 3944–3948 (2011)
53. D.H. Dalwadi, C. Canet, N. Roye, K. Hedman, Rheology: an important tool in ink development. Am. Lab. **37**(23), 18 (2005)
54. A. Davey, T. Guthrie, C. Haip, Flow properties of fluid systems. Surf. Coat. Int. **1991**(7), 329–339 (1991)
55. F. Coquel, E. Godlewski, N. Seguin, Relaxation of fluid systems. Math. Models Methods Appl. Sci. **22**(8), 1250014 (2012)
56. D.V. Boger, An introduction to rheology. J. Non-Newtonian Fluid Mech. **32**(3), 331–333 (1989)
57. H. Green, Rheological properties of paints, varnishes, lacquers, and printing inks. J. Colloid Sci. **2**(1), 93–98 (1947)
58. Z. Żołek-Tryznowska, *Printing on Polymers* (Elsevier, Amsterdam, 2016)
59. R. Buchdahl, J.E. Thimm, The relationship between the rheological properties and working properties of printing inks. J. Appl. Phys. **16**(6), 344 (1945)
60. A. Eich, *Visco Handbook* (SI Analytics GmbH, Mainz, 2015)
61. H.A. Barnes, J.F. Hutton, K. Walters, *An Introduction to Rheology* (Elsevier, Amsterdam, 1989)
62. T.I. Gudkova, L.A. Kozarovitskii, The relation between rheological properties of printing inks and their behavior in the printing process. Nauch. Trudy Moskov. Poligraf. Inst. **11**, 219–245 (1959)
63. T. Hartford, Rheological measurement of printing ink. Vehicles at low shear conditions. Am. Ink Maker **72**(11, Pt. 1), 42–49 (1994)
64. P. Oittinen, O. Perila, Rheological properties of printing inks. I. Flow curve behavior and tack. Suomen Kemistilehti B **45**(3), 95–99 (1972)
65. P. Oittinen, O. Perila, Rheological properties of printing inks. II. Flow curve and tack behavior and the picking of paper. Suomen Kemistilehti B **45**(3), 100–104 (1972)
66. P.C. Mishra, S. Mukherjee, S.K. Nayak, A. Panda, A brief review on viscosity of nanofluids. Int. Nano Lett. **4**(4), 109–120 (2014)
67. H. Pahlke, Rheological behavior of printing inks. Farbe + Lack **75**(3), 236–243 (1969)
68. K. Watanabe, T. Amari, Rheological properties of coatings during drying processes. J. Appl. Polym. Sci. **32**(2), 3435–3443 (1986)
69. H.C. Hamaker, The London–van der Waals attraction between spherical particles. Physica **4**(10), 1058–1072 (1937)

70. B.N. Shakhkel'dyan, The rheological properties of printing inks. Kolloidn. Zh. **18**, 111–119 (1956)
71. M. Black, L. Lin, J. Guthrie, Electrochemical sensors, US5658444A (1997)
72. W. Herbst, T. Hofheim, A. Rudolphy, H. Peter Simson, US3950288 - Pigment Compositions in Paste or Powder Form. Wiesbaden (1998)
73. L. Lin, Mechanisms of pigment dispersion. Pigm. Resin Technol. **32**(2), 78–88 (2003)
74. D.L. Zhang, Processing of advanced materials using high-energy mechanical milling. Prog. Mater. Sci. **49**(3–4), 537–560 (2004)
75. I. Bratkowska, W. Zwierzykowski, Effect of raw materials on rheological properties of carbon black dispersion in mineral oil used for printing inks production. Przem. Chem. **66**(8), 393–395 (1987)
76. A. Carlson, A.M. Bowen, Y. Huang, R.G. Nuzzo, J.A. Rogers, Transfer printing techniques for materials assembly and micro/nanodevice fabrication. Adv. Mater. **24**(39), 5284–5318 (2012)
77. F.M. Fowkes (ed.), *Contact Angle, Wettability, and Adhesion*. Advances in Chemistry, vol. 43 (American Chemical Society, Washington, 1964)
78. R.J. Good, A thermodynamic derivation of Wenzel's modification of Young's equation for contact angles; together with a theory of hysteresis. J. Am. Chem. Soc. **74**(20), 5041–5042 (1952)
79. L.R. White, On deviations from Young's equation. J. Chem. Soc. Faraday Trans. 1 F **73**, 390–398 (1977)
80. J.D. Berry, M.J. Neeson, R.R. Dagastine, D.Y.C. Chan, R.F. Tabor, Measurement of surface and interfacial tension using pendant drop tensiometry. J. Colloid Interface Sci. **454**, 226–237 (2015)
81. G.L. Robertson, *Food Packaging: Principles and Practice*, 3rd edn. (CRC Press, Boca Raton,2012)
82. D.Y. Kwok, A.W. Neumann, Contact angle measurement and contact angle interpretation. Adv. Colloid Interface Sci. **81**(3), 167–249 (1999)
83. W.C. Wake, Theories of adhesion and uses of adhesives: a review. Polymer **19**(3), 291–308 (1978)
84. R.C. Tolman, The effect of droplet size on surface tension. J. Chem. Phys. **17**(3), 333 (1949)
85. D. Zhang, L.C. Wadsworth, Corona treatment of polyolefin films - a review. Adv. Polym. Technol. **18**(2), 171–180 (1999)
86. Y. Rotenberg, L. Boruvka, A.W. Neumann, Determination of surface tension and contact angle from the shapes of axisymmetric fluid interfaces. J. Colloid Interface Sci. **93**(1), 169–183 (1983)
87. O.I. del Río, A.W. Neumann, Axisymmetric drop shape analysis: computational methods for the measurement of interfacial properties from the shape and dimensions of pendant and sessile drops. J. Colloid Interface Sci. **196**(2), 136–147 (1997)
88. M.D. Pashley, J.B. Pethica, D. Tabor, Adhesion and micromechanical properties of metal surfaces. Wear **100**(1–3), 7–31 (1984)
89. P.E. Thomas, G.E. Raley, Corona treatment of perforated film, US4351784A (1982)
90. L.-H. Lee (ed.), *Fundamentals of Adhesion* (Springer, New York, 1991)
91. H. Krupp, Theory of adhesion of small particles. J. Appl. Phys. **37**(11), 4176 (1966)
92. C.Y. Kim, D.A.I. Goring, Surface morphology of polyethylene after treatment in a corona discharge. J. Appl. Polym. Sci. **15**(6), 1357–1364 (1971)
93. J. Gassan, Effects of corona discharge and UV treatment on the properties of jute-fibre epoxy composites. Compos. Sci. Technol. **60**(15), 2857–2863 (2000)
94. J. Ferrante, J.R. Smith, Metal interfaces: adhesive energies and electronic barriers. Solid State Commun. **20**(4), 393–396 (1976)

95. P. Cheng, D. Li, L. Boruvka, Y. Rotenberg, A.W. Neumann, Automation of axisymmetric drop shape analysis for measurements of interfacial tensions and contact angles. Colloids Surf. **43**(2), 151–167 (1990)
96. J.U. Brackbill, D.B. Kothe, C. Zemach, A continuum method for modeling surface tension. J. Comput. Phys. **100**(2), 335–354 (1992)
97. M. Bousmina, Comparing the effect of corona treatment and block copolymer addition on rheological properties of polystyrene/polyethylene blends. J. Rheol. **39**(3), 499 (1995)
98. J.N. Anand, Re: contact theory of adhesion. J. Adhes. **5**(3), 265–267 (1973)
99. W. Chen, R.H.W. Lam, J. Fu, Photolithographic surface micromachining of polydimethylsiloxane (PDMS). Lab. Chip **12**(2), 391–395 (2012)
100. K.K. Hockman, D. Berengut, Design of experiments. Chem. Eng. **102**(11), 142 (1995)
101. J.A. Jacquez, Design of experiments. J. Frankl. Inst. **3358**(2), 259–279 (1998)
102. M.J. Anderson, P.J. Whitcomb, Design of experiments, in *Kirk-Othmer Encyclopedia of Chemical Technology* (Wiley, London, 2010)
103. G. Vicente, A. Coteron, M. Martinez, J. Aracil, Application of the factorial design of experiments and response surface methodology to optimize biodiesel production. Ind. Crops Prod. **8**(1), 29–35 (1998)
104. T.P. Ryan, *Modern Experimental Design* (Wiley, London, 2007)

Chapter 5
Printing Technologies

Abstract Printing is a mature, ubiquitous industry and is used extensively across the globe for mass producing a wide range of decorative products; from books and magazines through to packaging, advertising posters and even automobile dials. This chapter will provide an in-depth description of four of the most widely used methods: inkjet, screen, gravure and flexographic printing. In addition, 3D printing as a deposition method will also be discussed. The chapter includes an overview of the vital parameters of note for each of the different printing methods and gives a quantitative description of thc different interactions within individual printing systems. Finally, generic starting formulations of 2D material inks unique to each printing method are also provided.

The print process itself has a fundamental role in realising functional applications. The understanding and control of variables within the print process are important to ensure repeatability of performance. The four predominant printing technologies relevant to the deposition of 2D materials are inkjet, screen, gravure and flexographic printing. All these processes are widely used within the graphics printing industry. and are capable of handling narrow (<0.5 m in breadth) through to wide-format (>3 m breadth) substrates. They differ in their application method, printing speed, ink viscosity, resolution and whether or not there is any contact with the substrate. A comparison of these four processes is given in Table 5.1

The selection of print process is based on a fit-for-purpose approach as to the main requirements of the final product. Substrate requirements are a significant part in the selection of the printing method and it should be expected that for any application, a period of time will be required to optimise the interaction between ink formulation and the process parameters. Figure 5.1a provides an overview of these four printing technologies in the context of print speed, cost, the ease of making a prototype and the level of research and development required, whereas Fig. 5.1b shows the difference in the print speed and resolution.

Inkjet printing, despite having the highest cost of print per device unit, is at present perhaps the most widely researched and exploited method of 2D material application development. This is primarily due to the ease of fabricating a single

L. W. T. Ng et al., *Printing of Graphene and Related 2D Materials*,
https://doi.org/10.1007/978-3-319-91572-2_5

Table 5.1 Comparison of print parameters across printing methods

Printing method	Ink viscosity (cP)	Minimum line width (μm)	Minimum print thickness (μm)	Speed (m min^{-1})	Contact
Inkjet	(Very low) 4–30	30–50	~1	Slow 10 (m s^{-1})	No
Gravure	(Low) 100–1,000	10–50	~1	Fast ~1,000	Yes
Flexo	(Low-medium) 1,000–2,000	45–100	<1	Fast ~500	Yes
Screen	(High) 1,000–10,000	30–50	~ 5–100	Medium ~70	Yes

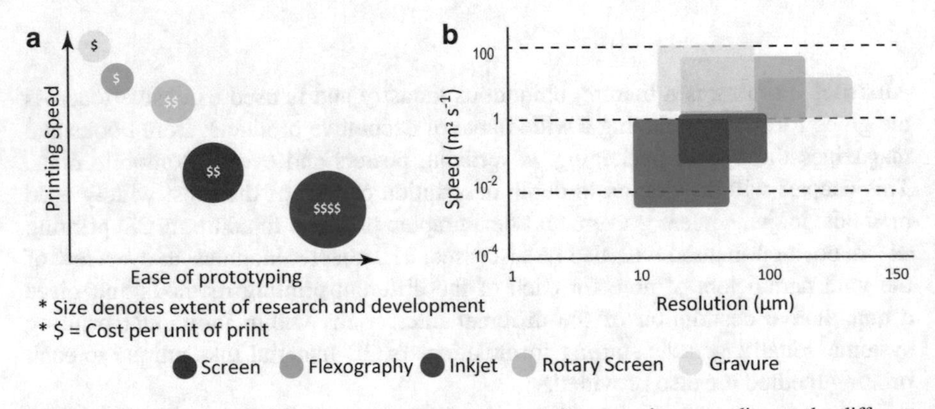

Fig. 5.1 (**a**) Comparison of price, print speed and ease of prototyping according to the different print methods. (**b**) Comparison of print speed and best achievable resolution ranges according to print technology

device without the need to produce large physical templates, and the relative ease of ink formulation from liquid phase exfoliated (LPE) 2D crystals. Inkjet printing has two key drawbacks, in having the highest cost per unit and the slowest printing speed. This makes it the least scalable of all the methods discussed in this chapter. However, the process does offer high resolution which is required for fine pitch applications such as transistors. This, when combined with the maskless patterning capability, makes inkjet printing well-suited to 2D material device prototyping.

Screen and rotary screen-printing are closely related as printing methods and offer a good balance between ease of prototyping, speed and cost. For prototyping purposes, flat-bed screen printing is particularly ideal, as it is a low cost, simple process. Results obtained using the flatbed process can easily be translated onto the rotary screen process for scale-up. Rotary screen is more suited to scale up to large volume than conventional flat-bed screen printing, although the transition entails larger physical stencils and more complicated machinery, which introduces additional control parameters.

Screen and rotary screen printing in general are not as capable in printing fine resolution prints as inkjet printing. Although rotary screen-printing is capable of reaching reasonable speeds, it does not offer as high a throughput as flexography or gravure printing.

Flexography and gravure printing are closely related in terms of print speed and their low viscosity of ink. Both processes offer the greatest potential for scale-up production due to their ability to achieve very low costs per unit of print. However, as we will discuss later in the chapter, flexography and gravure printing are the least researched of the methods due to their relatively high prototyping costs and the need to produce large amounts of ink.

This chapter offers an overview of these printing processes, a summary of the work done to date on 2D material deposition and provides starting ink formulation guidelines. This chapter also addresses 3D printing, in particular fused deposition modelling (FDM), as a practical manufacturing process.

5.1 Common Printing Parameters

Although each of the printing methods have their own unique properties, they share certain parameters and common issues. The important parameters to note when printing 2D material devices include:

- *Printing accuracy and resolution*: For printed electronics such as printed TFTs and printed display applications, a resolution of a few micrometres with $\pm 5\ \mu m$ registration accuracy will be required. Multilayer printing accuracy or registration between layers is important to manage to ensure minimal device defects.
- *Uniformity of print*: This is typically controlled by ink composition, design and the drying process.
- *Wetting control and interface formation*: Sharpness at pattern edges and bonding with substrates are strongly dependent on the surface tensions of the material/substrate and its design.
- *Ink compatibility*: The compatibility of inks with printing components such as rollers, masks, doctor blades and inkjet heads. This has a significant effect on yield and quality in mass production.

In functional printing, the viscosity and wettability of the ink in film formation and uniform deposition is vital. We will therefore discuss in detail how viscosity directly affects functional print quality via the coffee ring effect and how substrate wetting can be achieved through manipulation of the liquid components of an ink.

5.1.1 Viscosity Effects on Ink

There are many potential defects when printing (e.g. pinholes and voids). However, perhaps the greatest concern when printing 2D materials is the coffee ring effect as it represents a critical non-uniform deposition. Such deposition strongly affects the characteristics of the final printed device or coating and print-to-print consistency. This is especially so in electronic applications at fine pitch resolutions where overprinting (i.e. ink on ink) and registration (alignment between print layers) are crucial parameters.

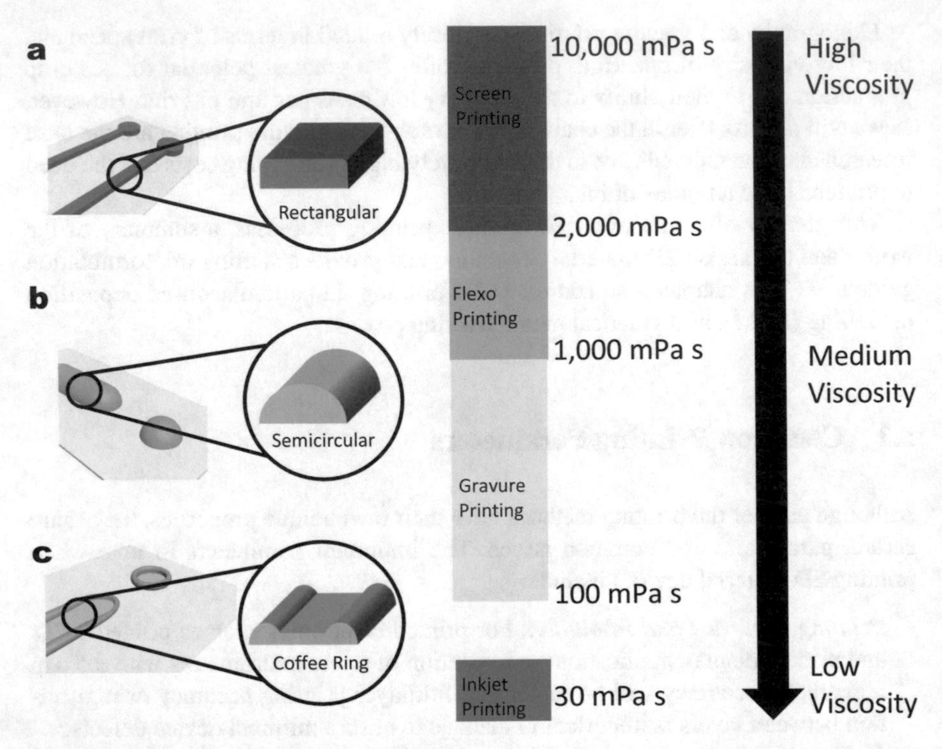

Fig. 5.2 Typical profiles of printed lines of (**a**) high and (**b**) medium and (**c**) low viscosity inks. As the viscosity of the ink utilised by the printing process is reduced, the greater is the likelihood of the coffee ring formation

Figure 5.2a–c schematically presents the typical cross sectional profiles of printed lines at different ink viscosities. Typically, a rectangular shape is most desired, as this represents a uniform deposition of material and hence, a uniform morphology across the printed structure [1]. However, this usually occurs only in the case of highly viscous inks, such as screen inks. With a decrease in ink viscosity, less polymeric binder content remains present in the ink to hold a stable dispersion. Eventually, very low viscosities lead to an undesired coffee-ring effect which deposits material unevenly on the periphery of the print; Fig. 5.2c. A rectangular profile is easy to achieve on screen printing systems and can be readily achieved on flexo and gravure systems by increasing the binder content (and hence the viscosity of the ink). Inkjet-printable inks, which are typically of far lower viscosity, are unlikely to achieve a rectangular profile. Therefore, the target for inkjet printing applications is a semicircular cross-sectional profile.

The coffee ring effect is caused by a range of factors that include the ink viscosity, wettability of the substrate and solvent vapourisation uniformity [1]. However, the greatest contributing factor for the coffee ring effect is related to the ink (or dispersion) drying processes [2, 3]. Deegan et al. proposed that the coffee ring effect arose from an unbalanced evaporation in a droplet during the drying process [2, 3].

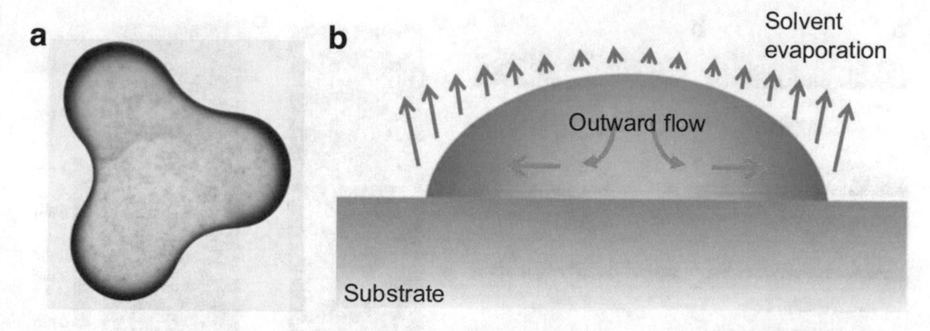

Fig. 5.3 (**a**) Photograph showing how a drop of coffee containing 1 wt% solids forms a perimeter ring with the solids. Reproduced with permission from Ref. [2]. Copyright (1997) Nature Publishing Group; (**b**) Schematic showing high evaporation at the droplet edges during coffee ring formation. Adapted from Ref. [1]

A droplet deposited on to a substrate first forms contact lines at the droplet–substrate interface. The higher surface area to volume ratio at the contact edges causes a more rapid solvent evaporation than in the droplet centre which leads to an outward solvent flow from the droplet centre to the edges to replenish the evaporated solvents [2, 3]. This flow carries the dispersed material to the droplet edges. If there is no (or weak) inward Marangoni flow, the material remains deposited at the edges which forms a material ring, leaving little to no material deposited at the centre of the dried droplet; Fig. 5.3b.

One promising way to overcome this is to induce a secondary inward flow to balance the evaporation-induced outward flow. In absence of any inward flows in the primary ink carrier solvent, preceding studies show that creating a solvent composition variation across the droplet may give rise to a surface tension and composition gradient driven Marangoni flow [4–10].

For example, in an ink composed of a binary solvent mixture, the differing solvent properties can lead to variations in the solvent proportions across the droplet during drying, such that a surface tension, composition and temperature gradient is generated to drive an overall Marangoni flow. Figure 5.4a, b presents a simplified schematic of the kinetic solvent drying of a binary solvent ink with and without Marangoni flow [10]. Hu et al. demonstrated the effect of such induced Marangoni flows in a black phosphorus (BP) ink with a binary solvent of isopropyl alcohol (IPA) and 2-butanol. The authors proposed that a combination of the surface tension gradient and composition gradient drove the formation of a Marangoni flow; Fig. 5.4c, d [5–7, 10]. This inward Marangoni flow redistributed BP flakes back to the droplet centre, rendering a significantly improved deposition uniformity [10]. This contrasts with the majority of single solvent ink systems demonstrated for 2D materials, such as those based on N-methyl-2-pyrrolidone (NMP) or IPA, where obvious coffee rings can be seen; Fig. 5.4c, d. Such effective control of solvents is a viable strategy to improve the deposition uniformity in low viscosity ink systems.

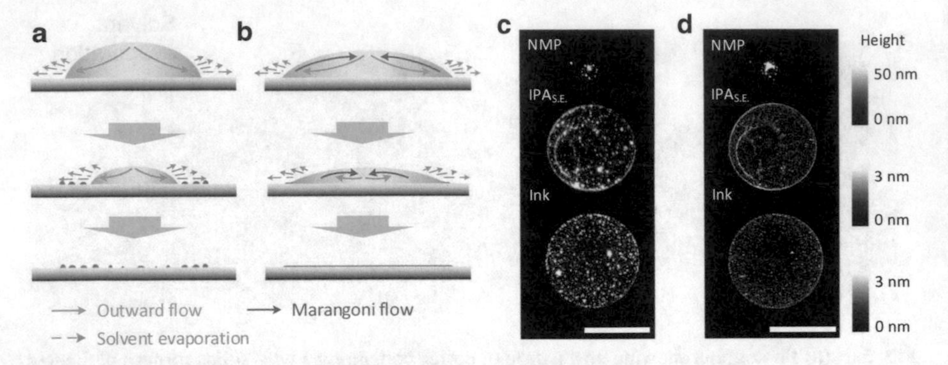

Fig. 5.4 Droplet drying (**a**) without and (**b**) with a Marangoni flow induced to prevent coffee ring effects. (**c**) Optical micrographs and (**d**) AFM images of the dried droplets, scale bar 50 μm. Reproduced with permission from Ref. [10]. Copyright (2017), American Chemical Society

5.1.2 Ink Wetting of Surface

Another important parameter to consider when depositing 2D materials is ink wetting of the substrate surface. As discussed in the previous chapter, the surface energy of non-absorbent substrates such as PET and coated papers which affects the spread of any liquid (in the context of this book, a 2D material ink). In general, good wetting is achieved when the surface tensions of the liquid are lower than that of the surface tension of the substrate. A guideline given in a packaging manual by Robertson states that the surface tension of the ink should be at least 7–10 mN m^{-1} lower than that of the critical surface tension (CST) of the substrate to ensure good, printable wetting [11].

The wetting ability of a liquid can be characterised using simple methods such as the sessile drop test using a contact angle goniometer [12–14].

However, when doing predictive formulation, it is also possible to measure the individual liquid's surface tension and the surface energy of a substrate separately. The surface energy of a liquid can be measured using the pendant drop method where the drop is suspended from a needle and the surface tension of the liquid can be ascertained from the shape of the liquid drop.

The pendant drop method is based on the Young–Laplace equation, which relates the Laplace pressure across an interface with the curvature of the surface and the interfacial tension (γ_{l-v}) ([12])

$$\gamma_{l-v}\left(\frac{1}{R_1}+\frac{1}{R_2}\right)=\Delta P\equiv\Delta P_0-\Delta\rho gz \tag{5.1}$$

where R_1 and R_2 are the principal radii of curvature; $\Delta P \equiv P_{in} - P_{out}$ is the Laplace pressure across the interface; $\Delta\rho = \rho_d - \rho$ where $\Delta\rho$ is the density difference between the drop phase density, ρ_d and the continuous phase density ρ.

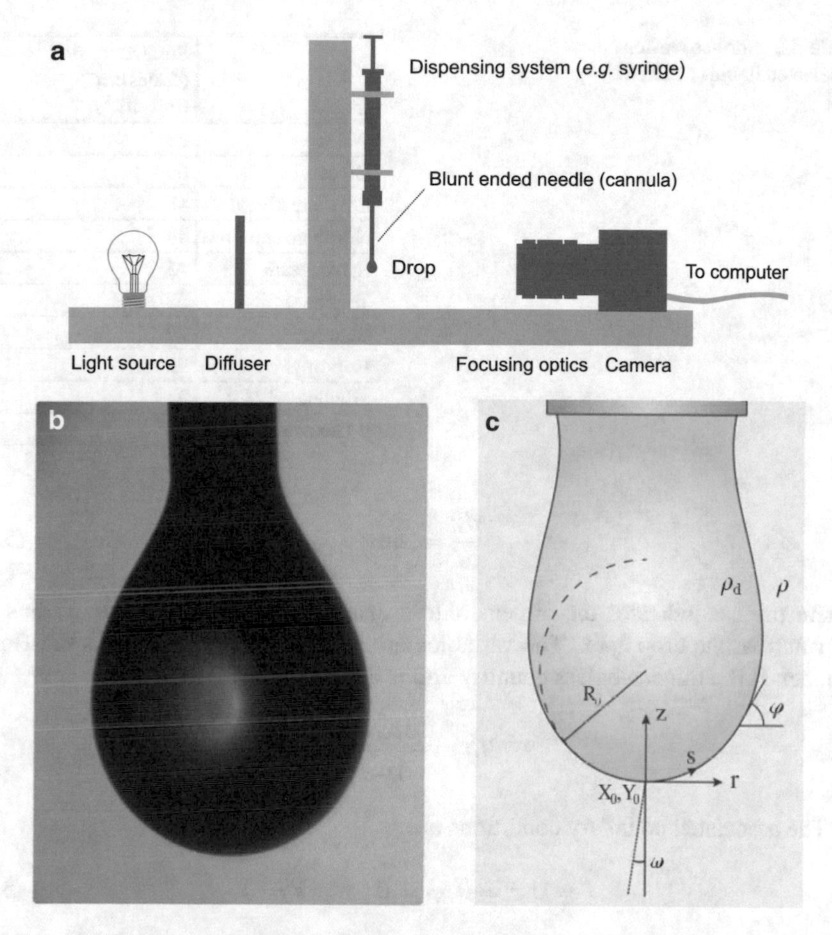

Fig. 5.5 (**a**) A basic goniometer set-up for droplet tensionometry. (**b**) A photograph of a typical droplet acquired by a camera. (**c**) Parameters measured during droplet tensionometry. Adapted with permission from Ref. [12]. Copyright (2015) Elsevier Inc

This can be written in terms of a reference pressure ΔP_0 at $z = 0$ and a hydrostatic pressure $\Delta \rho g z$. As a droplet is asymmetrical, Eq. (5.1) can be expressed in terms of the cylindrical coordinates r and z, together with the tangent angle φ, as shown in Fig. 5.5c. Thus, the Young-Laplace equation can be obtained as a coupled set of dimensionless differential equations in terms of the arc length s measured from the drop apex.

$$\frac{d\varphi}{d\bar{s}} = 2 - Bo\bar{z} - \frac{\sin\varphi}{\bar{r}} \tag{5.2}$$

$$\frac{d\bar{r}}{d\bar{s}} = \cos\varphi \tag{5.3}$$

Table 5.2 Surface tensions of selected liquids at 20 °C [15]

Liquid	Surface tension (dynes cm^{-1}) (mN m^{-1})
Water	73
Glycerol	63
Ethylene glycol	48
Nitrobenzene	44
Epoxy resin	43
Terpineol	33
Benzene	29
Isopropyl alcohol	22
Silicone oil	21
n-Hexane	18

$$\frac{d\bar{z}}{d\bar{s}} = \sin\varphi \tag{5.4}$$

where the bar indicates the dimensionless quantities scaled by R_0, the radius of curvature at the drop apex. The variables are defined in Fig. 5.5c. Bo is the Bond number. It is a dimensionless quantity and is defined by

$$Bo \equiv \frac{\Delta\rho g R_0^2}{\gamma_{l-v}} \tag{5.5}$$

The associated boundary conditions are

$$\bar{r} = 0, \bar{z} = 0, \bar{\varphi} = 0 \quad \text{at} \quad \bar{s} = 0 \tag{5.6}$$

The shape of the pendant drop is therefore dependent on Bo. If Bo associated with a pendant drop can be determined together with the drop radius R_0 at the apex, the interfacial tension γ_{l-v} is then readily obtained from Eq. (5.5) [12–14]. Modern goniometers are typically supplied with software that can automatically compute and derive the γ_{l-v} of any liquid, removing the need for complex computation on the part of the user. The surface tensions of some sample liquids are given in Table 5.2.

A good indicator of the surface energy of a substrate is its CST (γ_c), the surface tension where ideal spreading occurs [12–14, 16]. This can be ascertained using the Zisman method, as discussed in the previous chapter [17]. In the Zisman method, liquids of known surface tensions are dropped on a surface and their γ_{l-v} is plotted against the $\cos\theta$ measured; Fig. 5.6a. If no complete wetting occurs, a best-fit line is drawn through the points up to $\cos\theta = 1$. This then gives the γ_c of the particular material.

A demonstration of this on four different surfaces conducted by Kabza et al. is given in Fig. 5.6b–e using four different liquids, *n*-propanol, acetic acid, dioxane and acetonitrile. The Teflon surface measured the lowest γ_c of the four materials, making it the most difficult to wet. Table 5.3 gives the γ_c of some materials that may potentially act as substrates for printing.

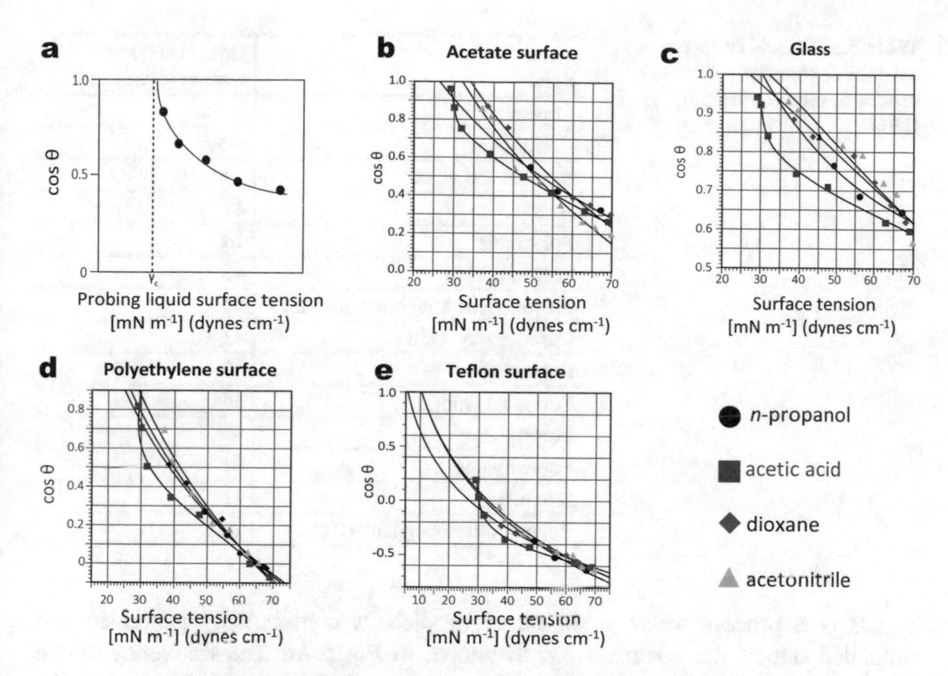

Fig. 5.6 Determination of the γ_c of different surfaces using different liquids. (**a**) Illustration of a Zisman plot and determination of γ_c at $\cos\theta$. Zisman plots of (**b**) acetate surface, (**c**) glass, (**d**) polyethylene and (**e**) Teflon surface determined using *n*-propanol, acetic acid, dioxane and acetonitrile as test liquids. Adapted with permission from Ref. [16]. Copyright (2000) American Chemical Society

In predicting the spreading behaviour of inks on the surface, it may be advantageous to find out the γ_{l-v} of a solvent before using it as a diluent for ink systems, as well as the γ_c of a substrate. Appropriate spreading or wetting of a surface ensures a uniform deposition of 2D materials, which in turn ensures device quality and printing consistency.

The following sections will discuss the printing methods mentioned above, beginning with the lower viscosity ink systems.

5.2 Inkjet Printing

Inkjet printing is a digital, non-impact printing technique in which ink droplets are propelled from an ink reservoir and deposited in a rapid succession onto the substrate to generate a designed image [9, 18–26]. The word 'digital' comes from the fact that both the printing pattern and deposition of ink droplets are controlled electronically. Figure 5.7 presents the schematics of the two prevalent droplet jetting mechanisms: continuous inkjet (CIJ) and drop-on-demand inkjet (DOD).

Table 5.3 Critical surface tensions of common substrates used for printing [15]

Solid	Critical surface tension (dynes cm^{-1})
Copper	~1000
Aluminium	~500
Sandblasted glass	290
Acetal	47
Polyamide	46
Polycarbonate	46
Polyethylene terephthalate (PET)	43
Cured epoxy resin	43
Polyimide	40
Polyvinyl chloride	39
Polystyrene	33
Polyethylene	31
Silicone	21
Polytetrafluoroethylene (PTFE)	18

CIJ is a process where a stream of droplets is continuously generated and propelled out of the ink reservoir, as shown in Fig. 5.7a. The frequency of the droplets is controlled by regulating the pressure applied to the ink within the reservoir. The droplets then pass through an electrostatic field, such that the individual droplets are either charged or not charged. The charged droplets are then selectively deflected to deposit onto substrate while the uncharged ones are collected by the ink receiver.

DOD is a process where the ink droplets are only generated and jetted on demand. As shown in Fig. 5.7b, c, the ink droplets are generated either through a piezoelectric inkjet process or a thermal inkjet process. In a piezoelectric inkjet process, a voltage pulse is applied to the piezoelectric material to generate a shape change of the reservoir. This exerts a pressure pulse on the ink that forces it out of the ink reservoir as ink droplets. In a thermal inkjet process, the ink is rapidly heated up to generate bubbles in the ink. The bubbles propel the ink out of the ink reservoir as ink droplets. The jetted ink droplets then impact, spread, merge with other printed droplets and dry on the substrate to form the desired image.

Inkjet printing offers several advantages over flexographic, gravure and screen printing: firstly, because the jetted droplets usually have a volume of 1–100 pL, inkjet printing achieves high resolution patterning, typically ~50–80 μm [19, 21, 26–28]; secondly, as the printing pattern is digitally defined, there is no need to design physical masks. Inkjet printing is therefore more flexible, faster and cheaper as a process in early prototyping of first devices in comparison to other printing processes with large set-up costs [9]. Finally, unlike other printing technologies, inkjet printing is a low loading, low viscosity ink printing process, and it does not require a large amount ink for printing trials. These features make it particularly suitable for prototyping [9]. For instance, the as-produced liquid phase exfoliated

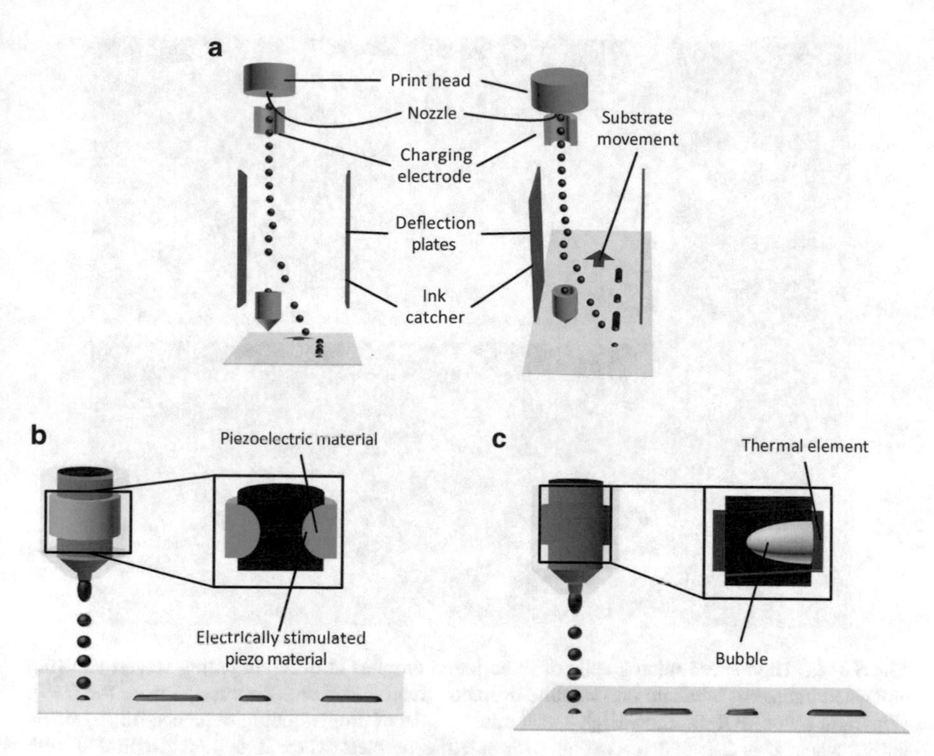

Fig. 5.7 (**a**) A continuous inkjet (CIJ) printer (**b**) Drop-on-demand (DOD) inkjet printer with piezoelectric head (**c**) DOD printer with thermal head.

(LPE) 2D material dispersions are readily adapted to inkjet printing for device fabrication [22, 29–37].

However, as the inkjet printing head needs to scan over the substrate to deposit ink droplets, inkjet printing is limited to low printing speeds, typically well below $1\ \mathrm{m\,s^{-1}}$ [38]. Among these two inkjet printing technologies, CIJ offers a number of advantages over its counterpart DOD.

CIJ allows a higher jetting speed and the use of volatile inks with low boiling point solvents as there is less risk of nozzle clogging resulting from ink drying. However, the complexity of CIJ (e.g. control of droplet jetting, deflecting and recycling) often limits the application of CIJ. Despite the limitations of nozzle clogging, DOD inkjet is the most widely researched and employed technique in the printing of 2D material inks. The similarities in viscosities between LPE dispersions and inkjet inks also means that inkjet printing has become an ideal method of printing 2D materials.

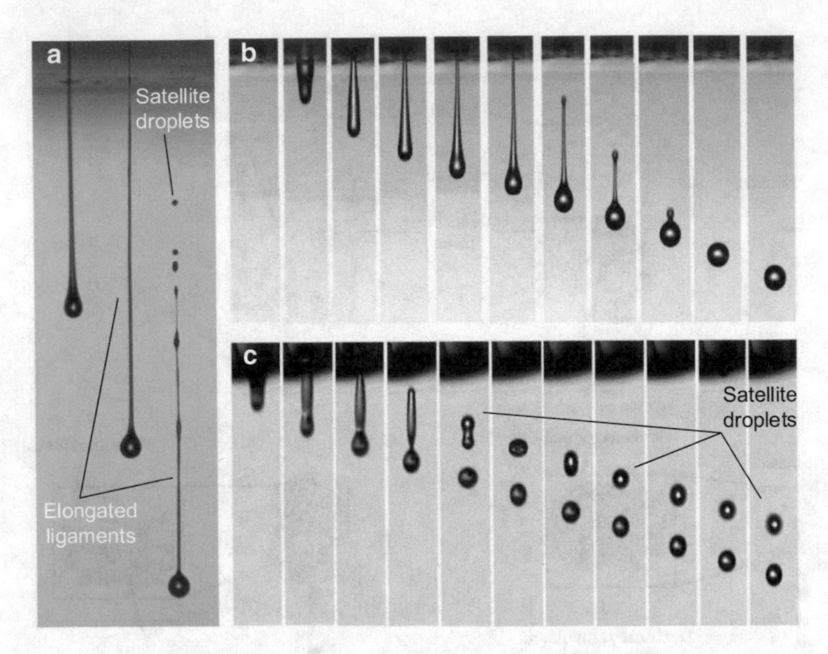

Fig. 5.8 (**a**) High-speed micrographs of three jetted droplets at different jetting stages, showing elongated ligaments break up into satellite droplets. Reproduced with permission from Ref. [39], Copyright (2008) IOP Science. High-speed micrographs of droplet jetting sequence for (**b**) stable jetting when Z is 2.2, and (**c**) jetting with satellite droplets when Z is 17.3. Reprinted with permission from Ref. [40]. Copyright (2009) American Chemical Society

5.2.1 *Inkjet Printing Principles*

Stable Droplet Formation and Jetting

A key requirement of DOD is the generation of stable droplets. DOD requires the formation of a single droplet and successful jetting under each electrical impulse without the formation of satellite droplets or secondary droplets, as shown in Fig. 5.8. Unstable jetting may lead to deviation from the droplet jetting trajectory or even deposition onto untargeted areas [9, 19, 21, 22, 26, 27, 40].

Stable droplet formation can theoretically be predicted by the Ohnesorge (*Oh*) number, which is characterised by the Reynolds (*Re*) and Weber (*We*) numbers [19, 22, 26]:

$$Re = \frac{\upsilon \rho a}{\eta} \tag{5.7}$$

$$We = \frac{\upsilon^2 \rho a}{\gamma_{l-\upsilon}} \tag{5.8}$$

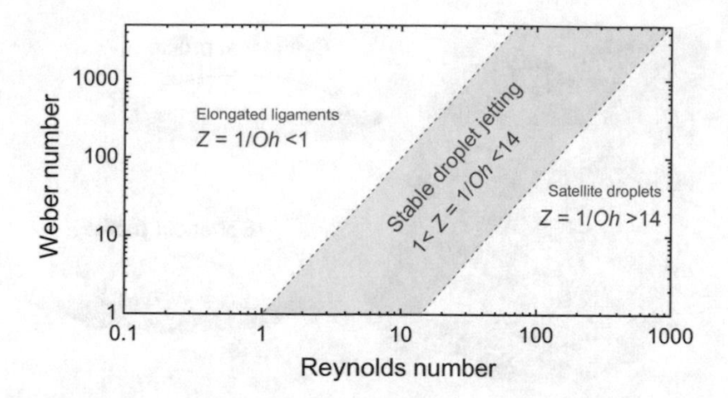

Fig. 5.9 Map of Reynolds (*Re*) and Weber (*We*) numbers, showing the optimal region of inverse Ohnesorge (*Oh*) number (1–14) for stable droplet jetting, adapted from Ref. [26]

The *Oh* number is defined by:

$$Oh = \frac{\sqrt{We}}{Re} = \frac{\eta}{\sqrt{\gamma_{l-v}\rho a}} \tag{5.9}$$

where η, γ_{l-v} and ρ are the viscosity (mPa s), surface tension (mN m^{-1}) and density (g cm^{-3}) of the ink, respectively, υ is the ejection velocity (m s^{-1}) of the droplet, and a is the diameter (μm) of the jetting nozzle. Hence, it is possible to achieve an *Oh* number by manipulation of an ink's viscosity, surface tension and density.

A 1984 publication on DOD printing by Fromm suggested the use of the inverse of the *Oh* number, $Z = 1/Oh$, to determine whether an ink could form a stable jetting [41]. As a rule of thumb, it is now commonly accepted that Z should be <14 to avoid formation of satellite droplets [19, 22, 26, 40]. Meanwhile, Z should be >1 to avoid elongated ligaments, which or even jetting failure [19, 22, 26, 40]. Figure 5.8b is a demonstration of stable droplet jetting when Z is within 1–14, while Fig. 5.8c shows generation of satellite droplets when Z is over 14. The above investigation allows the generation of a *Re* and *We* map (Fig. 5.9), defining the optimal value region of *Re* and *We* for stable droplet jetting. We note that these conditions are optimised for printing inks with particle sizes significantly smaller than the nozzle diameter.

Droplet Impact and Spreading

Following the impact of a droplet on a substrate, subsequent ink spreading and drying defines the printed image and morphology. This section considers the spreading process, while the drying process is discussed in Sect. 5.2.1. The dominant forces during droplet impact and spreading are mainly inertial and capillary forces. As droplets are of 1–100 pL volume with low density, the gravitational forces can typically be neglected after the impact [26]. It has been suggested that the spreading

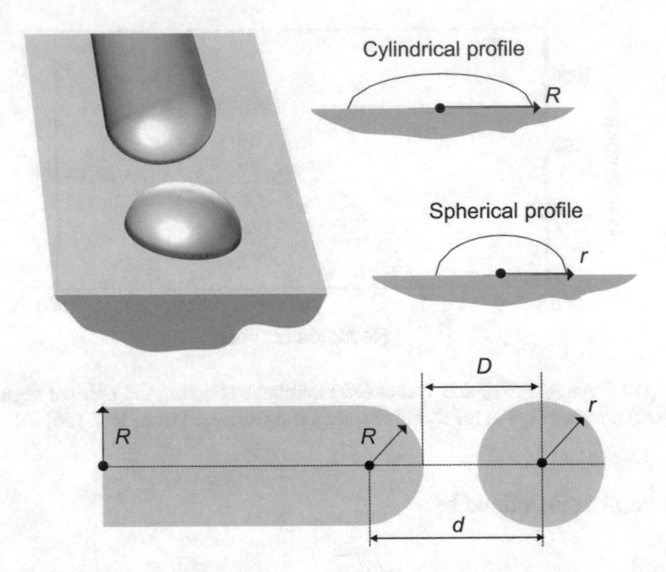

Fig. 5.10 Schematic figure with top view showing a droplet of a radius of r deposited to a defined cylindrical line of a radius of R with a droplet spacing of d

after impact can be divided into two stages: the impact driven stage and the capillary driven stage [24]. The impact driven stage is the spreading process immediately after the droplet comes in contact with the substrate, where inertial force dominates. This stage leads to a maximum spreading of the droplet until a capillary-force driven stage dominates. This can be further divided into two separate situations depending on the interaction between the droplet and the substrate.

For inks with sufficient wettability, the droplet continues a capillary spreading [11, 42]. For inks with insufficient wettability, the droplet retracts and 'beads up', leading to discontinuous deposition of ink [11, 42]. Therefore, to ensure a continuous printing feature, the ink surface tension has to be sufficiently low.

Upon deposition and subsequent wetting of the substrate, the impingement of a droplet onto a predefined pattern needs to be controlled in a way that it does not cause either overspreading or insufficient merging [43]. Figure 5.10 further depicts the schematic geometry of a droplet deposited onto a substrate [43]. Before solvent evaporation, the droplet has a spherical cap geometry with radius of r, while the line has a cylindrical geometry with radius of R. Thus, the spherical cap geometry can be defined as:

$$V = \frac{4}{3}\pi r^3 \delta \tag{5.10}$$

where δ is a volume correction factor. If d is optimal, the impingement is then ideal such that the droplet extends the line by d while the radius of the line remains R. The cylindrical geometry is then defined as:

$$\frac{V}{d} = \pi R^2 \delta' \tag{5.11}$$

where δ' is a volume correction factor. Note that δ is expected to be very small when the droplet wets the substrate well (i.e. forms a small contact angle). On the other hand, δ approaches 1 for contact angles close to 180°. The relationship between R and r can be written as:

$$\frac{R}{r} = \sqrt{\frac{4}{3d/r}\frac{\delta}{\delta'}} \tag{5.12}$$

In addition, as shown, the impinging distance from the droplet to the end of the line (defined herein as the 'bead') is:

$$D = d - R \tag{5.13}$$

Therefore, the relationship between D and r is:

$$\frac{D}{r} = \frac{d}{r} - \frac{R}{r} = \frac{d}{r} - \sqrt{\frac{4}{3d/r}\frac{\delta}{\delta'}} \tag{5.14}$$

The morphologies of the printed patterns are governed by the droplet impinging behaviour, which is essentially defined by the above parameters.

As depicted in Fig. 5.11a, b, when d is small ($d < R$, i.e. in Eq. (5.13): $D < 0$), the droplet lands on the bead of the line and expands around the bead rather than creating its own contact line, forming 'stacked coins' or 'bulging' lines. However,

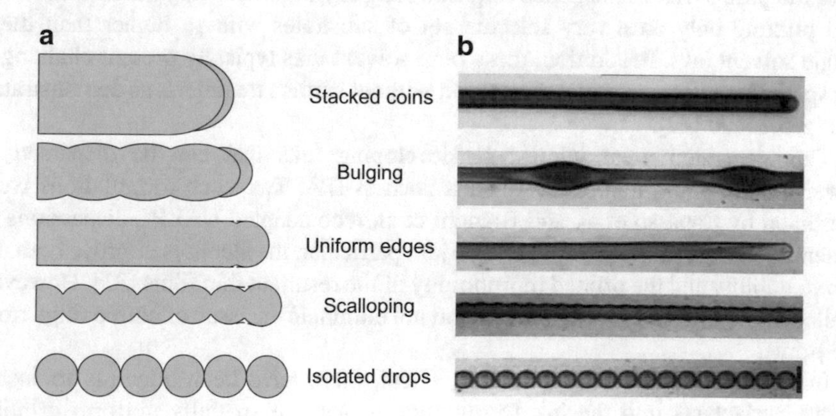

Fig. 5.11 (**a**) Schematic figure for droplets impinged onto a printed line, showing the formation of different morphologies of printed structures. The dashed lines indicate the contact lines; (**b**) Photographs showing different printed morphologies. Reproduced with permission from Ref. [43]. Copyright (2009) American Chemical Society

when d is large ($2R < d < 2r$, i.e. in Eq. (5.13): $D > R$), the impinging of the droplet with the bead is restrained, forming 'scalloped' lines. For excessively large d ($d > 2r$), the droplet does not impinge onto the line, forming 'isolated drops'. For d values between the bulging and scalloping scenarios (i.e. $0 < D < R$), the droplet impinges onto the bead and forms contact lines with uniform edges [43]. This simple geometrical calculation can be used as a guide to optimise the printing patterns of an inkjet ink.

5.2.2 *Current 2D Material Inkjet Ink Formulations*

The current formulations for 2D material inkjet inks have gone through three different generations starting with the pure solvent inks (i.e. the as-produced LPE dispersions in pure solvents), surfactant inks (i.e. the as-produced LPE dispersions with surfactants) and inks with binders (i.e. exfoliated 2D materials formulated into inks with the addition of polymers) [27].

As mentioned previously, Torrisi et al. first demonstrated inkjet printing ultrasound assisted LPE (UALPE) graphene dispersions in N-Methyl-2-pyrrolidone (NMP) as a pure solvent ink to print graphene transistors [22]. These UALPE dispersions generally have a viscosity of $\sim$2 mPa s with flake dimension of <1 µm, making the dispersions suitable for low viscosity printing and coating technologies for device fabrication, in particular inkjet printing [22, 29–31, 34, 35]. This method was later extended to other 2D materials, including transition metal dichalcogenides (TMDs; e.g. MoS_2, WS_2, WSe_2) and hexagonal boron nitride (h-BN) [29–31]. However, these stable UALPE dispersions usually require high boiling point, toxic organic solvents, with high surface tensions ($\sim$40 mN m^{-1}) such as NMP. Printing with the pure solvent inks hence required long drying times, very careful handling and printing only on a very selected set of substrates with γ_c higher than these single solvent inks. In addition, these pure solvent inks typically present challenges for spatially uniform material deposition without surface treatment, as demonstrated in Fig. 5.12c, d [22].

This generated great interest in developing inks that can be dispersed in lower boiling point, non-toxic solvents such as IPA. Two such formulations were presented by Capasso et al. and Bianchi et al. who adapted UALPE dispersions in water/alcohol for inkjet printing [33, 45]. In particular, the alcohols improve both the ink wettability and the printed morphology of the resultant thin films [33]. However, while wetting was improved, this still did not eliminate the issue of coffee rings from the prints.

In order to negate the coffee ring effect, there have been attempts to incorporate surfactants into the ink formulation to achieve spatially uniform printing [34, 35]. For instance, Ref. [32, 46] adapted as-produced UALPE dispersions of 2D crystals materials with surfactants and small molecules for inkjet printing. In particular, McManus et al. argued that pyrenesulphonic acid sodium salt (PS1)—small molecule surfactant—could lead to improved printed morphologies without

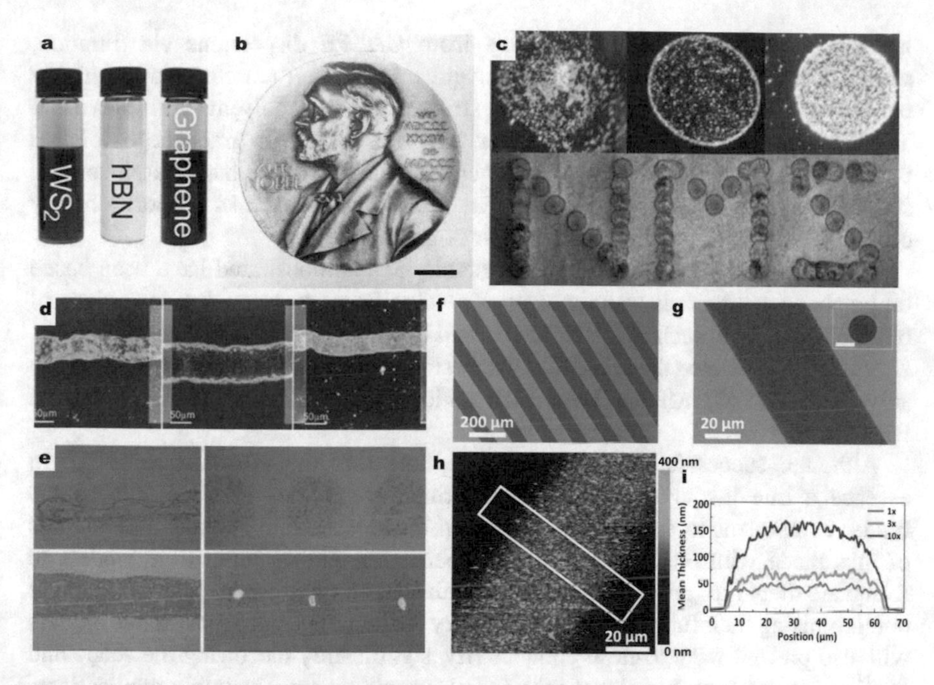

Fig. 5.12 Inkjet printing with current 2D material inks: (**a**) Photograph of representative NMP inks. Reprinted with permission from Ref. [30]. Copyright (2014) American Chemical Society; (**b**) Inkjet-printed graphene pattern on paper, scale bar 5 cm. Reprinted with permission from Ref. [32]. Copyright (2017) Nature Publishing Group; (**c**) Inkjet-printed droplets of graphene NMP ink on (top row) Si/SiO_2 treated with O_2 plasma, untreated Si/SiO_2, Si/SiO_2 treated with hexamethyl disiloxane (HMDS), and (bottom row) paper, respectively. Reprinted with permission from Ref. [22]. Copyright (2012) American Chemical Society. (**d**) Inkjet-printed lines of graphene NMP ink on untreated Si/SiO_2, Si/SiO_2 treated with O_2 plasma and Si/SiO_2 treated with HMDS, respectively, showing change in coffee ring effect. Reprinted with permission from Ref. [22]. Copyright (2012) American Chemical Society; (**e**) Inkjet-printed graphene water/PS1 ink on untreated Si/SiO_2. Reprinted with permission from Ref. [32] Copyright (2017) Nature Publishing Group; (**f**) Inkjet-printed graphene/EC ink on untreated Si/SiO_2. (**g**) A single printed line of the same inkjet-printed graphene (inset, droplet of graphene, scale bar corresponds to 40 µm) illustrating the uniformity of the printed features. (**h**) An atomic force microscopy (AFM) image of a single line after 10 printing passes that show no coffee ring features. (**i**) Averaged cross-sectional profiles of printed lines after 1, 3 and 10 printing passes, which demonstrate the reliable increase in thickness obtained after multiple printing passes. The cross-sectional profiles are obtained from the averaged AFM profile over ~20 µm as indicated by the boxed region in (**h**). (**f–i**) Reproduced with permission from Ref. [44]. Copyright (2013) American Chemical Society

compromising the functionalities (e.g. electrical conductivity, optoelectronic properties) of the 2D materials [32]. However, Fig. 5.12e indicates that the improvements are somewhat limited compared to pure solvent inks (Fig. 5.12c, d). In addition, incorporating a surfactant into the formulation also introduces unwanted impurities that may affect the performance of the fabricated device.

Eventually, it was found that formulating inks with binders such as ethyl cellulose (EC) offered a more effective approach. This is usually done as follows: (1) extracting the exfoliated 2D materials flakes from UALPE dispersions via filtration, sedimentation, solvent exchange, or solvent evaporation; (2) formulating highly concentrated inks by re-dispersing the extracted flakes in solvents with polymers. The first demonstration of such binder inks was in 2013: graphene exfoliated in dimethylformamide (DMF) was solvent exchanged into ethanol/terpineol and stabilised by EC [47]. As depicted in Fig. 5.12f–i, the such inks support a highly controllable, spatially uniform deposition of graphene [44].

To date, most 2D material inkjet printable inks demonstrated have been based on graphene and there have only been a few significant reported demonstrations of other 2D materials, such as *h*-BN, TMDs and BP [10, 27, 29–32, 48]. Most recently, Hu et al. demonstrated the successful dispersion and deposition of BP using a binary solvent mixture of NMP and IPA which achieved a spatially uniform deposition of BP [10].

Although successful in deposition, we note that this formulation is not in essence a true ink compared to inkjet printable graphics inks, as it lacks the binder component to ensure good adhesion to the substrate surface. The success of inks made with other 2D materials depends on finding compatible binders to formulate them into printable inkjet inks with spatially uniform deposition, without compromising their functionality. This is very challenging, since most stable binders will also prevent flake-to-flake connectivity, significantly narrowing the scope and applicability of such functional inks for electronic or optoelectronic applications. Table 5.4 summarises the demonstrated ink formulations to date in the literature and applications enabled by them.

5.3 Screen Printing

Screen printing or serigraphy is a stencil process whereby ink is transferred to the substrate through a stencil made of a fine fabric mesh of silk, synthetic fibres or metal threads. The image carrier is called a screen and is essentially a porous mesh. The screen is tightly stretched over a frame that is either made of wood or metal. The pores of the mesh are blocked up in the non-imaging areas by a photopolymerised resin. The remaining pores in the imaged areas are left open to allow ink to flow through [25, 53].

During printing, ink is first flooded over the screen. A squeegee is then drawn across it, applying shear forces to force the ink through the open pores of the screen which have not been covered by the photopolymerised resin. At the same time the substrate is held in close proximity the screen to facilitate the ink transfer [25, 53]. Many screen printing systems still consist of a simple hand-operated unit; Fig. 5.13a. These can be particularly useful when printing on very thick or thin substrates that cannot be automatically fed for printing or where a test run of a new image is required. Flatbed screen printers have the additional advantage of being able to register precise prints directly on top of each other [54]. This is very important during functional printing where precisely registered contacts and overlapping layers of functional materials are required to ensure device efficiency [55].

Table 5.4 Current inkjet-printable 2D material inks and their applications

Year	Inks	Substrates	Treatment	Applications	Ref.
Pure solvent inks					
2012	Graphene in NMP	Si/SiO_2; Glass	HMDS; O_2; 170 °C	Thin film transistors (TFTs)	[22]
2014	Graphene, MoS_2 in NMP	Coated PET	200 °C	Photodetectors	[29]
2014	Graphene, TMDs, *h*-BN in NMP	Si/SiO_2	–	TFTs Photodetectors	[30]
2015	Graphene in water, ethanol	PET	–	Conductor	[45]
2017	Graphene, TMDs, *h*-BN in NMP	Coated PET	–	TFTs	[31]
2017	Graphene in water, ethanol	Si/SiO_2	–	Non-linear optical device	[33]
2017	BP in Binary solvent of isopropanol, NMP	Si/SiO_2	–	Functional ink platform	[10]
Surfactant inks					
2014	Graphene in SDBS/PANI, water	Carbon	80 °C	Electrode	[46]
2017	Graphene, TMDs in PS1, water	Si/SiO_2	300 °C	Photodetectors; ROM	[32]
Binder inks					
2013	Graphene in EC,	Si/SiO_2; Glass	400 °C	TFTs	[47]
2013	Graphene in EC, ethanol	Si/SiO_2; Kapton	Up to 450 °C	Conductor	[44]
2014	Graphene in Plasdone S-630, isopropanol, butanol	Paper	–	Conductor	[49]
2014	MoS_2 in EC, terpineol	Si/SiO_2	HMDS	TFTs	[50]
2015	Graphene in EC, cyclohexanone and terpineol	PET; PEN; PI; Glass	HMDS; Pulsed light; Up to 300 °C	Conductor	[51]
2015	Graphene in PVP, isopropyl alcohol	Si/SiO_2	–	Humidity sensor	[37]
2016	Graphene in isopropanol–PVP	Si/SiO_2; FTO glass	–	Counter electrodes for solar cells	[36]
2017	Graphene in nitrocellulose, IPA, acetone, ethyl lactate, octyl acetate, ethylene glycol diacetate	Glass; Kapton	350 °C	Conductor	[52]
2018	Graphene in IPA-PVP	Kapton	–	Thermoelectrics	[28]

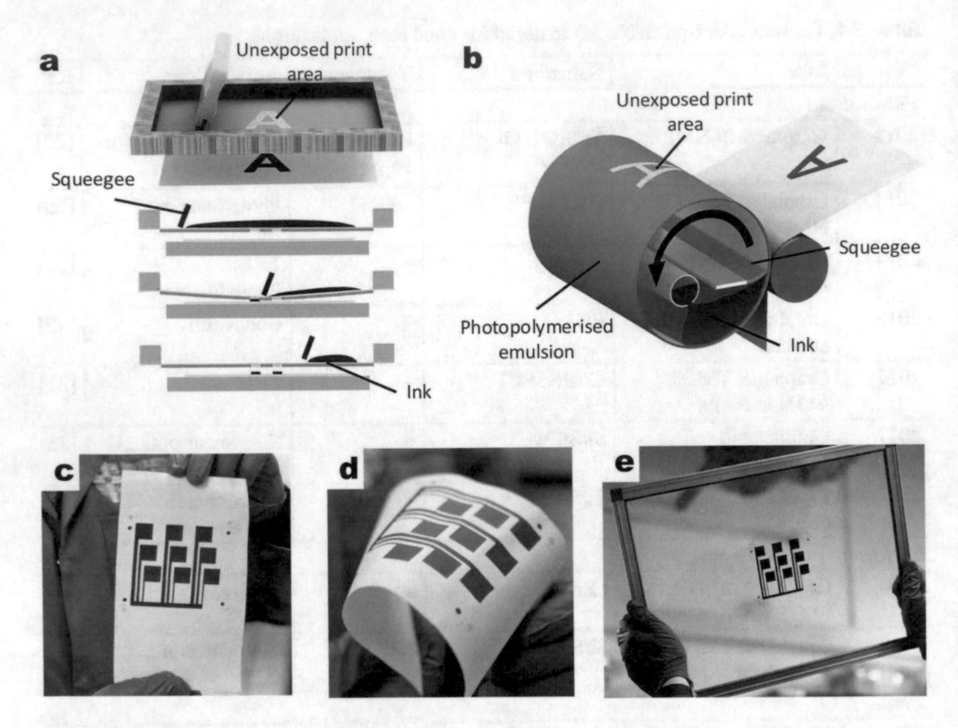

Fig. 5.13 Schematic of a (**a**) flatbed and (**b**) rotary screen printer. Screen printed conductive graphene ink on (**c**) paper when (**d**) bent and on (**e**) glass

Semi-automatic machines have the substrates fed and taken off by hand but utilise a mechanised squeegee. Fully automatic flat-bed presses are also available where the substrate is fed in and taken off by automatic feed and delivery systems. After printing, the sheets are taken through an air drier to evaporate any solvent. Prints emerge dry and can be stacked ready for further processing [56].

Highly efficient cylinder presses can also be employed for roll-to-roll (R2R) and higher speed sheet-fed operations; Fig. 5.13b. With these presses, the squeegee remains stationary and the screen, cylinder and substrate all move in unison. This permits faster operation, since the substrate does not have to be brought to a halt and fed into a vacuum base as with flat-bed presses. Speeds of up to $\sim$100 m s^{-1} can be achieved with this process. However, the rotary process can prove challenging when producing printed structures requiring particularly precise registration. This is because automatic registration control has been developed for the less critical application of graphics printing which may have typically greater tolerance than functional printing [57]. Because of their intrinsic simplicity, screens can be produced inexpensively, making it an attractive process for short-run work. Screen printing has been used to print 2D materials directly on substrates such as paper, plastic and glass; Fig. 5.13c, d [46, 58, 59].

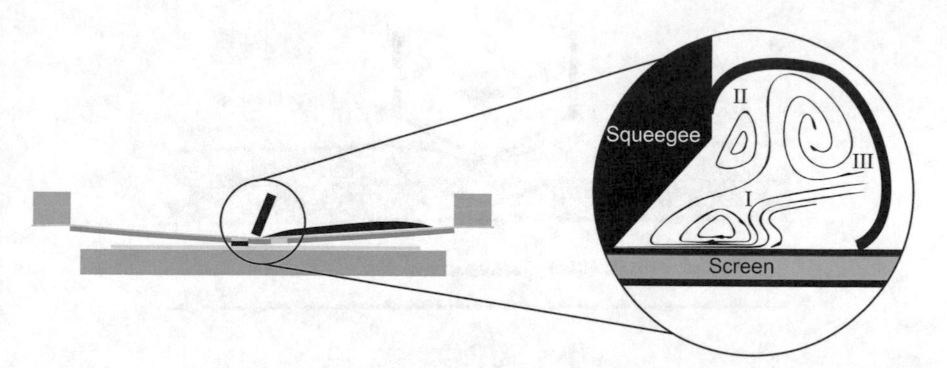

Fig. 5.14 Owczarek and Howland's tri-region model of screen printing flow where region (I) is the *pressurisation* region, (II) is the *downward screen cross-flow* region and (III) the *ink-collection* region

5.3.1 Screen Printing Principles

In screen printing, the thixotropic (shear-thinning) behaviour of the ink plays a critical role in printability [60, 61]. As the ink is drawn along the screen via the squeegee, the shear stress ($\dot{\gamma}$) acting upon the ink changes from very low ($\sim$1 s^{-1}) to very high ($\sim$1,000 s^{-1}) [62] while viscosity (η) decreases [60]. Upon deposition, $\dot{\gamma}$ decreases as shear stress from the squeegee is removed while η increases to form a solid, unbroken image on the substrate surface.

Owczarek and Howland modelled the flow of screen printing in 1990 [63–65] where they identified three distinct regions of the flow field in front of a squeegee as it is drawn across the screen; Fig. 5.14. Region I is the *pressurisation* region in which the flow is mainly affected by the viscous and pressure forces exerted on the screen. It includes the screen cross-flow region from the substrate towards the squeegee. On its left boundary, the ink escapes from this region under the squeegee in the masked regions and is forced through the mesh in the unmasked regions. Region II is the *downward screen cross-flow* region which is affected mainly by the viscous, gravitational and pressure forces. Region III is the *ink collection* region which is affected mainly by the inertial and gravitational forces.

The volume of ink deposited is governed by several factors; firstly, the thread count of the screen (i.e. the number of threads per unit distance and thread diameter), which in turn defines the open area percentage of the screen; secondly, the thread diameter which defines the thickness of the screen (D) and hence the depth of the ink column at each open hole in the mesh; thirdly, the pressure and angle (which in turn defines the area of the squeegee in contact with the screen), and finally, the speed of the blade with respect to the screen during ink deposition [25, 53, 55, 66].

The thickness of the ink under the squeegee, H_{SC}, can be theoretically calculated by subtracting the volume of the wires, V_W, and the volume placed by the squeegee tip, V_{DS}, from the volume of the screen filled with ink and adding the equivalent

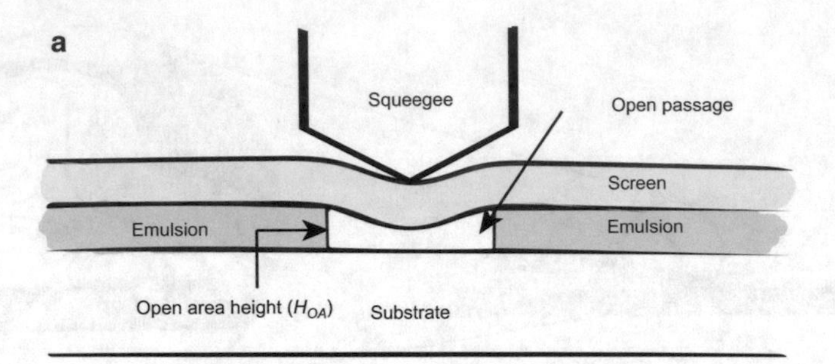

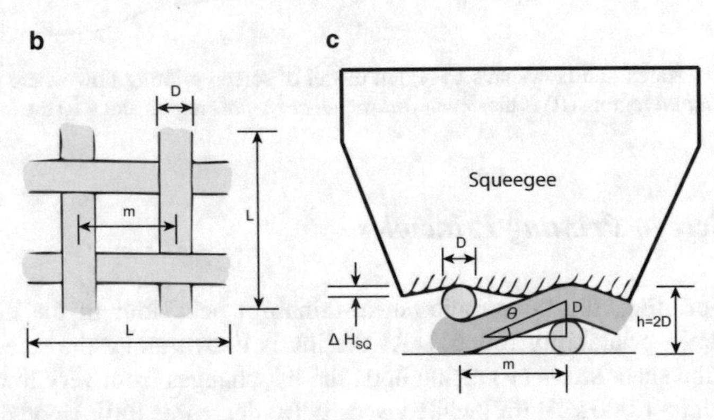

Fig. 5.15 Cross section and labels of derivation of screen printing thicknesses. (**a**) Cross section of the point where the squeegee makes contact with the mesh. (**b**) Magnified schematic of the screen mesh. (**c**) Magnified schematic of point of contact of the squeegee with the mesh

open area height H_{OA} [63], as indicated in Fig. 5.15a. Hence, referring to Fig. 5.15b, c we can derive the following relation

$$\begin{aligned}(H_{SC} - H_{OA})L^2 &= hL^2 - (V_W) - (V_{DS}) \\ &= 2 \times DL^2 - \frac{\pi D^2 L}{\cos\theta} - (V_{DS})\end{aligned} \tag{5.15}$$

from which we get

$$H_{SC} = 2 \times D - \frac{\pi D^2}{L\cos\theta} - \frac{(V_{DS})}{L^2} + H_{OA}. \tag{5.16}$$

where $\frac{(V_{DS})}{L^2} = \Delta H_{SQ}$ represents the average squeegee penetration height into the screen. The mesh count, M, (typically expressed per inch) is given by

$$M = \frac{1}{m} \tag{5.17}$$

where m is the distance between the centre lines of two parallel wires; Fig. 5.15b, and

$$L = 2m = \frac{2}{M} \tag{5.18}$$

As a result

$$\begin{aligned} H_{SC} &= D\left[2 - \frac{\pi}{2} DM\sqrt{1 + (DM)^2}\right] \\ &= \Delta H_{SQ} + H_{OA} \end{aligned} \tag{5.19}$$

Hence, the thickness of the ink deposited beneath the squeegee is primarily based on the size of the apertures of the screen and the thickness of the mesh.

However, due to ink pressure build-up in the ink ahead of the screen in Region I as the squeegee is drawn across the screen, more ink is usually forced beneath the squeegee tip leading to a thicker deposited wet ink, H_{WP}, than the ink thickness under the squeegee, H_{SC}. It can be therefore assumed that in all circumstances, $H_{WP} > H_{SC}$, with the thickness of the deposited ink increasing for higher squeegee speeds over the surface. The relationship between H_{WP} and H_{SC} is depicted in Fig. 5.16.

The continuity equation for the ink flow in time (Δt) per unit area of squeegee in contact with the screen can be given as:

$$V_{SQ}(\Delta t)(H_{WP} - H_{SC} + H_{PR}) = V_P(\Delta t)H_{SC} \tag{5.20}$$

where H_{PR} represents the equivalent thickness of the ink residue which will be left on the screen after the squeegee is drawn.

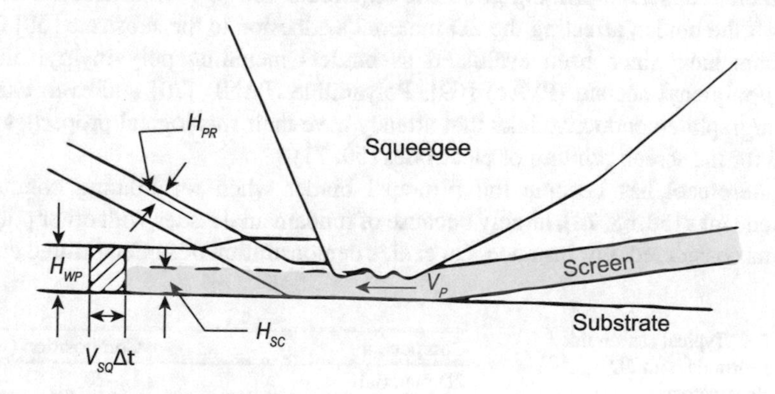

Fig. 5.16 Deposited wet ink thickness (H_{WP}) in relation to ink thickness under squeegee (H_{SC})

$$H_{WP} = H_{SC}\left(1 + \frac{V_P}{V_{SQ}}\right) - H_{PR} \quad (5.21)$$

It can be seen that the velocity of the print has a direct influence over the final wet film thickness of the ink, H_{WP}, with a greater V_{SQ} corresponding to a lower wet film thickness. Therefore, to control the final thickness of the ink deposited by screen printing, it is important to accurately define the thickness of the diameter of the mesh, the mesh aperture size, the angle and pressure of the squeegee and in turn, the speed at which the squeegee is drawn across the screen.

5.3.2 *Current 2D Material Screen Ink Formulations*

Screen printing inks are typically higher in viscosity and lower in volatility than liquid inks. The lower volatility allows for a longer dwell time on the screen (i.e. the time between spreading the ink on the screen and printing the pattern). The ink must remain stable during this time to avoid drying in the screen and clogging up the mesh. The higher viscosity prevents the ink from flowing through the screen prior to printing. Screen printing inks therefore typically have a non-shear starting viscosity of 1,000–10,000 mPa s (1–10 Pa s) [56, 66]. Due to their inherently high starting viscosity, screen inks are usually processed using a three-roll mill [25, 66]. A typical starting formulation of 2D material screen printing ink is given in Table 5.5.

In contrast to the development of 2D material inkjet inks, early 2D material based screen printable inks, owing to their inherent higher viscosity, included polymeric binders in their formulations. Zhang et al. were the first to demonstrate this by dispersing reduced graphene oxide (rGO) in ethyl cellulose (EC)/terpineol to screen print counter-electrodes for dye-sensitised solar cells (DSSC) [68]. However, this first attempt at screen printing graphene required a 400 °C treatment to effectively burn off the binder, affecting the 2D material's adhesion to the substrate [68] Other polymers have since been evaluated as binders including polyvinylpyrrolidone (PVP)/polyvinyl acetate (PVAc) [69], Polyaniline (PANI) [70] and even existing carbon/graphite-conductive inks that already have their rheological properties optimised for the screen printing of electrodes [59, 71].

EC/terpineol has become the principal binder when formulating conductive graphene inks [58, 68, 72], largely because of fundamental issues with other polymer systems considered. For instance, Cu et al.'s demonstration of screen printed current

Table 5.5 Typical screen ink starting formula with 2D materials as active pigments[67]

Component	Composition (wt%)
2D material	12–20
Binder	45–65
Solvents	20–30
Additives (defoamers/thickeners)	1–5

Table 5.6 Current screen-printable 2D material inks and their applications

Year	Type of ink	Substrates	Applications	Ref.
2011	rGO in EC, terpineol	FTO-glass	Dye-sensitised solar cells	[68]
2013	Graphene in PANI in H_2O, EtOH	PET/glass	Supercapacitors	[70]
2015	Graphene in EC, terpineol	Glassine paper	Paper-based organic TFTs	[58]
2016	Graphene in PVP/PVAc	Paper/PET	Paper-based organic TFTs	[69]
2017	rGO in water, acrylate resin	Textile	Graphene supercapacitors	[73]
2017	MoS_2 mixed in graphite ink	Polyester film	Screen-printed electrodes for electrocatalysis	[71]
2017	CuO decorated rGO in EC, terpineol	Graphite	Non-enzymatic glucose sensor	[72]
2017	Electrolytic graphene mixed in carbon ink	Glass	Graphene electrode	[59]

collectors, though surviving multiple (~200) bend cycles, required lamination between two plastic sheets to be mechanically robust [70]. Similarly, Arapov et al.'s demonstration of a gelated PVP/PVAc binder in IPA required very controlled heating up to 75 °C to achieve pseudoplasticity [49]. This pseudoplasticity, although eventually shear thinning, is not ideal for screen printing due to the lack of flow when in an unsheared state and can be prone to damaging screens. Abdelkader et al. successfully used an acrylate binder with water as a diluent in order to disperse rGO [73]. Though a viable strategy, this was only possible due to the hydrophilic nature of rGO and may not be possible to use with other hydrophobic 2D materials [73]. The EC/terpineol combination, on the other hand, has exhibited superior electrochemical responses when compared to other binders that could be screen-printed [71].

At present, graphene is the most researched 2D material to be incorporated into a screen ink. There have been some attempts to incorporate other 2D materials such as *h*-BN [74] and MoS_2 [71]. However, much work still needs to be done to optimise these and other 2D materials for screen printing deposition. Table 5.6 gives the most recently reported formulations of 2D material inks for screen printing, their respective substrates and intended purposes.

5.4 Gravure and Flexographic Printing

Gravure and flexographic printing are two common technologies used to print numerous packaging applications including boxes, cartons, plastics and foils. With regard to 2D material deposition, however, little work has been done, mainly due to the high prototyping costs and large ink volume requirements of both printing processes. Both gravure and flexography rely on large metal or ceramic rolls (called a gravure cylinder in gravure printing and an anilox roller in flexographic printing) to meter and control the amount of ink deposited. Gravure printing is a 'direct' process whereby ink is carried directly from the ink trough to the substrate via the gravure cylinder which doubles up as the image carrier. Gravure printing often relies on inks with viscosities of 100–1,000 mPa s. Flexography, whilst similar to gravure, is a form of 'indirect' printing which involves the transfer of the metered ink film from the anilox roller to a printing plate that carries the image before transferring onto the substrate. Flexography can handle inks with slightly higher viscosities 1,000–2,000 mPa s (and potentially higher 2D material content). Both printing processes are particularly suited to mass fabrication of devices due to their extremely high throughput. As the gravure and flexographic processes share essential similarities this section will cover them both.

5.4.1 Gravure Printing

In gravure printing, the desired pattern is directly engraved into a metal cylinder in the form of cells which are designed to fill with ink as the cylinder is passed through an ink trough [25, 66, 75–77]. Ink transfer is achieved by passing the substrate between the gravure cylinder and an impression roller. The printing unit of a gravure press consists of an ink duct in which the etched cylinder rotates in a fluid ink and a metal doctor blade, which spans the width of the gravure cylinder and scrapes excess ink from the cylinder surface; Fig. 5.17a, b [25, 56, 76, 79, 80].

Nguyen et al. developed a model for the ink-transfer mechanism of the gravure printing process that would allow the effective control of parameters for fine line prints [79, 80]. In this model, the gravure printing process can be divided into four distinct phases: (1) the inking phase at the ink trough, (2) doctoring phase at the doctor blade where the ink is metered out, (3) printing phase in the area where the substrate meets the gravure roller and the (4) setting phase once ink has been successfully released from the cylinder cell and the print has been made.

In the ideal scenario, all the ink is transferred at each stage, and set on the substrate. However, the reality and imperfections within rollers and ink design means that imperfect ink transfer occurs at all the stages of the process. These imperfections in transfer can be described in a series of ratios given as:

- Ink transfer ratio (δ_T), which is the ratio of printed pattern volume (V_P) to cell volume (V_C): $\delta_T = V_P/V_C$.
- Inking ratio (δ_I), which is the ratio of filled volume (V_F) to cell volume (V_C): $\delta_I = V_F/V_C$.
- Doctoring ratio (δ_D), which is the ratio of remained volume (V_R) to filled volume (V_F): $\delta_D = V_R/V_F$.
- Printing ratio (δ_P), which is the ratio of printed volume (V_P) to remained volume (V_R): $\delta_P = V_P/V_R$.

The three-stage ink transfer process (δ_T) can hence be described as:

$$\delta_T = \delta_I \cdot \delta_D \cdot \delta_P \tag{5.22}$$

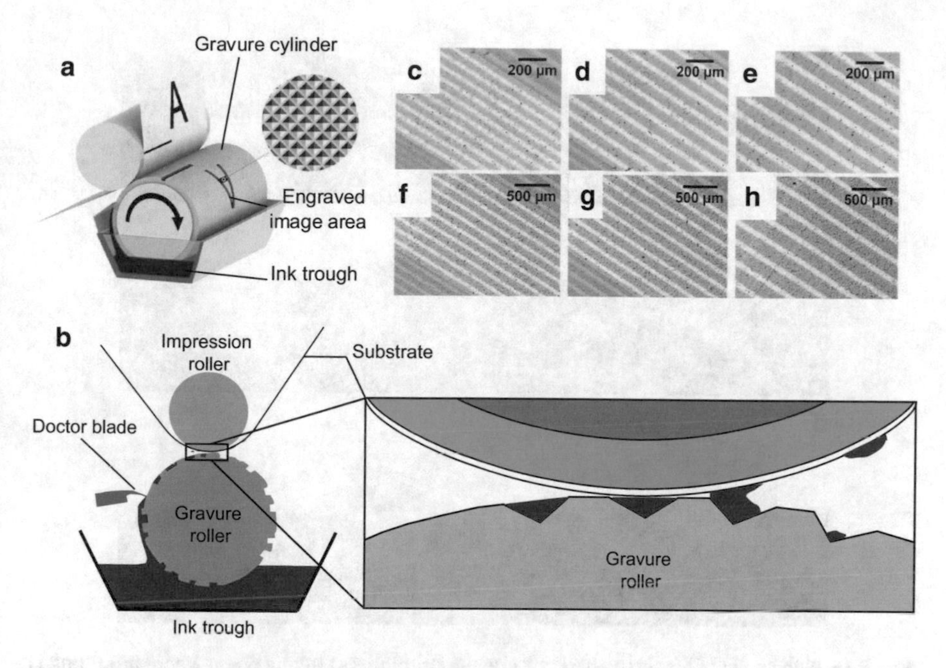

Fig. 5.17 (**a**) Schematic of gravure printing (**b**) 2D schematic of gravure printing. Inset: Magnified schematic of ink transfer from gravure roller to substrate. Optical microscopy images of printed graphene lines using cell sizes of 15, 20, 25, 30, 35 and 50 μm for (**c–h**), respectively. Lines printed without the specified cell size are shaded gray. Figure from E.B. Secor and M.C. Hersam

The schematic of the four phases and the corresponding volume ratio and area ratio is shown in Fig. 5.18. In the inking phase, the aspect ratio (AR = width/depth) and contact angle between the ink and cell (θ_{IC}) play an important role. The inking ratio decreases as contact angle θ_{IC} increases and AR decreases; Fig. 5.18a [79, 80]. In the doctoring phase, the bending stiffness, pressure, angle, tip shape of doctor blade as well as cell geometry and printing speed determine doctoring ratio (δ_D); Fig. 5.18b. Issues related to these parameters in this phase are often due to misshapen, blunt or damaged doctor blades.

During the printing phase, the ink contained in the cells is brought into contact with the substrate. The ink is then transferred to the substrate to create the printed pattern; Fig. 5.18c. In reality however, the printing phase is far more complicated and incomplete emptying of cells have different characteristics.

The nuances of the ink transfer phase are shown in Fig. 5.19a. Assuming the inking and doctoring phase occurred perfectly, incomplete printing ($0 \leq \delta_P < 1$) can occur due to stretching of the ink from forces acting on the ink.

Three distinct forces act on the ink. The cohesive force (F_I), the adhesive forces on the cell surface (F_{IC}) and the adhesive force on the substrate (F_{IS}). If $F_{IC} > F_{IS}$ and $F_I > F_{IS}$ occurs over the entire interface, no ink will transfer ($\delta_P = 0$). If $F_I < F_{IS}$ and F_{IC}, the ink will ideally be fractured according to Mode I. This is termed as the cohesive fracture. When $F_{IS} > F_{IC}$, most of the ink is transferred to

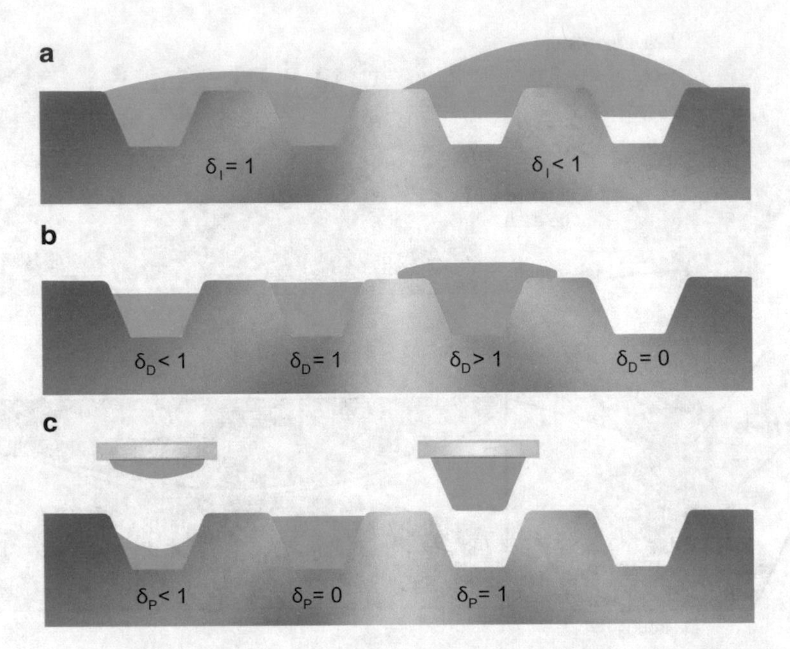

Fig. 5.18 Schematic of the three phases of gravure printing. (**a**) Inking phase where full filling is given as η_I and partial filling as $0 \leq \delta_D < 1$. (**b**) Doctoring phase $0 \leq \delta_D < 1$ where doctoring is $0 \leq \delta_D < 1$, ideal is $\delta_D = 1$ and over doctoring $\delta_D < 0$. (**c**) Printing phase whereby complete printing is $\delta_P = 1$ and incomplete emptying of cells is $0 \leq \delta_P < 1$

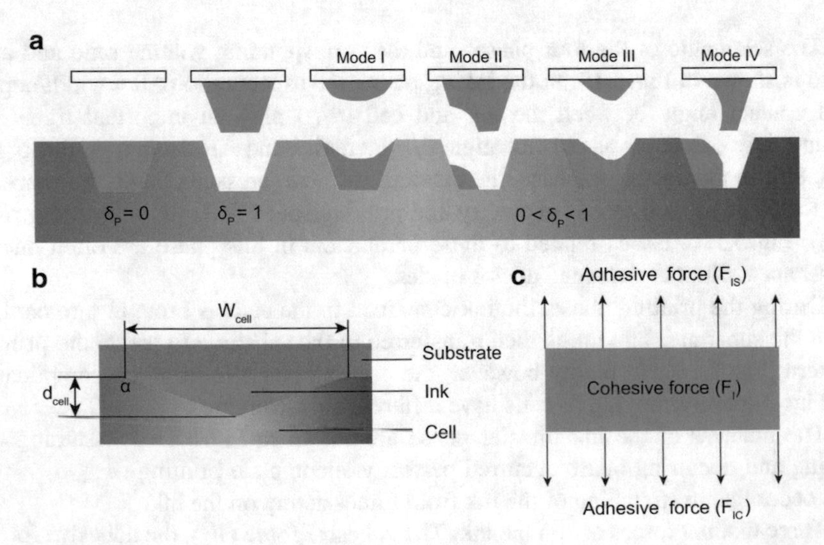

Fig. 5.19 Ink transfer in the gravure process. (**a**) Ink transfer fracture Modes in gravure printing. (**b**) Typical geometry of a gravure cell, indicating the key parameters. (**c**) Summary of forces acting on the ink during transfer from roller to substrate

the substrate without leaving pinholes or voids in the printed pattern. A summary of the fracture modes and their associated forces are given in Table 5.7.

In reality, the non-uniformity of these forces means that the ink in the cell can be prone to fracture according to Modes II, III and IV, also called adhesive-cohesive fractures; Fig. 5.19a [79, 80]. Some ink is left in the cell in Mode II when $F_{IC} > F_I$, even though most of the ink is transferred because $F_{IS} > F_{IC}$. Similarly in Mode III, some ink is transferred onto the substrate when $F_{IS} > F_I$. This is however not ideal as it represents suboptimal transfer of ink to the substrate.

Mode IV can be considered the least ideal of all the printing scenarios when both F_{IS} and $F_{IC} < F_I$ for certain areas in the cells and the substrate. This causes a mixture of cohesive and cohesive-adhesive fractures at both the substrate and the cell, leaving pinholes and voids in the printed patterns.

The adhesive forces can be estimated using the work of adhesion. The work of adhesion is the force required to separate two phases, which can be either liquid–liquid (ink–ink) or solid–liquid (ink–roller/ink–substrate), and can also be seen as a measure of the strength of the contact between two phases. The following analysis which investigated the ink-transfer mechanism during the printing phase of roll-to-roll gravure printing was derived in Ref. [79].

$$F_{IC} = w_{IC} \cdot L_{IC} \tag{5.23}$$

$$F_{IS} = w_{IS} \cdot L_{IS} \tag{5.24}$$

where w_{IC}, w_{IS}, L_{IC} and L_{IS} are the works of adhesion (w) and contact lengths (L) at the ink-cell (IC) and ink-substrate (IS) interfaces. For a given set of extrinsic conditions (e.g. the printing speed and the pressure of the plate on the substrate), L_{IS} can be considered a constant and it is assumed that $L_{IS} = W_{cell}$, where W_{cell} is the width of the cell. Assuming a triangular cell as in Fig. 5.19b, L_{IC} can be calculated via

$$L_{IC} = \frac{W_{cell}}{\cos(\tan^{-1}[2d_{cell}/W_{cell}])} \tag{5.25}$$

As ink transfer is optimal when $F_{IS} > F_{IC}$, we can derive the difference between the adhesive forces in terms of cell geometry: (cell depth (d_{cell}) width (W_{cell}) and works of adhesion (w_{IC} and w_{IS}) at the interface as shown by combining Eqs. (5.23)–(5.25)

$$\Delta F = F_{IS} - F_{IC} = W_{cell}\left[w_{IS} - \frac{w_{IC}}{\cos(\tan^{-1}[2d_{cell}/W_{cell}])}\right] > 0 \tag{5.26}$$

According to Eq. (5.26), ΔF is mainly dependent on the geometry of the cell and the works of adhesion at the IC and IS interfaces for a given γ_{ink}. The thermodynamic work of adhesion (w_{IS} or w_{IC}) can be interpreted as the work required to separate a unit area of the IS or IC surfaces while leaving a clean substrate or cell and an unfractured 'block' of ink.

Table 5.7 Ink-transfer modes during the printing phase depending on the adhesive and cohesive forces acting on the ink [79]

Conditions		Fracture mode	Ink transfer	Mode
$F_{IS} > F_{IC}$	For entire cell surface	Adhesive fracture at cell	Complete printing	
$F_{IC} < F_I$				
$F_{IC} > F_{IS}$	For entire substrate surface	Adhesive fracture at substrate	No printing	
$F_{IC} > F_I$				
$F_{IC} > F_I$	For entire cell surface	Cohesive fracture	Partial printing	Mode I (no defects)
$F_{IS} > F_I$	For entire substrate surface			
$F_{IS} > F_{IC}$	For partial area of substrate surface	Cohesive-adhesive fracture at cell	Partial printing	Mode II (no defects)
$F_{IC} < F_I$				
$F_{IC} > F_I$	For partial area of cell surface	Cohesive-adhesive fracture at substrate	Partial printing (with pinholes and voids)	Mode III (pinholes and voids)
$F_{IS} < F_I$				
$F_{IC} < F_I$	For partial area of cell surface	Cohesive-adhesive fracture at both cell and substrate	Partial printing (with pinholes and voids)	Mode IV (pinholes and voids)
$F_{IS} < F_I$	For partial area of substrate surface			

The work of adhesion (w) can be calculated from the Young–Dupré equations which relate the surface tension of a liquid to the work of adhesion [81].

$$w_{IC} = \gamma_{ink}(1 + \cos\theta_{IC}) \tag{5.27}$$

$$w_{IS} = \gamma_{ink}(1 + \cos\theta_{IS}) \tag{5.28}$$

where θ_{IC} and θ_{IS} are the contact angle of the ink on the cell and substrate, respectively.

Neumann et al. [82] proposed an equation of state with an empirical constant $\beta = 0.0001247\,\mathrm{m}^2\,\mathrm{mN}^{-1}$ which allows the relation to the γ_c of a solid surface. Hence, for a given surface tension of ink (γ_{ink}) and surface energy of a substrate (γ_{sub}) or cell (γ_{cell}).

$$\begin{aligned} w_{IC} &= \gamma_{ink}(1 + \cos\theta_{IC}) \\ &= 2\sqrt{\gamma_{ink}\gamma_{cell}}e^{-\beta(\gamma_{ink}-\gamma_{cell})^2} \end{aligned} \tag{5.29}$$

$$\begin{aligned} w_{IS} &= \gamma_{ink}(1 + \cos\theta_{IS}) \\ &= 2\sqrt{\gamma_{ink}\gamma_{sub}}e^{-\beta(\gamma_{ink}-\gamma_{sub})^2} \end{aligned} \tag{5.30}$$

It is hence stated that ΔF is a function of γ_{ink}, γ_{cell} and γ_{sub}. This scenario assumes perfect inking and doctoring in the preceding phases of printing. However, this is often not the case in high-speed gravure printing where imperfect inking and doctoring occurs. For instance, an air bubble trapped in a cell could reduce L_{IC}, thereby increasing the value of ΔF. Also, the leveling phenomenon of ink within the cavity after doctoring might subsequently increase or decrease L_{IS}, which would also affect the value of ΔF. However, ΔF can in general be used to represent the degree of ink transferred to the substrate and for estimating whether printed patterns might have defects such as pinholes or voids. Hence, adjusting the cell geometry, cell materials substrates and surface tension of inks will allow for optimised printing.

Despite carefully controlling these parameters it should be noted that surface area of the gravure cylinder, the speed of the press and the thinning out of ink with the doctor blade means that solvent evaporation can be quite rapid, leading to an increase in the viscosity of ink during printing. This can lead to further uncontrollable cohesive-adhesive fractures in printing as F_I increases due to increase in viscosity. Effective viscosity control is required to ensure good ink circulation. This can be aided by computer monitoring systems or by a press operative who constantly monitors the viscosity of the ink over the production run. [25, 66].

In the final ink setting stage, it is important to note the behaviour of the ink on the substrate. The deposited ink usually takes a shape of the cell as shown in Fig. 5.20a immediately after printing and reaches an equilibrium (Fig. 5.20b) that

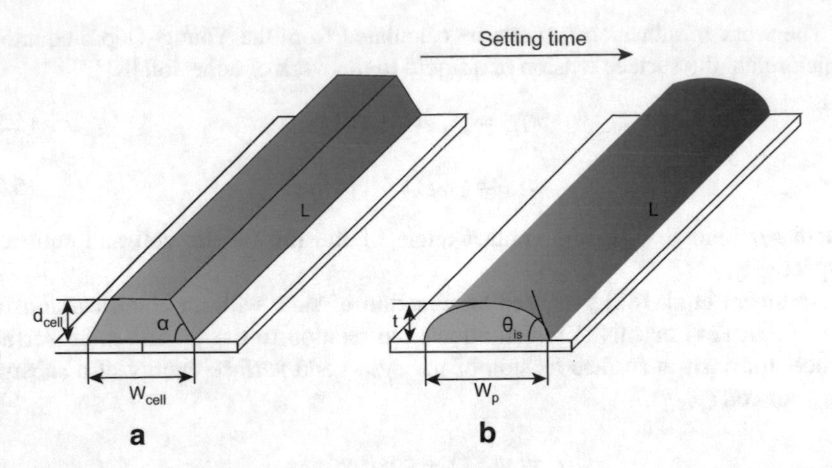

Fig. 5.20 Geometry of printed patterns after the printing phase in gravure printing when ink is (**a**) released and (**b**) when thermo-dynamic equilibrium is attained

can be defined also by the Young-Dupré equation [81]. This was derived in Ref. [80] and is given by:

$$\gamma_{sub} = \gamma_{sub\text{–}ink} + \gamma_{ink} \cos\theta_{IS} \tag{5.31}$$

where γ_{sub} and γ_{ink} are the surface tensions of the ink and substrate, $\gamma_{sub\text{–}ink}$ is the interfacial tension between the substrate and the ink, and θ_{IS} is the contact angle of the ink on the substrate as shown in Fig. 5.20b [80].

Assuming that the ink transfer from the cell onto the substrate remains constant and the ink transfer ratio (δ_T) as shown in Eq. (5.22) is variable, the extent of the spreading over the substrate could be determined. For a given cell geometry (i.e. fixed volumes: cell cross section, A_{cell} and L) the width of the printed patterns (W_P) is a function of the contact angle (θ_{IS}) as shown in Fig. 5.20b and is given by:

$$W_p = 2\sin\theta_{IS}\sqrt{\frac{\delta_T \cdot A_{cell}}{(\theta_{IS} - \sin\theta_{IS} \cdot \cos\theta_{IS})}} \tag{5.32}$$

The resulting W_P is hence dependent not only on cell geometry and wetting behaviour of the ink on the substrate (θ_{IS}) but also on δ_T. Nguyen et al. [80] also found that increasing the volume of a gravure cell ($A_{cell} \times d_{cell} \times W_{cell}$) increases the width of the final printed line and decreases the contact angle (θ_{IS}).

By ensuring effective wetting, doctoring and inking and maximising ink transfer ratio (δ_T) allows a greater spread of the ink on the substrate after deposition. Conversely, altering any of these parameters allows effective control of final, desired line-width.

5.4.2 Flexographic Printing

Flexography is a relief printing process in which the impression stands proud of the printing plate as opposed to being 'flat' on a plate, like screen printing, or recessed as in gravure printing [25, 56, 66, 83, 84].

To form an image, soft and flexible relief printing plates are mounted and registered on a plate cylinder; Fig. 5.21a. Prints are usually made at low pressure due to the combination of very fluid inks and soft, flexible printing plates [56, 66, 85]. Ink is applied to the surface of the printing plate using an anilox roller. This is rolled through the ink trough, filling the cells with ink. Excess ink is removed via a doctor blade, ensuring an even metering of the ink. The ink is transferred to a plate cylinder with the soft relief plate attached; Fig. 5.21g. Typically flexographic printing is a reel fed process [56, 66, 85].

Following Nguyen et al.'s model as with gravure printing [79, 80], it is possible to break down flexography into a series of stages.

- Stage (i)—Inking, whereby the ink fills the cells of an anilox roller
- Stage (ii)—Doctoring, where the wet ink film thickness is controlled
- Stage (iii)—Ink transfer, whereby the ink is transferred from the anilox roll to the printing plate
- Stage (iv)—Printing stage, whereby the ink is transferred from printing plate to the substrate
- Stage (v)—Ink setting stage once ink has been successfully deposited on the substrate and the print has been made

Similar to gravure printing, the success of the print begins with the inking and doctoring of the anilox roll, which is a hard engraved cylinder of millions of cells; Fig. 5.21a, inset. Unlike gravure printing, however, these cells are not the main image-forming region and are responsible purely for metering the amount of ink transferred. The amount of ink that a single cell (V_C) can hold is given as [27, 86]

$$V_C = \frac{h}{3}[a_T + a_B + \sqrt{(a_T a_B)}] \quad (5.33)$$

where h is the height of the cell, a_T is the area of the opening of the cell and a_B is the area of the base of the cell; Fig. 5.21g. The volume obtained can then be multiplied by the total number of cells on the roller to obtain the total volume of ink the roller can hold. The ink transfer process in stages (iii) and (iv) of flexography is governed by the cell volume, the cell's emptying behaviour, mechanical throw-on and rolling as well as the behaviour and adjustment of the surface properties of the printing plate and substrate [38].

However, using Nguyen et al.'s model at the printing stage of gravure printing, it can be assumed that the ink transfer from metal anilox roller to printing plate and from printing plate to substrate is a function of the cohesive force of the ink

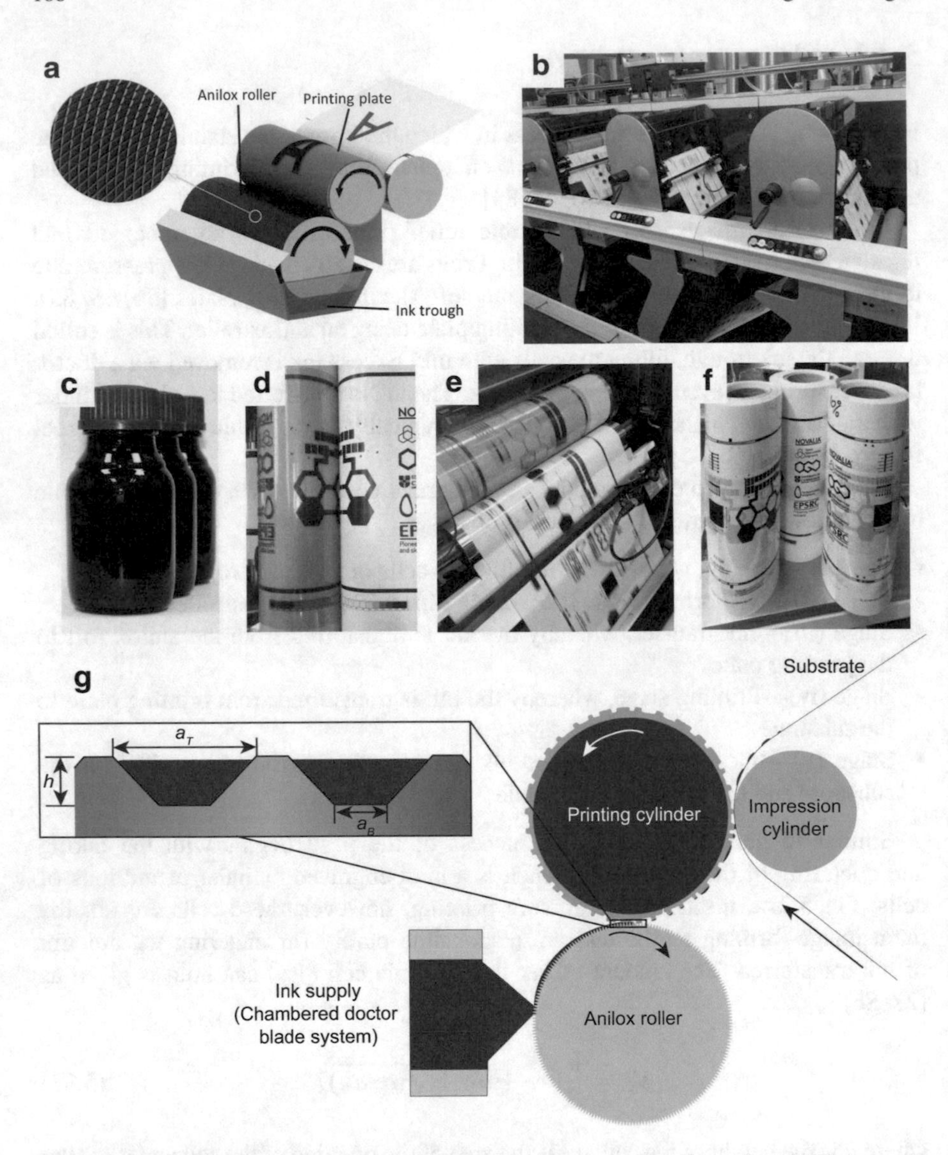

Fig. 5.21 (**a**) Schematic of flexographic printing; Inset: close-up of anilox roller; (**b**) Photographs of a roll-to-roll web flexographic press (Nilpeter FA4) printing graphene. (**c**) Graphene ink formulated for flexographic printing, (**d**) Impression of the printing plate on the paper substrate. (**e**) Flexographic press during operation. (**f**) Printed patterns of graphene ink on paper and polymer rolls. (**g**) Detailed 2D schematic of a flexographic printing system. Inset: dimensions of an anilox roller cell

(F_I) which is directly influenced by its viscosity (higher viscosities would lead to higher cohesive forces) and the CST of the roller ($\gamma_{c,cell}$), the printing plate ($\gamma_{c,plate}$) and the surface tension of the ink (γ_{ink}) [79, 80]. Adjusting the viscosity and surface tension of the inks can reduce the chance of cohesive-adhesive failure in between the anilox roller and the printing plate and maximise the transfer ratio (δ_T) and hence maximise the amount of ink that is deposited on the substrate. However, as compared to gravure printing, few studies have been conducted on the hydrodynamics of ink transfer and understanding of the relationship between ink viscosity, surface tension and ink transfer in flexographic printing is limited.

5.4.3 Gravure and Flexography Printing Ink Formulations

Gravure and flexographic printing requires inks to be mobile, low in viscosity and be rapidly drying. This is because inks must fill the cells of the gravure cylinder and anilox roller fully and also be fluid enough to transfer from roller to substrate to achieve a good δ_T [25, 56, 66, 79, 85]. Yet, viscosity must also be controlled to allow sufficient (F_I) to exist to prevent unwanted cohesive-adhesive fractures within the inks [25, 66, 79].

The range of ink viscosities utilised in gravure printing is generally ~100–1,000 mPa s while flexographic inks are usually in the range of ~1,000–2,000 mPa s. When control over the materials used in printing machinery (i.e. the metals/ceramics used for the gravure or anilox roller) and the substrate is not possible, the only parameter ink formulators have available to adjust are the solvent compositions to control the γ_{ink} and viscosity by changing the amount of solvent.

At the same time, due to the modest viscosities of flexographic and gravure inks, particular attention must be paid to the coffee ring effect as discussed in Sect. 5.1.1. A typical starting formulation of a flexographic ink and gravure ink is given in Table 5.8. Other factors to consider on press to ensure good quality of the print include:

- The rheological characteristics of the ink
- Speed of printing
- Rate of evaporation of the solvent system
- Shape and range of cell depths
- Wiping characteristics of the ink by the doctor blade
- Print design parameters
- Nature of the substrate

Table 5.8 Typical flexographic and gravure starting formulae with 2D material as the active pigment [67]

Component	Gravure ink (wt. %)	Flexographic ink (wt. %)
2D Material	12–17	12–17
Binder	20–35	40–45
Solvent mixture	60–65	25–45
Additives (defoamers/surfactants)	1–2	1–5

Table 5.9 Current gravure and flexographic 2D material inks and their applications

Year	Printing process	Type of ink	Substrates	Applications	Ref.
2014	Gravure	Graphene in EC, terpineol	Kapton	Printed electrodes	[78]
2014	Flexography	GNP in Na-CMC	ITO-PET	Dye-sensitised solar cells	[87]
2015	Gravure	MoS_2, rGO in PVA	Polyimide	Flexible microsupercapacitors	[88]
2015	Flexography	Graphene in carbon ink	Paper/PET	Printed electrodes	[89]

Recently, Secor et al. demonstrated gravure printing of high-resolution patterns of LPE graphene sheets onto a flexible substrate (Kapton) by using a graphene-EC ink [78]. A stable graphene/terpineol polymeric ink with a viscosity (η) varying in the range 0.2–3 Pa s was obtained from a homogeneous dispersion of graphene-EC powder (5–10 wt%) in a low viscosity ethanol/terpineol mixture (2.5:1 vol%) followed by ethanol removal [78]. The authors pointed out that the use of small lateral size graphene sheets ($\sim$50 × 50 nm with typical thickness of $\sim$2 nm), is critical for high-resolution gravure printing where sub-micrometer particles are needed [78]. Secor et al. hence demonstrated continuous, electrically conductive printed stripes in the micrometer range; Fig. 5.17c–g.

A graphene nanoparticle (GNP)-based flexo ink was developed by Baker et al. by optimising the GNP/binder (i.e., sodium carboxymethylcellulose (Na-CMC)) ratio in water/IPA solutions, reaching a η of $\sim$20 mPa s. The ink was printed onto a flexible ITO substrate at a speed of 0.4 m s^{-1}, maximising the optical transmission of the printed layer to replace platinum as a counter-electrode catalyst in flexible dye-sensitised solar cells [87].

In 2015, the Hybrid Nanomaterials Engineering group at the University of Cambridge demonstrated a complete flexographic press run of a hybrid conductive carbon-graphene ink on both paper and PET; Fig. 5.21b–f. The press run achieved speeds of $\sim$100 m/min, allowing the printing of hundreds of circuits in 1 min. The measured R_s from this demonstration was $\sim$16.5 k$\Omega\,\square^{-1}$ and $\sim$11.5 k$\Omega\,\square^{-1}$ for paper and PET, respectively, which is a significant improvement from conventional flexo-printed carbon inks with R_s of $\sim$40 k$\Omega\,\square^{-1}$.

Table 5.9 presents the notable work done to date on 2D material inks for flexographic and gravure printing. Compared to inkjet and screen printing, less work has been done in this area. This is largely due to the high set-up and prototyping costs of having to produce large gravure rollers or that large quantities of ink (>$\sim$10 L) usually have to be produced to achieve stable runs. However, both flexography and gravure printing remain highly viable candidates for high-throughput, low-cost printing as compared to inkjet and screen due to their inherent ability to fabricate larger numbers of printed devices at higher speeds.

5.5 3D Printing

As already discussed in Chap. 3, 2D materials such as graphene, with its exceptional electronic, mechanical and thermal properties, make excellent candidates for applications such as high-speed electronics, energy storage devices and electrochemical sensors that function well as 2D-printed devices. More recently, it has also been used as a biocompatible material [90] for drug delivery [91], stem-cell differentiation [92], biosensors [93, 94] and osteo, cardiac and neuro-tissue engineering [95].

3D printing is a method of manufacturing in which materials, such as plastic or metal, are deposited onto one another in layers to produce a three-dimensional object [96]. Inks used in 3D printing are generally solids, powders or liquid polymers that are polymerised in-situ. This is in contrast with earlier discussed printing methods which produce two-dimensional products. 3D printing has primarily been used to create engineering prototypes with an emphasis on strong mechanical properties with few biomedical and electrical applications. However, incorporating 2D materials into a 3D-printed architecture/polymer scaffold has only started recently. It is hence only natural that the first significant applications for 3D printing are mostly biomedical related. However, other electronic applications [97] such as for batteries and supercapacitors have also been explored [98].

Although there have been many forms of 3D printing developed, the main deposition method of note for 2D materials is fused deposition modelling (FDM).

5.5.1 *Fused Deposition Modelling*

The fused deposition modeling FDM is a process where a thin filament of plastic feeds a heated extruding print head; Fig. 5.22. The print head melts the plastic filament and extrudes directly onto a printed bed that may or may not be heated. This produces a print with a thickness typically of 0.25 mm per layer. Some materials that can be extruded in this manner are polycarbonate (PC), acrylonitrile butadiene styrene (ABS), polyphenylsulphone (PPSF), PC-ABS blends, and PC-ISO, which is a medical grade PC. With FDM, no chemical post-processing is required. Furthermore, the simplicity of the process means less expensive machinery [99].

FDM is a slow process which sometimes takes days to build large complex parts. To save printing time, some FDM printing machines are capable of operating in two modes: a fully dense mode which produces a filled out, solid printed piece and a sparse mode that leaves the printed piece hollow. However, the latter may reduce the mechanical strength of the final printed piece [100–106].

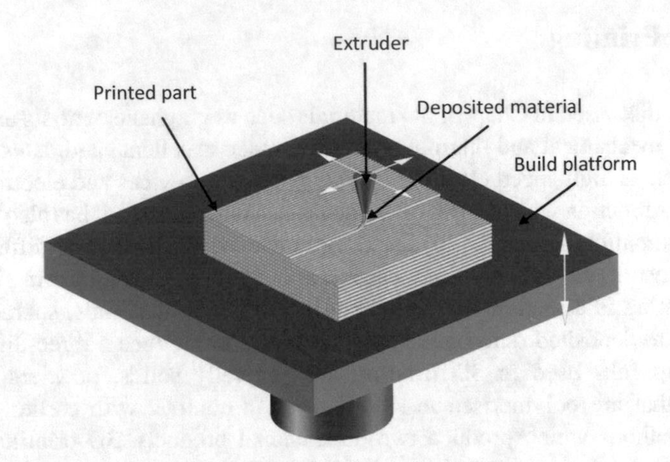

Fig. 5.22 Schematic illustrating the fused deposition modelling process

Table 5.10 Recent demonstrations of 3D printing of 2D materials

Year	2D material	3D printing process	Applications	Ref.
2015	Graphene	Modified FDM	Scaffolds for electronic and biomedical applications	[91]
2015	Graphene	FDM	3D graphene structures	[108]
2016	rGO	FDM	Flexible circuits	[97]
2016	Graphene	Modified FDM	Graphene aerogels	[109]
2016	GO	Modified FDM	Printed batteries	[98]

5.5.2 *Recent Demonstrations of 3D Printing of 2D Materials*

FDM has been used to deposit 2D materials such as rGO [97, 107], GO [98] and graphene [91] for a variety of applications as shown in Table 5.10. However, due to the inherent liquid nature of many 2D material dispersions, often a modified FDM method is utilised in which a liquid is deposited from the extruder instead of a melted plastic filament.

The first significant demonstration of graphene heterostructure printing was by Jakus et al. when graphene ink was extruded to create stable graphene-polymer composites that were capable of being used for bioelectronic applications.

As well as biomedical applications, 3D printing of graphene has also been used in electrical applications. Garcia-Ton et al. utilised FDM to deposit a GO/BCS ink, while Zhang et al. demonstrated the 3D printing of graphene aerogels [108].

Compared to the other printing methods, 3D printing is very much in the early stage of development with most demonstrations having no particular application, despite successfully printing interesting heterostructures. However, early work indicates that 3D-printed 2D material structures, due to its biocompatibility, electrical conductivity and strength, might first find applications in biomedical implants and devices and as lightweight, mechanically strong composites.

5.6 Conclusion

This chapter has discussed multiple methods of printing as a deposition method for 2D materials. Four printing methods, screen, inkjet, gravure and flexography, have been presented as the main processes with the most potential for printing 2D materials.

Of these four methods, inkjet printing has been the most researched due to the ease of set-up and its ability to print single prototypes quickly and cheaply. However, it is also the most complicated method to optimise and offers limited opportunities for scale up. At the other end of the spectrum, flexography and gravure printing offer the greatest potential for industrial scale up, but are, at present, the least developed. This is largely due to the costs associated in prototyping new devices. Screen printing offers the greatest compromise between short print runs and scale-up ability; especially due to the ability to configure rotary screen presses which can operate at high speed. 3D printing, particularly FDM, as a 2D material deposition method is also in its early stages, but shows promise as a method of printing biocompatible electronics. It is limited by its printing speed, but ultimately could be attractive for custom 3D prints that are not intended for mass production.

References

1. K. Suganuma, *Printing Technology*. SpringerBriefs (Springer, Berlin, 2014)
2. R.D. Deegan, O. Bakajin, T.F. Dupont, G. Huber, S.R. Nagel, T.A. Witten, Capillary flow as the cause of ring stains from dried liquid drops. Nature **389**(6653), 827–829 (1997)
3. R.D. Deegan, O. Bakajin, T.F. Dupont, G. Huber, S.R. Nagel, T.A. Witten, Contact line deposits in an evaporating drop. Phys. Rev. E **62**(1), 756–765 (2000)
4. J.A. Lim, W.H. Lee, H.S. Lee, J.H. Lee, Y.D. Park, K. Cho, Self-organization of ink-jet-printed triisopropylsilylethynyl pentacene via evaporation-induced flows in a drying droplet. Adv. Funct. Mater. **18**(2), 229–234 (2008)
5. H. Hu, R.G. Larson, Marangoni effect reverses coffee-ring depositions. J. Phys. Chem. B **110**(14), 7090–7094 (2006)
6. H. Hu, R.G. Larson, Analysis of the effects of Marangoni stresses on the microflow in an evaporating sessile droplet. Langmuir **21**(9), 3972–3980 (2005)
7. H. Wang, Z. Wang, L. Huang, A. Mitra, Y. Yan, Surface patterned porous films by convection-assisted dynamic self-assembly of zeolite nanoparticles. Langmuir **17**(9), 2572–2574 (2001)
8. H. Liu, W. Xu, W. Tan, X. Zhu, J. Wang, J. Peng, Y. Cao, Line printing solution-processable small molecules with uniform surface profile via ink-jet printer. J. Colloid Interface Sci. **465**, 106–111 (2016)
9. M. Singh, H.M. Haverinen, P. Dhagat, G.E. Jabbour, Inkjet printing-process and its applications. Adv. Mater. **22**(6), 673–685 (2010)
10. G. Hu, T. Albrow-Owen, X. Jin, A. Ali, Y. Hu, R.C.T. Howe, K. Shehzad, Z. Yang, X. Zhu, R.I. Woodward, T.-C. Wu, H. Jussila, J.-B. Wu, P. Peng, P.-H. Tan, Z. Sun, E.J.R. Kelleher, M. Zhang, Y. Xu, T. Hasan, Black phosphorus ink formulation for inkjet printing of optoelectronics and photonics. Nat. Commun. **8**(1), 278 (2017)
11. G.L. Robertson, *Food Packaging: Principles and Practice*, 3rd edn. (CRC Press, Boca Raton, 2012)

12. J.D. Berry, M.J. Neeson, R.R. Dagastine, D.Y.C. Chan, R.F. Tabor, Measurement of surface and interfacial tension using pendant drop tensiometry. J. Colloid Interface Sci. **454**, 226–237 (2015)
13. C. Huh, R.L. Reed, A method for estimating interfacial tensions and contact angles from sessile and pendant drop shapes. J. Colloid Interface Sci. **91**(2), 472–484 (1983)
14. C.E. Stauffer, The measurement of surface tension by the pendant drop technique. J. Phys. Chem. **69**(6), 1933–1938 (1965)
15. Kruss Website Glossary, https://www.kruss.de/services/education-theory/glossary. Accessed 10 June 2017
16. K. Kabza, J.E. Gestwicki, J.L. McGrath, Contact angle goniometry as a tool for surface tension measurements of solids, using Zisman plot method. A physical chemistry experiment. J. Chem. Educ. **77**(1), 63–65 (2000)
17. W.A. Zisman, Relation of the equilibrium contact angle to liquid and solid constitution, in *Contact Angle, Wettability, and Adhesion*, vol. 43 (American Chemical Society, 1964), pp. 1–51
18. S. Magdassi, *The Chemistry of Inkjet Inks* (World Scientific, Singapore, 2009)
19. I.M. Hutchings, G.D. Martin (eds.), *Inkjet Technology for Digital Fabrication* (Wiley, Hoboken, 2012)
20. J.G. Korvink, P.J. Smith, D.-Y. Shin (eds.), *Inkjet-Based Micromanufacturing* (Wiley-VCH, Weinheim, 2012)
21. P. Calvert, Inkjet printing for materials and devices. Chem. Mater. **13**(10), 3299–3305 (2001)
22. F. Torrisi, T. Hasan, W. Wu, Z. Sun, A. Lombardo, T.S. Kulmala, G.-W. Hsieh, S. Jung, F. Bonaccorso, P.J. Paul, D. Chu, A.C. Ferrari, Inkjet-printed graphene electronics. ACS Nano **6**(4), 2992–3006 (2012)
23. E. Tekin, P.J. Smith, U.S. Schubert, Inkjet printing as a deposition and patterning tool for polymers and inorganic particles. Soft Matter **4**(4), 703 (2008)
24. B.-J. de Gans, P.C. Duineveld, U.S. Schubert, Inkjet printing of polymers: state of the art and future developments. Adv. Mater. **16**(3), 203–213 (2004)
25. R.H. Leach, R.J. Pierce, *The Printing Ink Manual* (Springer, Amsterdam, 1993)
26. B. Derby, Inkjet printing of functional and structural materials: fluid property requirements feature stability, and resolution. Annu. Rev. Mater. Res. **40**(1), 395–414 (2010)
27. G. Hu, J. Kang, L.W.T. Ng, X. Zhu, R.C.T. Howe, C. Jones, M.C. Hersam, T. Hasan, Functional inks and printing of two-dimensional materials. Chem. Soc. Rev. **47**, 3265–3300 (2018)
28. T. Juntunen, H. Jussila, M. Ruoho, S. Liu, G. Hu, T. Albrow-Owen, L.W.T. Ng, R.C.T. Howe, T. Hasan, Z. Sun, I. Tittonen. Inkjet printed large-area flexible few-layer graphene thermoelectrics. *Adv. Funct. Mat.* **28**(22), 1800480 (2018)
29. D.J. Finn, M. Lotya, G. Cunningham, R.J. Smith, D. McCloskey, J.F. Donegan, J.N. Coleman, Inkjet deposition of liquid-exfoliated graphene and MoS_2 nanosheets for printed device applications. J. Mater. Chem. C **2**(5), 925–932 (2014)
30. F. Withers, H. Yang, L. Britnell, A.P. Rooney, E. Lewis, A. Felten, C.R. Woods, V. Sanchez Romaguera, T. Georgiou, A. Eckmann, Y.J. Kim, S.G. Yeates, S.J. Haigh, A.K. Geim, K.S. Novoselov, C. Casiraghi, Heterostructures produced from nanosheet-based inks. Nano Lett. **14**(7), 3987–3992 (2014)
31. A.G. Kelly, T. Hallam, C. Backes, A. Harvey, A.S. Esmaeily, I. Godwin, J. Coelho, V. Nicolosi, J. Lauth, A. Kulkarni, S. Kinge, L.D.A. Siebbeles, G.S. Duesberg, J.N. Coleman, All-printed thin-film transistors from networks of liquid-exfoliated nanosheets. Science **356**(6333), 69–73 (2017)
32. D. McManus, S. Vranic, F. Withers, V. Sanchez-Romaguera, M. Macucci, H. Yang, R. Sorrentino, K. Parvez, S.-K. Son, G. Iannaccone, K. Kostarelos, G. Fiori, C. Casiraghi, Water-based and biocompatible 2D crystal inks for all-inkjet-printed heterostructures. Nat. Nanotechnol. **12**(4), 343–350 (2017)

33. V. Bianchi, T. Carey, L. Viti, L. Li, E.H. Linfield, A.G. Davies, A. Tredicucci, D. Yoon, P.G. Karagiannidis, L. Lombardi, F. Tomarchio, A.C. Ferrari, F. Torrisi, M.S. Vitiello, Terahertz saturable absorbers from liquid phase exfoliation of graphite. Nat. Commun. **8**, 15763 (2017)
34. F. Bonaccorso, A. Bartolotta, J.N. Coleman, C. Backes, 2D-crystal-based functional inks. Adv. Mater. **28**(29), 6136–6166 (2016)
35. R.C.T. Howe, G. Hu, Z. Yang, T. Hasan, Functional inks of graphene, metal dichalcogenides and black phosphorus for photonics and (opto)electronics. Proc. SPIE 9553, 95530R (2015)
36. D. Dodoo-Arhin, R.C.T. Howe, G. Hu, Y. Zhang, P. Hiralal, A. Bello, G. Amaratunga, T. Hasan, Inkjet-printed graphene electrodes for dye-sensitized solar cells. Carbon **105**, 33–41 (2016)
37. S. Santra, G. Hu, R.C.T. Howe, A. De Luca, S.Z. Ali, F. Udrea, J.W. Gardner, S.K. Ray, P.K. Guha, T. Hasan, CMOS integration of inkjet-printed graphene for humidity sensing. Sci. Rep. **5**(1), 17374 (2015)
38. H. Kipphan (ed.), *Handbook of Print Media* (Springer, Berlin, 2001)
39. G.D. Martin, S.D. Hoath, I.M. Hutchings, Inkjet printing - the physics of manipulating liquid jets and drops. J. Phys. Conf. Ser. **105**, 012001 (2008)
40. D. Jang, D. Kim, J. Moon, Influence of fluid physical properties on ink-jet printability. Langmuir **25**(5), 2629–2635 (2009)
41. J.E. Fromm, Numerical calculation of the fluid dynamics of drop-on-demand jets. IBM J. Res. Dev. **28**(3), 322–333 (1984)
42. Y. Aleeva, B. Pignataro, Recent advances in upscalable wet methods and ink formulations for printed electronics. J. Mater. Chem. C **2**(32), 6436 (2014)
43. D. Soltman, V. Subramanian, Inkjet-printed line morphologies and temperature control of the coffee ring effect. Langmuir **24**(5), 2224–2231 (2008)
44. E.B. Secor, P.L. Prabhumirashi, K. Puntambekar, M.L. Geier, M.C. Hersam, Inkjet printing of high conductivity, flexible graphene patterns. J. Phys. Chem. Lett. **4**(8), 1347–1351 (2013)
45. A. Capasso, A.E. Del Rio Castillo, H. Sun, A. Ansaldo, V. Pellegrini, F. Bonaccorso, Ink-jet printing of graphene for flexible electronics: an environmentally-friendly approach. Solid State Commun. **224**, 53–63 (2015)
46. Y. Xu, I. Hennig, D. Freyberg, A. James Strudwick, M. Georg Schwab, T. Weitz, K. Chih-Pei Cha, Inkjet-printed energy storage device using graphene/polyaniline inks. J. Power Sources **248**, 483–488 (2014)
47. J. Li, F. Ye, S. Vaziri, M. Muhammed, M.C. Lemme, M. Östling, Efficient inkjet printing of graphene. Adv. Mater. **25**(29), 3985–3992 (2013)
48. A.G. Kelly, D. Finn, A. Harvey, T. Hallam, J.N. Coleman, All-printed capacitors from graphene-BN-graphene nanosheet heterostructures. Appl. Phys. Lett. **109**(2), 023107 (2016)
49. K. Arapov, R. Abbel, G. de With, H. Friedrich, Inkjet printing of graphene. Faraday Discuss. **173**, 323–336 (2014)
50. J. Li, M.M. Naiini, S. Vaziri, M.C. Lemme, M. Östling, Inkjet printing of MoS_2. Adv. Funct. Mater. **24**(41), 6524–6531 (2014)
51. E.B. Secor, B.Y. Ahn, T.Z. Gao, J.A. Lewis, M.C. Hersam, Rapid and versatile photonic annealing of graphene inks for flexible printed electronics. Adv. Mater. **27**(42), 6683–6688 (2015)
52. E.B. Secor, T.Z. Gao, A.E. Islam, R. Rao, S.G. Wallace, J. Zhu, K.W. Putz, B. Maruyama, M.C. Hersam, Enhanced conductivity, adhesion, and environmental stability of printed graphene inks with nitrocellulose. Chem. Mater. **29**, 2332–2340 (2017)
53. S. Scherp, S.J.D. Ericsson, US4267773 - Silkscreen Printing Machine (1981)
54. S.J.D. Ericsson, US4226181 - Method and apparatus for adjusting the position of a stencil relative to a printing table (1980)
55. B. Kang, W.H. Lee, K. Cho, Recent advances in organic transistor printing processes. ACS Appl. Mater. Interfaces **5**(7), 2302–2315 (2013)
56. H. Lievens, Wide web coating of complex materials. Surf. Coat. Technol. **76–77**, 744–753 (1995)

57. S.J.D. Ericsson, US4485447 - Method and arrangement for registration of a print on a material (1977)
58. W.J. Hyun, E.B. Secor, G.A. Rojas, M.C. Hersam, L.F. Francis, C.D. Frisbie, All-printed, foldable organic thin-film transistors on glassine paper. Adv. Mater. **27**(44), 7058–7064 (2015)
59. C. Karuwan, A. Wisitsoraat, P. Chaisuwan, D. Nacapricha, A. Tuantranont, W.C. Hooper, V. Vaccarino, R.W. Alexander, D.G. Harrison, A.A. Quyyumi, Screen-printed graphene-based electrochemical sensors for a microfluidic device. Anal. Methods **9**(24), 3689–3695 (2017)
60. R.F. Rosu, R.A. Shanks, S.N. Bhattacharya, Shear rheology and thermal properties of linear and branched poly (ethylene terephthalate) blends. Polymer **40**, 5891–5898 (1999)
61. M.J. Barker, Screen inks, in *The Printing Ink Manual*, Chap. 10, ed. by R. Leach (Society of British Printing Ink Manufacturers, Edinburgh, 1999), pp. 599–635
62. L. Dybowska-Sarapuk, D. Janczak, G. Wróblewski, M. Słoma, M. Jakubowska, The influence of graphene screen printing paste's composition on its viscosity. Proc. SPIE **9662**, 966242 (2015)
63. J.A. Owczarek, F.L. Howland, A study of the off-contact screen printing process—part I: model of the printing process and some results derived from experiments. IEEE Trans. Compon. Hybrids Manuf. Technol. **13**(2), 358–367 (1990)
64. D. He, Modelling and computer simulation of the behaviour of solder paste in stencil printing for surface mount assembly. PhD thesis, University of Salford, 1998
65. N. Kapur, S.J. Abbott, E.D. Dolden, P.H. Gaskell, Predicting the behavior of screen printing. IEEE Trans. Compon. Packag. Manuf. Technol. **3**(3), 508–515 (2013)
66. A. Goldschmidt, H.-J. Streitburger, *BASF Handbook on Basics of Coating Technology* (William Andrew, Norwich, 2003)
67. E.W. Flick, Printing inks, in *Printing Ink and Overprint Varnish Formulations*, 2nd edn. (Elsevier, New York, 1999), pp. 1–61
68. D.W. Zhang, X.D. Li, H.B. Li, S. Chen, Z. Sun, X.J. Yin, S.M. Huang, Graphene-based counter electrode for dye-sensitized solar cells. Carbon **49**(15), 5382–5388 (2011)
69. K. Arapov, E. Rubingh, R. Abbel, J. Laven, G. de With, H. Friedrich, Conductive screen printing inks by gelation of graphene dispersions. Adv. Funct. Mater. **26**(4), 586–593 (2016)
70. Y. Xu, M.G. Schwab, A.J. Strudwick, I. Hennig, X. Feng, Z. Wu, K. Müllen, Screen-printable thin film supercapacitor device utilizing graphene/polyaniline inks. Adv. Energy Mater. **3**(8), 1035–1040 (2013)
71. S.J. Rowley-Neale, G.C. Smith, C.E. Banks, Mass-producible 2D-MoS_2 -impregnated screen-printed electrodes that demonstrate efficient electrocatalysis toward the oxygen reduction reaction. ACS Appl. Mater. Interfaces **9**(27), 22539–22548 (2017)
72. Z. Zhang, P. Pan, X. Liu, Z. Yang, J. Wei, Z. Wei, 3D-copper oxide and copper oxide/few-layer graphene with screen printed nanosheet assembly for ultrasensitive non-enzymatic glucose sensing. Mater. Chem. Phys. **187**, 28–38 (2017)
73. A.M. Abdelkader, N. Karim, C. Vallés, S. Afroj, K.S. Novoselov, S.G.Yeates, Ultraflexible and robust graphene supercapacitors printed on textiles for wearable electronics applications. 2D Mater. **4**(3), 35016 (2017)
74. A.M. Joseph, B. Nagendra, E. Bhoje Gowd, K.P. Surendran, Screen-printable electronic ink of ultrathin boron nitride nanosheets. ACS Omega **1**(6), 1220–1228 (2016)
75. K.I. Bardin, US4003311 - Gravure printing method (1977)
76. M. Lahti, S. Leppävuori, V. Lantto, Gravure-offset-printing technique for the fabrication of solid films. Appl. Surf. Sci. **142**(1), 367–370 (1999)
77. C. Deus, J. Salomon, U. Wehner, Roll-to-roll coating of flexible glass: equipment, layer stacks and applications. Vak. Forsch. Prax. **28**(4), 40–44 (2016)
78. E.B. Secor, S. Lim, H. Zhang, C.D. Frisbie, L.F. Francis, M.C. Hersam, Gravure printing of graphene for large-area flexible electronics. Adv. Mater. **26**(26), 4533–4538 (2014)
79. H.A.D. Nguyen, C. Lee, K.-H. Shin, D. Lee, An investigation of the ink-transfer mechanism during the printing phase of high-resolution roll-to-roll gravure printing. IEEE Trans. Compon. Packag. Manuf. Technol. **5**(10), 1516–1524 (2015)

80. H.A.D. Nguyen, J. Lee, C.H. Kim, K.-H. Shin, D. Lee, An approach for controlling printed line-width in high resolution roll-to-roll gravure printing. J. Micromech. Microeng. **23**(9), 095010 (2013)
81. M.E. Schrader, Young-Dupre revisited. Langmuir **11**(9), 3585–3589 (1995)
82. A.W. Neumann, R.J. Good, C.J. Hope, M. Sejpal, An equation-of-state approach to determine surface tensions of low-energy solids from contact angles. J. Colloid Interface Sci. **49**(2), 291–304 (1974)
83. J.A. Martens, US5172072 - Flexographic printing plate process (1992)
84. R.N. Fan, US5719009 - Laser ablatable photosensitive elements utilized to make flexographic printing plates (1990)
85. T. Smith, Flexographic inks. Pigm. Resin Technol. **15**(3), 11–12 (1986)
86. Printwiki, Anilox roller, http://printwiki.org/Anilox_Roller. Accessed 25 May 2018
87. J. Baker, D. Deganello, D.T. Gethin, T.M. Watson, Flexographic printing of graphene nanoplatelet ink to replace platinum as counter electrode catalyst in flexible dye sensitised solar cell. Mater. Res. Innov. **18**(2), 86–90 (2014)
88. Y. Xiao, L. Huang, Q. Zhang, S. Xu, Q. Chen, W. Shi, Gravure printing of hybrid MoS_2-rGO interdigitated electrodes for flexible microsupercapacitors. Appl. Phys. Lett. **107**(1), 013906 (2015)
89. New graphene based inks for high-speed manufacturing of printed electronics, http://www.cam.ac.uk/research/news/new-graphene-based-inks-for-high-speed-manufacturing-of-printed-electronics. Accessed 10 June 2017
90. D. Bitounis, H. Ali-Boucetta, B.H. Hong, D.-H. Min, K. Kostarelos, Prospects and challenges of graphene in biomedical applications. Adv. Mater. **25**(16), 2258–2268 (2013)
91. A.E. Jakus, E.B. Secor, A.L. Rutz, S.W. Jordan, M.C. Hersam, R.N. Shah, Three-dimensional printing of high-content graphene scaffolds for electronic and biomedical applications. ACS Nano **9**(4), 4636–4648 (2015)
92. G.Y. Chen, D.W.P. Pang, S.M. Hwang, H.Y. Tuan, Y.C. Hu, A graphene-based platform for induced pluripotent stem cells culture and differentiation. Biomaterials **33**(2), 418–427 (2012)
93. Y. Shao, J. Wang, H. Wu, J. Liu, I.A. Aksay, Y. Lina, Graphene based electrochemical sensors and biosensors: a review. Electroanalysis **22**(10), 1027–1036 (2010)
94. E.K. Wujcik, C.N. Monty, Nanotechnology for implantable sensors: carbon nanotubes and graphene in medicine. Wiley Interdiscip. Rev. Nanomed. Nanobiotechnol. **5**(3), 233–249 (2013)
95. A. Fraczek-Szczypta, Carbon nanomaterials for nerve tissue stimulation and regeneration. Mater. Sci. Eng. C **34**(1), 35–49 (2014)
96. C. Schubert, M.C. van Langeveld, L.A. Donoso, Innovations in 3D printing: a 3D overview from optics to organs. Br. J. Ophthalmol. **98**(2), 159–61 (2014)
97. D. Zhang, B. Chi, B. Li, Z. Gao, Y. Du, J. Guo, J. Wei, Fabrication of highly conductive graphene flexible circuits by 3D printing. Synth. Met. **217**, 79–86 (2016)
98. K. Fu, Y. Wang, C. Yan, Y. Yao, Y. Chen, J. Dai, S. Lacey, Y. Wang, J. Wan, T. Li, Z. Wang, Y. Xu, L. Hu, Graphene oxide-based electrode inks for 3D-printed lithium-ion batteries. Adv. Mater. **28**(13), 2587–2594 (2016)
99. X. Yan, P. Gu, A review of rapid prototyping technologies and systems. Comput. Aided Des. **28**(4), 307–318 (1996)
100. P.M. Pandey, N.V. Reddy, S.G. Dhande, Real time adaptive slicing for fused deposition modelling. Int. J. Mach. Tools Manuf. **43**(1), 61–71 (2003)
101. R. Anitha, S. Arunachalam, P. Radhakrishnan, Critical parameters influencing the quality of prototypes in fused deposition modelling. J. Mater. Process. Technol. **118**(1–3), 385–388 (2001)
102. A.K. Sood, R.K. Ohdar, S.S. Mahapatra, Improving dimensional accuracy of Fused Deposition Modelling processed part using grey Taguchi method. Mater. Des. **30**(10), 4243–4252 (2009)
103. D.T. Pham, R.S. Gault, A comparison of rapid prototyping technologies. Int. J. Mach. Tools Manuf. **38**(10–11), 1257–1287 (1998)

104. S.H. Masood, W.Q. Song, Development of new metal/polymer materials for rapid tooling using Fused deposition modelling. Mater. Des. **25**(7), 587–594 (2004)
105. J.-P. Kruth, M.C. Leu, T. Nakagawa, Progress in additive manufacturing and rapid prototyping. CIRP Ann. Manuf. Technol. **47**(2), 525–540 (1998)
106. A.K. Sood, R.K. Ohdar, S.S. Mahapatra, Parametric appraisal of mechanical property of fused deposition modelling processed parts. Mater. Des. **31**(1), 287–295 (2010)
107. J.H. Kim, W.S. Chang, D. Kim, J.R. Yang, J.T. Han, G.-W. Lee, J.T. Kim, S.K. Seol, 3D printing of reduced graphene oxide nanowires. Adv. Mater. **27**(1), 157–161 (2015)
108. E. García-Tuñon, S. Barg, J. Franco, R. Bell, S. Eslava, E. D'Elia, R.C. Maher, F. Guitian, E. Saiz, Printing in three dimensions with graphene. Adv. Mater. **27**(10), 1688–1693 (2015)
109. Q. Zhang, F. Zhang, S.P. Medarametla, C. Zhou, H. Li, D. Lin, 3D printing of graphene aerogels. Small **12**(13), 1702–1708 (2016)

Chapter 6
Applications of Printed 2D Materials

Abstract The primary goal of this book is to comprehensively review 2D materials that have gained research interest and present the use of printing as a low-cost, high-throughput method of exploiting 2D materials in mass produced devices and components. This chapter covers the existing demonstrations of printed 2D materials in a wide variety of devices and summarises their current status. In addition, this chapter also discusses the state-of-the-art literature on the trends and development stages, future technology directions and their likely convergence for next generation of applications, devices and systems.

Throughout this book we have discussed methods for the exfoliation and production of 2D materials and detailed the ink production and printing techniques required to translate these 2D materials into applications. This chapter will discuss the relevant printable applications of 2D materials.

There have been numerous demonstrations and device prototypes based on solution processed 2D materials over the last 10 years. However, applications based on printing 2D material inks with optimised rheological properties specific to target applications have only recently begun to emerge [1].

2D material applications relevant to functional printing can be divided into five broad categories—conductive inks, including transparent conductive electrodes (TCEs), (opto)electronics and photonics, printed sensors, printed energy storage, and barrier, shielding and membrane applications; Fig. 6.1.

We note that whilst not all these demonstrations have been proven to have been 'printed', their methods are highly relatable to the ink production process and show potential in being fabricated using one of the mainstream printing methods discussed in this book.

L. W. T. Ng et al., *Printing of Graphene and Related 2D Materials*,
https://doi.org/10.1007/978-3-319-91572-2_6

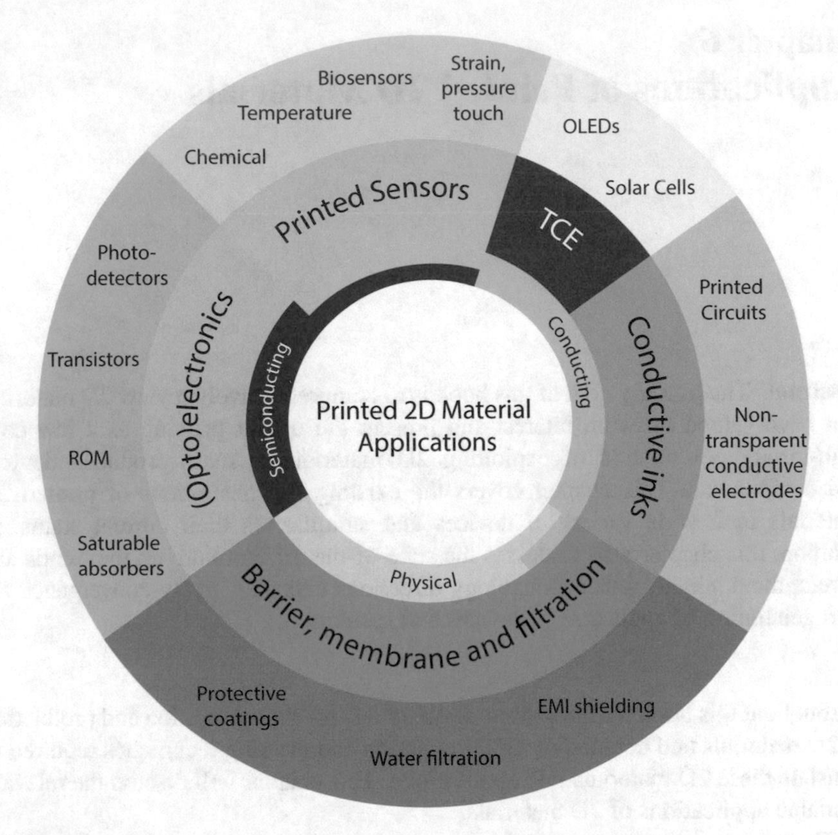

Fig. 6.1 Overview of printed 2D material applications and required properties

6.1 Conductive Inks

Graphene is a promising material for both transparent and non-transparent conductive electrodes for various devices. The properties of solution-processed graphene in particular have been widely proven in the formulation of low-cost, highly conductive ink for flexible and printable electronics and interconnects [2–9]. As well as solution-processed graphene, the graphene used in such inks can also be produced by other production methods described in earlier chapters, such as dry exfoliation and direct synthesis from carbon precursors.

The key to achieving a conductive film is to ensure the uniform presence of graphene in the dried printed film is at a concentration above its percolation threshold [10]. In a standard conducting ink formulation, the dried printed film may consist of randomly positioned and oriented flakes embedded in a polymer binder. These flakes are required to form a continuous pathway to allow current to flow, forming a conductive network (i.e. percolated pathway); Fig. 6.2. The behaviour of this conductive network for different concentrations of graphene in the dried film

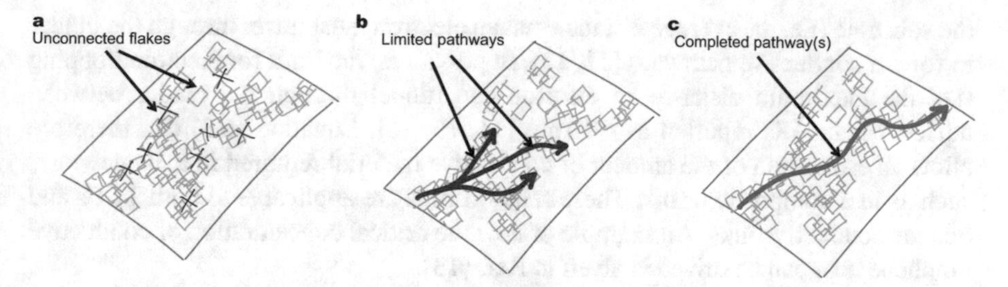

Fig. 6.2 Formation of a conductive network in a printed film, showing the effect of (**a**) insufficient 2D material concentration, (**b**) partially connected pathways between 2D materials, and (**c**) complete, connected pathways leading to conductivity

can be predicted via percolation theory, widely employed to characterise conductive thin films of nanomaterials [10–12]. When the material concentration is too low, a continuous network cannot form, and the conductivity is zero. Above the percolation threshold conductive pathways start building up. Completely percolated networks represent numerous continuous or connected networks between two points. The electrical conductivity gradually increases with an increasing number of conductive pathways in a two-dimensional network following the power-law relationship [10, 11, 13]:

$$\phi \propto (X - X_0)^i \tag{6.1}$$

where X and X_0 are the density and critical density of conducting objects, respectively, above which the density of objects X results in electrical conductivity. X_0 is also termed the percolation threshold. i is termed the percolation coefficient. For a three-dimensional network, such as a composite with conductive filler materials distributed in an insulating matrix, the above relationship may still be used. This equation may also be rewritten in terms of volume fractions:

$$\phi \propto (v - v_0)^i \tag{6.2}$$

where v and v_0 are the volume fraction and the critical volume fraction of the conducting objects, respectively [11, 14].

Since this relation does not take into account particle size, shape, orientation their distribution uniformity inside a composite matrix, i and v_0 are empirically derived for a given system [11, 14]. Modelling the graphene flakes as thin, circular, 2D conductive platelets and considering their 3D random distribution within a composite, it can be shown that [11, 14]:

$$v_0 = \frac{27\pi d^2 t}{4(d + d^*)^3} \tag{6.3}$$

where d and t are the diameter and thickness of the graphene flakes, respectively, and d^* is the average separation between graphene flakes in the dried film formed on the substrate (i.e. the average distance that an electron must travel through the binder to form a conductive pathway) [11, 13]. In polymers, the limit for electron hopping (i.e. the maximum distance an electron can tunnel through the binder between adjacent flakes) is reported as $\sim$10 nm [11, 15, 16]. Equation (6.3) can therefore allow an estimation of the amount of conductive material required in a 3D network, such as in a composite or ink. These relationships are applicable to both TCEs and opaque conductive inks. An example of how the critical concentration of conductive graphene inks can be drived is given in Ref. [13].

When calculating the conductivity of conductive material for TCEs with regard to Eq. (6.1), it is important to note that the percolation exponent i is dependent on the dimensionality of the space involved. The theoretical values for i are $\sim$1.33 for a 2D percolation network and $\sim$1.94 for a 3D percolation network [10, 17, 18]. An example of how the critical exponent can be derived and how it indicates the dimensionality of the conductive material is given in Ref. [17].

6.1.1 Transparent Conductive Electrodes

TCEs are widely used within applications such as light-emitting devices, photovoltaics, and touch-screens. The materials require a combination of low sheet resistance, R_s, and high optical transmission, T [9, 19–30].

Current TCE technology is mostly semiconductor-based [31], including indium oxide [32], zinc oxide [33] and tin oxide [34]. The current market is dominated by indium tin oxide (ITO) due to its low R_s ($\sim 15\,\Omega\,\square^{-1}$) and high transparency (>80%) [19, 35]. However, the relative brittleness of ITO with 2–3% crack onset strain means that it has limited use in applications that specifically require flexibility such as flexible displays [19]. In addition, the supply constraints of indium, general ITO processing requirements and the difficulties in its patterning pose additional obstacles [31].

For next generation electronics with mechanical flexibility, new TCE materials are therefore needed. As discussed in Chap. 2, optical and electrical properties of graphene suggest its potential as a flexible TCE. Indeed, as shown in Fig. 6.3 where T versus R_s is plotted for ITO, silver nanowires (AgNW), single-walled-nanotubes (SWNTs) and graphene, in practice, graphene (intentionally or unintentionally doped) can achieve the same R_s as ITO and other alternatives with a similar or even higher T.

Different strategies based on liquid phase exfoliated (LPE) graphene deposition have been explored including via drop and dip casting [36], rod coating [9], spray coating [37], vacuum filtration [38], and Langmuir–Blodgett assembly [39] and inkjet printing [40].

In 2008, Cote et al. demonstrated Langmuir–Blodgett assembly of graphene oxide (GO) thin films by dip-coating glass, quartz and mica substrates into a GO dispersion. Figure 6.4a–c shows the resulting flakes on the substrates after one, two and three dips. Despite having a heavily percolated network in Fig. 6.4e, this early

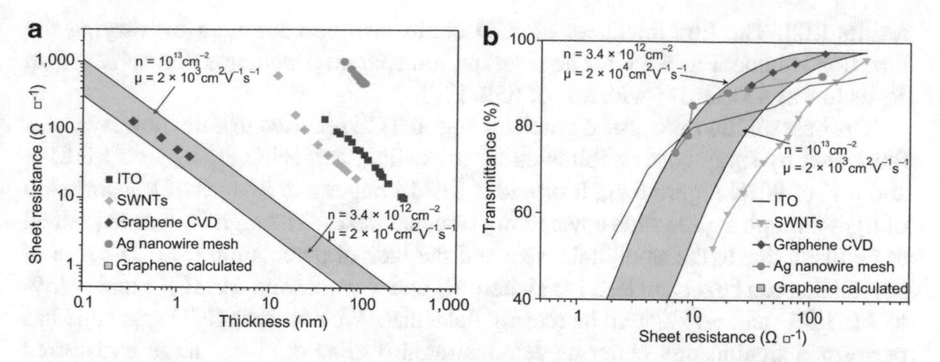

Fig. 6.3 (**a**) Thickness dependence of sheet resistance. (**b**) Transmittance versus sheet resistance for commercially available and newly emerging transparent conductors. Reproduced with permission from Ref [19]. Copyright (2010) Nature Publishing Group

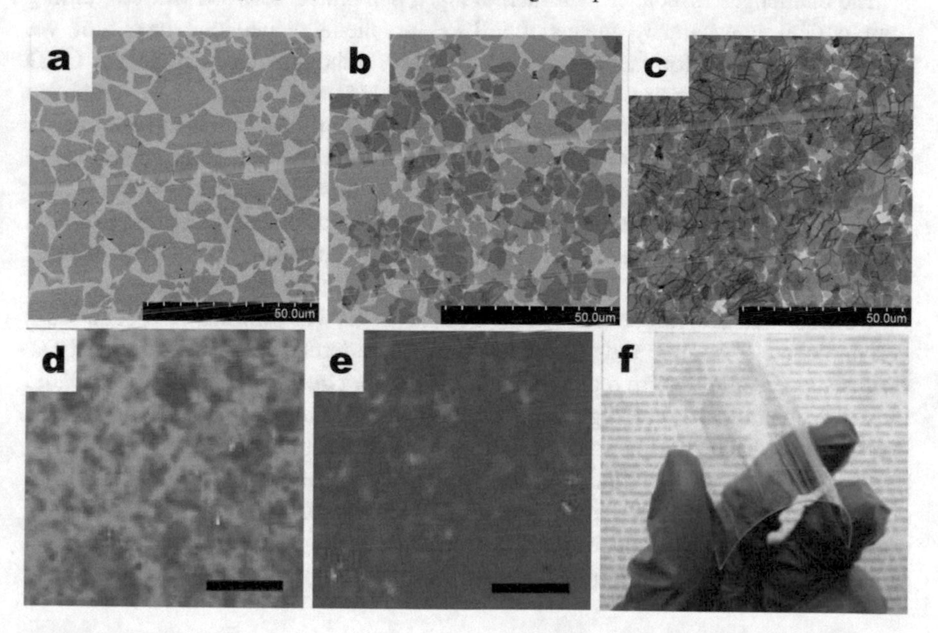

Fig. 6.4 SEM images showing layer-by-layer assembly of GO layers of similar sizes via Langmuir–Blodgett assembly. (**a**) Closely packed single layer GO as the first assembled layer. (**b**) Assembled double layers, with diluted top layer. (**c**) Assembled double layers, with high density top layer. Adapted with permission from Ref. [39]. Copyright (2009) American Chemical Society. (**d, e**) Optical micrographs of films of different densities of overlapped regions (darker regions) between graphene sheets assembled via vacuum filtration. (**f**) Image of TCE fabricated by Eda et al. Adapted with permission from Ref. [27]. Copyright (2008) Nature Publishing Group

demonstration achieved a relatively poor R_s of $19\,\mathrm{M}\Omega\,\square^{-1}$ and a transmission (T) of 95% [39]. Other strategies, as demonstrated by Eda et al., by spin coating reduced GO (rGO) dispersions directly on to glass and plastic substrates achieved better results [27]. The film thickness of rGO could also be controlled by varying the filtration volume; Fig. 6.4d–f. The resultant transparent conductive film achieved an R_s as low as $43\,\mathrm{k\Omega}\,\square^{-1}$ with a T of 65% [27].

The best solution-processed graphene based TCE reported to date, however, was fabricated by spray coating followed by annealing, achieving an $R_s = 5\,\mathrm{k\Omega}\,\square^{-1}$ and a T of 90%. Figure 6.4g, h provides SEM images and images of transmission of light through a glass slide covered in graphene flakes. The high R_s was explained to be likely due to the small flake size and the lack of percolation [37]. The role of percolation can be seen in Ref. [41] where R_s and T went from $6\,\mathrm{k\Omega}\,\square^{-1}$ and $\sim$75% to $2\,\mathrm{k\Omega}\,\square^{-1}$ and $\sim$77% with increasing flake size. We note that CVD graphene has performed significantly better as demonstrated by Bae et al. for large transferred graphene films. The authors achieved an R_s of $30\,\Omega\,\square^{-1}$ and T of $\sim$90% via wet chemical doping of the CVD-grown graphene films [22].

The challenges associated with achieving a percolated network while retaining high optical transparency means that LPE graphene dispersions have not yet matched ITO in performance. Looking forward, although doped, intrinsic CVD graphene theoretically presents itself as a viable ITO replacement, LPE graphene dispersions are an unlikely potential competitor to completely replace metal oxide based TCE, especially for flexible device applications. However, as an alternative to CVD graphene, LPE graphene is cheaper and easier to scale despite having a higher R_s at $T = 90\%$. They have already been demonstrated in a range of applications, including in organic light emitters (OLEDs) [42, 43] and solar cells [36]. It must hence be considered for applications in which cost reduction is critical.

We note that the large-scale fabrication of TCEs via printing also requires a certain degree of control and patterning. Custom-patterned solar cells and OLEDs, for instance, would require more control in patterning over large areas as compared to spray coating, bar coating or other 'flooding' techniques. Torrisi et al. first demonstrated such transparent conductive inkjet-printable ink from N-Methyl-2-pyrrolidone (NMP) based liquid phase exfoliation graphene dispersion in 2012 [40]. Although the reported electrical conductivity was far from suitable for practical applications, it is now well-recognised that there are significant opportunities yet to be explored not only with inkjet, but also with other forms of printing [44].

6.1.2 *Opaque Conductive Inks*

As mentioned in the previous section, Torrisi et al. first demonstrated conductive inkjet-printable ink from NMP based LPE graphene dispersion in 2012 which achieved an electrical conductivity of $100\,\mathrm{S\,m^{-1}}$ [40]. Whilst this was a successful printed demonstration, the conductivity of the ink is still far from $5.96 \times 10^7\,\mathrm{S\,m^{-1}}$, the conductivity of bulk copper which is used for current electrical applications [45]. Since then, much work has been devoted to enhancing the conductivities of the inkjet-printable graphene inks through improved material production, processing and engineering, ink formulation and print post-processing. As shown in Fig. 6.5a, a value of up to 40,000 S m^{-1} was reported by Secor et al. in 2017 with a graphene ink using nitrocellulose binder [46]. Although the conductivity of these graphene inks has already far exceeded the initial demonstrations, the conductivities are still over three orders of magnitude lower than those of existing metal-based inks (>100,000 S m^{-1}).

Compared to inkjet printing, for high throughput applications, techniques such as flexography come to the fore due to the high production speeds (>500–600 m min^{-1}), high resolution (~100 μm), and low ink weight (dry ink thickness <1 μm) [47–49]. Flexography is already a widely used technique for functional printing [48–50]. It should be noted, however, that the lower ink weight (i.e. thickness), when compared to techniques such as screen printing, can lead to higher resistances for similar or equivalent ink formulations.

Current functional inks for such large-area printing typically fall into two categories—low cost (~£50/kg), but high (>20 kΩ $\square^{-1}$) R_s inks containing carbon, and high cost (>£1,500/kg), low R_s (<1 Ω $\square^{-1}$) inks containing metals such as copper or silver. Some device manufacturers therefore typically print in a two-stage process to balance cost/performance. Using metal inks introduces several limitations to functional printing, even if they only form a small proportion of the coating. Firstly, after printing, metal inks often require high-temperature sintering or other curing procedures, which requires specialist equipment not usually found on a standard printing press. The two-stage printing process, meanwhile, can add complexity as additional printing plates are required. These must remain in alignment throughout the printing process to avoid either short circuits or inadequate contact. The high cost is also significant, even though only a small volume of metal ink is required for each device, as >1 kg of ink is required simply to operate a standard flexographic press in a commercial setting, regardless of how many devices are printed. Finally, when it comes to disposal of the printed device, metal inks are challenging to de-ink from the substrate, and do not biodegrade. With these factors in mind, it is highly desirable to remove metal inks entirely from the printing process.

Figure 6.5a gives an overview of the conductivity levels of graphene inks demonstrated to date. Graphene inks are highly promising for their low-cost and high conductivity. This coupled with deposition methods such as inkjet-, screen- and gravure-printing allow the printing of conductive patterns in a low-cost, high-throughput manner.

In this context, results from various research groups, including the authors of this work, have indicated that graphene-based inks can bridge the gap between carbon and metal based inks, with the potential for an ink of comparable cost to existing carbon-based inks, but that could be printed in a single pass with the need for metal ink overprints. An example of this is shown in Fig. 6.5b, c, showing photos from the first reported industrial scale press run for a flexographic graphene ink by the authors. Printed at ~100 m min^{-1}, the graphene ink was sufficiently conductive to be used as a capacitive touchpad; Fig. 6.5d. Such inks are likely to be used for low-cost flexible and disposable devices such as interactive packaging or biosensors.

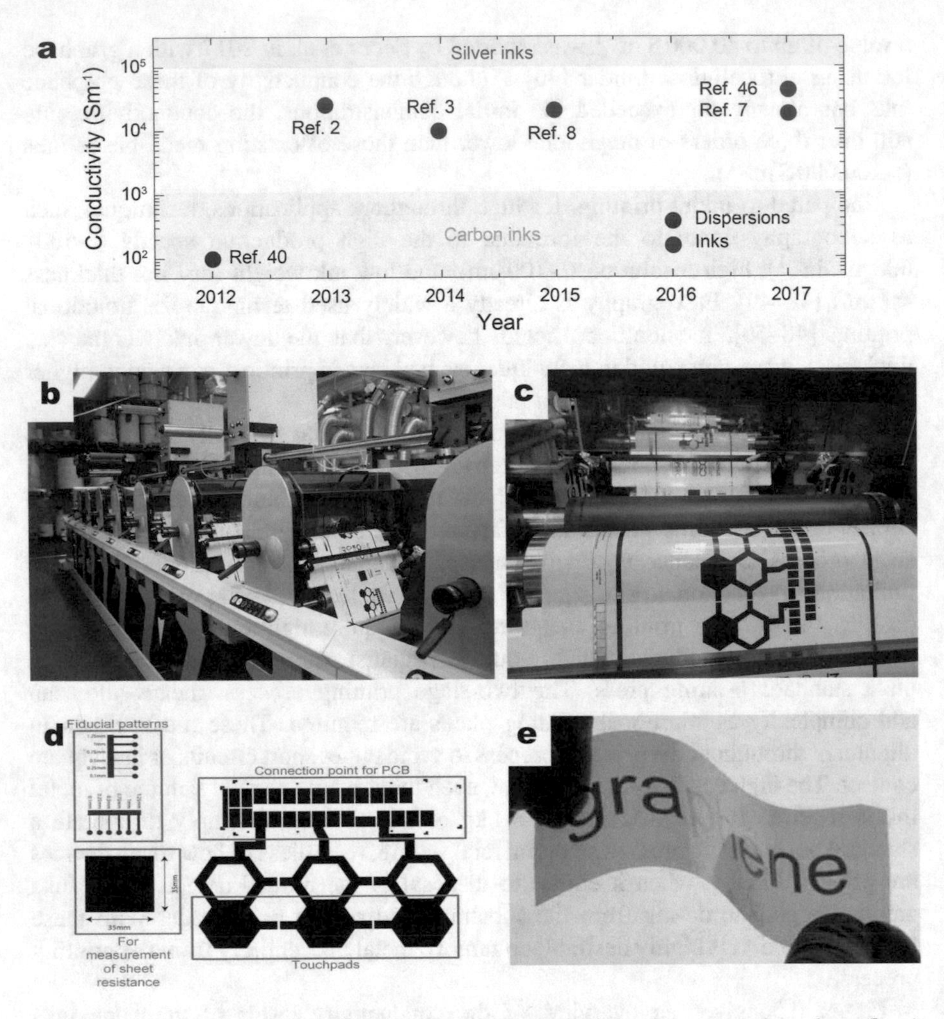

Fig. 6.5 (**a**) Electrical conductivity of selected graphene inks demonstrated to date. (**b**, **c**) Flexographic press running at 100 m min^{-1} with graphene ink and (**d**) a schematic of a printed touchpad that was produced via flexographic printing. (**e**) Inkjet printed graphene ink with a nitrocellulose (NC) binder. Reproduced with permission from Ref. [46]. Copyright (2017) American Chemical Society

The lowest electrical resistances reported from printed graphene in general have been demonstrated with screen printing as a deposition process due to the thick coatings it produces [5, 6, 8]. For instance, Hyun et al. demonstrated a graphene ink screen printed using a modified stencil process which achieved an R_s of ~30 $\Omega\square^{-1}$ at 25 µm thickness [8]. Such films can additionally be post-processed using methods such as physical compression to improve the inter-flake connectivity and hence decrease the electrical resistance [5, 6].

However, we note these aforementioned reports typically use an ink with minimal binder [6, 8] or a high-temperature curing step to remove the binder [5]. While this lowers the R_s, it also greatly reduces the durability of the printed film, which now consists only of loosely bound graphene flakes, meaning it has limitations for the majority of applications. A more realistic strategy for press-ready inks would appear to be an innovation within the binder systems that are capable of producing better adhesion or conductivities. For instance, the conductivity of 40,000 S m^{-1} reported by Secor et al. was achieved with a nitrocellulose binder, which was proven to improve the mechanical durability of the printed graphene even after binder removal via photonic annealing [46].

6.2 Printed (Opto)Electronics and Photonics

Although there have been many demonstrations from mechanically exfoliated and CVD graphene in (opto)electronics and photonics [19, 51, 52], low yields and high production costs still pose the greatest barriers toward scalable device development and real-world applications [53, 54]. Therefore, recent years have seen increasing efforts towards the development of scalable and low cost printable 2D material (opto)electronics and photonics applications [40, 55–57].

This was first demonstrated by Torrisi et al. in 2012 [40]. As schematically illustrated in Fig. 6.6a, the authors inkjet-printed graphene as a channel for transistors. The graphene ink used was NMP based LPE dispersions. NMP was used as the solvent to avoid residual impurities in the device due to the presence of surfactants in aqueous dispersions. The authors showed that the fabricated transistors exhibited a carrier mobility of up to 95 cm^2 V^{-1} s^{-1} [40], greater than that of typical printed organic transistors (usually $<$1 cm^2 V^{-1} s^{-1}) [59, 60]. However, the I_{ON}/I_{OFF} from the printed transistors was limited to only 10 (compared to $>10^5$ for organic transistors) due to the lack of a bandgap in graphene [40]. To address this low I_{ON}/I_{OFF} ratio, the authors then deposited poly[5,5'-bis(3-dodecyl-2-thienyl)-2,2'-bithiophene] on top of graphene, as presented in Fig. 6.6a. This strategy proved viable and increased the I_{ON}/I_{OFF} ratio to 4×10^5. However, the field effect mobility of the device decreased to 0.1 cm^2 V^{-1} s^{-1} as a result, still representing an order of magnitude of improvement compared to inkjet printed PQT-12 devices.

Recently, Carey et al. reported inkjet-printed graphene based heterostructure transistors [58]. Figure 6.6b depicts the photographs of the devices on textiles, where the inkjet-printed graphene was exploited as the channel and contacts and inkjet-printed h-BN was used as the dielectric layer [58]. The authors showed that the devices exhibited a significantly increased carrier mobility of up to 204 cm^2 V^{-1} s^{-1} (average mobility: $\sim$150 cm^2 V^{-1} s^{-1} on PET and $\sim$91 cm^2 V^{-1} s^{-1} on polyurethane textile) [58]. The authors also showed that the textile devices possessed a high mechanical durability, withstanding up to $\sim$4% strain and even more than 20 washing cycles [58]. However, as shown in Fig. 6.6c, the typical output characteristics of the devices on textiles exhibited an I_{ON}/I_{OFF}

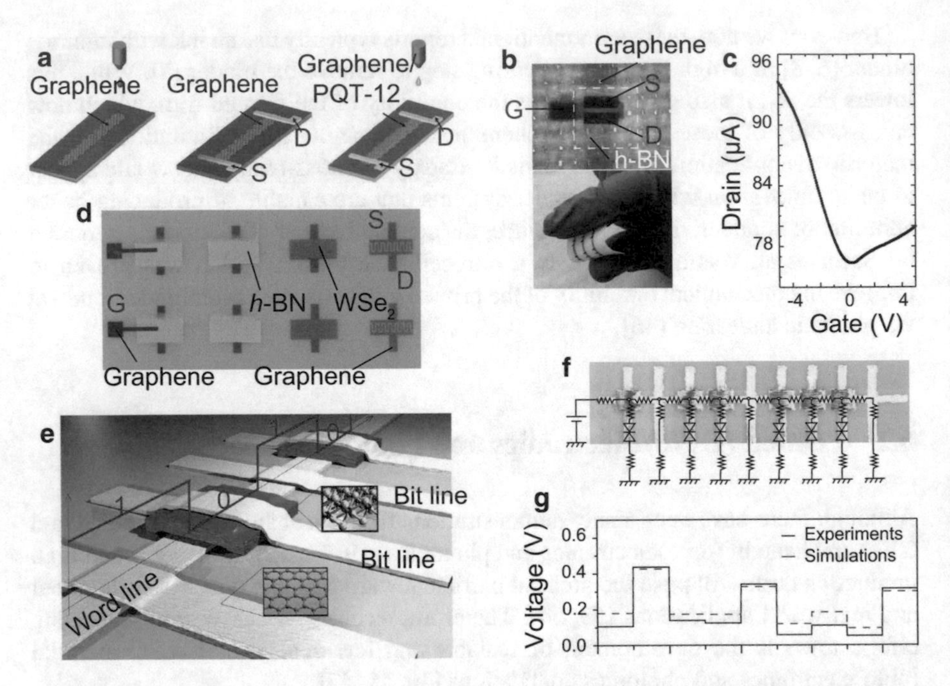

Fig. 6.6 Printed (opto)electronics and photonics: (**a**) Schematic of the first demonstration of inkjet-printed graphene transistors. Reproduced with permission from Ref. [40]. Copyright (2012), American Chemical Society. (**b**) Photographs of all inkjet-printed 2D material transistors on textile, and (**c**) associated transfer characteristics. Reproduced with permission from Ref. [58]. Copyright (2017), Nature Publishing Group. (**d**) All inkjet-printed 2D material transistors. Figure from A.G. Kelly and J.N. Coleman. (**e**) Schematic all inkjet-printed read-only memory, (**f**) the corresponding circuit designs and (**g**) the electrical response. Adapted with permission from Ref. [56]. Copyright (2017), Nature Publishing Group

of only ~1.2 [58]. We note that state-of-the-art organic field effect transistors have I_{ON}/I_{OFF} ratios of 10^6–10^{10} [61].

In contrast to graphene, semiconducting TMDs show promise for the development of electronics with high I_{ON}/I_{OFF} ratio [51, 62]. Kelly et al. reported inkjet-printed heterostructure transistors employing WSe_2 instead of graphene as the channel layer, as shown in Fig. 6.6d [55]. The transistors exhibited an I_{ON}/I_{OFF} ratio of up to ~600, significantly higher than the aforementioned graphene transistors. However, the carrier mobility was limited to $0.22\,cm^2\,V^{-1}\,s^{-1}$. To date, an effective method to increase the I_{ON}/I_{OFF} ratio whilst retaining a high mobility remains elusive and is a critical factor in the adoption for realistic applications in electronics.

Besides transistors, printed electronic circuits are also beginning to emerge. Figure 6.6e shows a schematic of a printed read-only memory (ROM) based on 2D material junctions [56], while Fig. 6.6f shows the printed ROM with the corresponding circuit design [56]. As shown, in this ROM, the graphene/WS_2/graphene junction with higher electrical resistance was interpreted as logic ‘0’, while the graphene/graphene junction was interpreted as logic ‘1’ [56]. This printed ROM output experimental results in a good agreement with simulation; Fig. 6.6g [56]. For practical interest, the ROM could be used in RFID tags to store identification information or integrated with additional electronics for more complex circuits and functions [56].

2D materials such as semiconducting TMDs and BP also offer potential for optoelectronic applications, such as photodetectors [51, 63]. A printed 2D material photodetector was first reported by Finn et al. in 2014 [64]; Fig 6.7a. The device consisted of inkjet-printed interdigitated graphene electrodes with an inkjet-printed MoS_2 active photodetection channel, with a calculated photoresponsivity of $<1\,\mu A\,W^{-1}$ (at 532 nm).

Similarly, Withers et al. demonstrated a heterostructure device using graphene, *h*-BN, and WS_2; Fig. 6.7b [65]. This demonstration involved drop-casting and inkjet printing WS_2 and *h*-BN inks to fabricate a photodetector with WS_2 acting as the active photodetection channel, achieving a photoresponsivity of $\sim 0.1\,mA\,W^{-1}$.

Different from this planar structure, the heterostructure graphene/WS_2/graphene junction was also utilised for photodetection [56]. This device exhibited a photoresponsivity of $>1\,mA\,W^{-1}$ (at 514 nm). To further improve the photoresponsivity, Hu et al. proposed a hybrid photodetector structure where inkjet-printed BP was integrated with a graphene/Si Schottky junction [57].

Figure 6.7c shows a photograph of Hu et al.’s printed devices on a silicon wafer, while the inset schematically depicts the structure of this hybrid device [57]. As BP has the tendency to degrade due to oxidation under ambient conditions, the hybrid device was encapsulated with parylene-C [57]. The device exhibited a photoresponsivity of up to $164\,mA\,W^{-1}$ at 450 nm [57], significantly higher than those of the previous photodetectors. More importantly, as presented in Fig. 6.7d, due to the layer-dependent bandgap of BP (0.3–2.0 eV), the device responded to 1550 nm light with a photoresponsivity of $1.8\,mA\,W^{-1}$, beyond the bandgap of Si (1.1 eV) [57].

Beyond (opto)electronics, the nonlinear optical absorption and ultrafast carrier dynamics of 2D materials (e.g. graphene, MoS_2 and BP) make them attractive for nanomaterial based nonlinear optical devices such as saturable absorbers (SAs) [9, 57, 66, 67]. SAs may be used for ultrafast optical pulse generation and are an underpinning technology in a wide variety of applications, ranging from materials processing, time-resolved spectroscopy and industrial micromachining to biomedical imaging [66, 68]. One dominant, well-developed technology for 2D material based SA fabrication is through developing polymer composites from solution-processed dispersions [9, 69–72]. However, it can be challenging to precisely control the optical parameters of the fabricated SA devices. The composite preparation can also be time consuming due to slow solvent evaporation, especially when high boiling point solvents (e.g. NMP) are used. To address these challenges, Hu et al. employed inkjet-printed BP as SA devices to mode-lock ultrafast lasers; Fig. 6.7e [57]. The issue of BP degradation and operation stability was again addressed by encapsulation with parylene-C [57]. Figure 6.7f, g shows stable

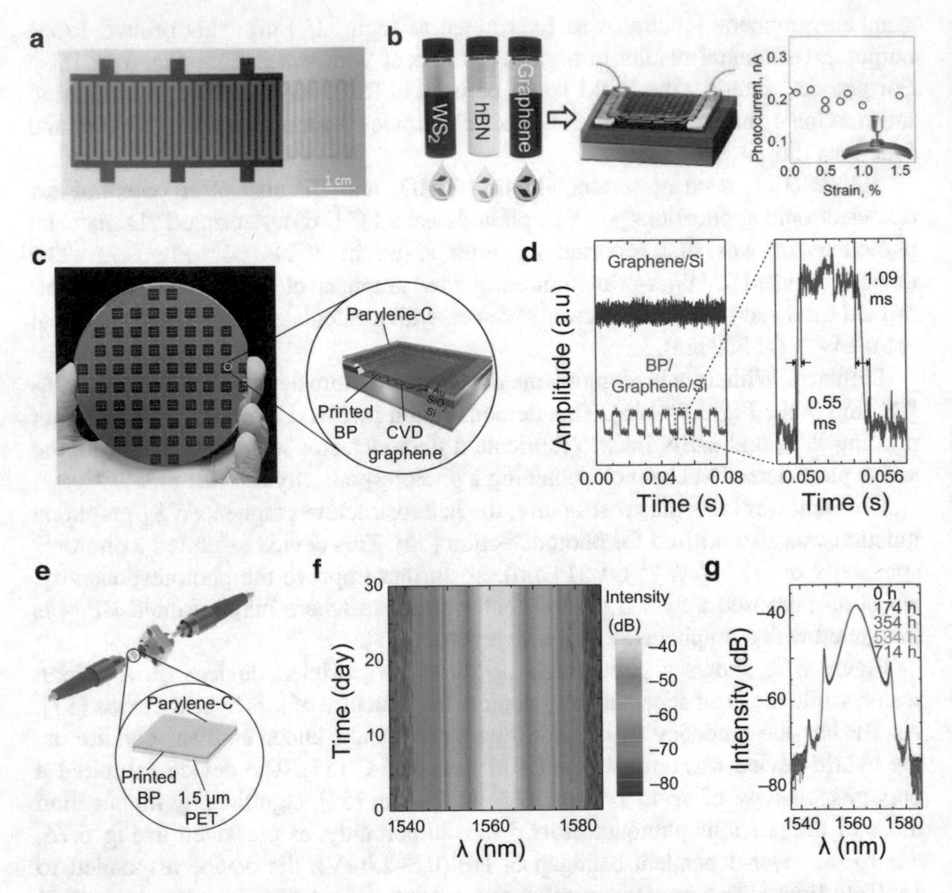

Fig. 6.7 (**a**) Photograph of printed graphene-MoS_2 photodetector. The black regions are the graphene interdigitated electrodes. Reproduced with permission from Ref. [64]. Copyright (2014) Royal Society of Chemistry. (**b**) Schematic of a general heterostructure device fabrication process using 2D-material inks. The last panel shows the final device and the cross-sectional image of the WS_2 thin film. The scale bar is 35 nm. Reproduced with permission from Ref. [65]. Copyright (2014) Nature publishing group. (**c**) Graphene/Si Schottky junction/inkjet-printed BP photodetector arrays in a silicon wafer, and (**d**) the corresponding measured response at 1,550 nm light. (**e**) Schematic showing integration of inkjet-printed BP nonlinear optical devices into a laser cavity to generate ultrafast lasers, and (**f**) the corresponding output laser spectrum across 30 days and (**g**) overlay of the spectrum acquired after 0, 174, 354, 534 and 714 h of operation, highlighting the stability of printed and encapsulated BP. Reproduced with permission from Ref. [57]. Copyright 2017, Nature Publishing Group

ultrafast pulse generation from the correspondingly printed BP-SAs for over 30 days which far exceeded the performance of prior demonstrations using BP [57]. Functional inks and inkjet printing therefore could be a very attractive approach for both discrete and hybrid and integrated photonic and optoelectronic device applications including photonic integrated circuits enabled by 2D materials.

6.3 Printed Sensors

Sensors are a major application field for printed 2D materials. Their unique photonic, (opto)electronic and mechanical properties as well as the large specific surface area make them particularly capable of responding to changes in the ambient conditions (e.g. gas, moisture, stress or other environmental elements) through contact/non-contact chemical/physical reactions [53, 54, 73]. These responses can be converted into signals that can be read out and interpreted.

High performance, proof-of-concept sensors based on 2D materials from mechanical exfoliation and CVD have already been widely reported. Examples include sensors for individual gas molecules [74], in-vitro bacterial sensors [75], and even environmental sensors for internet of things (IoT) applications [76].

Printing is beginning to emerge as a promising route to producing sensors. In particular, resistive sensors, with their simple and flexible device design and easy read-out capabilities, appear to be viable architecture for printable 2D material sensors. Generally, fabrication of resistive sensors involves printing of 2D materials-based sensing materials over parallel, interdigitated or fractal electrodes.

6.3.1 Chemical Sensors

Chemical sensors find widespread use in healthcare, environmental monitoring, industrial processes, agriculture and smart buildings. The properties of 2D materials lend themselves to some of these applications. For example, Yao et al. demonstrated inkjet-printed MoS_2 sensors for NH_3 detection [77]. The sensors were fabricated by printing LPE MoS_2 over metal microelectrodes deposited on SiO_2. The sensors exhibited resistance changes when exposed to NH_3, with a detection capability down to 5 ppm [77]. However, the reported long detection time (~1,500 s) and recovery time would need an order of magnitude improvement in order to be effective for use in real-world applications. This might require better selection of the 2D materials, the underlying materials processing and engineering, and/or an optimised device design.

Besides semiconducting MoS_2, Cho et al. showed that other 2D materials such as graphene and BP produced from LPE exhibited sensitivity to chemicals such as NO_2, NH_3, ethanol and acetaldehyde [78]. When compared with MoS_2 and graphene, the authors demonstrated that the sensitivity from BP was up to 20 times higher, and that the response time was ~40 times faster. This suggests a strong potential for BP in sensing applications.

In addition, BP has also been reported as responding to humidity [79]. He et al. deposited a mixture of GO and BP via inkjet printing as a highly sensitive sensing layer for humidity. Like BP, GO is known to be a highly hydrophilic 2D material. The electrical response of the GO and BP sensors was demonstrated by the authors for humidity levels ranging from 11 to 97% relative humidity, which revealed high capacitance sensitivity of 4.45×10^4 times for the GO sensor and 5.08×10^3 times for the BP sensor at 10 Hz operation frequency. Response/recovery times of the GO and BP sensor were measured to be 2.7/4.6 s and 4.7/3.0 s, respectively. These sensors also showed sensitive and fast response to a proximal human fingertip, giving potential applications in contactless switching [79]. However, the development of printable chemically unmodified BP gas and humidity sensors for long-term use could be limited due to BP's degradation in the ambient environment.

MXenes have also been proposed for gas sensing [80]. This is due to the adsorption of hydrogen-bonded gases such as ammonia (NH_3) directly on the MXene surface. Kim et al. demonstrated this by fabricating MXene gas sensors capable of detecting volatile organic compounds (VOCs) at the part per billion (ppb) level [81]. The authors utilised $Ti_3C_2T_x$ as a sensing layer for acetone, ethanol and ammonia, with demonstrated detection levels as low as 1,000 ppb [81]. The $Ti_3C_2T_x$ sensor had a lower limit of detection for acetone (50 ppb) as compared to MoS_2 (500 ppm) and graphene (300 ppm). The demonstrated sensor also outperformed WS_2 (1 ppm), trilayer MoS_2 (10 ppm), MoS_2 (100 ppm) and BP (10 ppm) in the detection of ammonia. Also, the signal-to-noise ratio was up to 2 orders of magnitude higher compared to other 2D materials.

Despite demonstrating good sensitivities, such sensors can typically have poor selectivity (i.e. the ability to uniquely respond to specific analytes). This has prompted the design of 2D material-based functionalised hybrid materials. Functionalisation of 2D materials via chemical doping or decoration with other functional materials (e.g. metal/metal oxide nanoparticles) has been demonstrated to be a viable strategy for enhancing sensitivity, response/recovery times, and selectivity [82, 83].

For instance, the functionalisation of graphene with groups such as =O, $-NH_2$, –OH, –F, $-CH_3$ and $-SO_3H$ can significantly expand its sensing capabilities to a wide range of chemicals, including NO, NO_2, Cl_2, SO_2, $CHCl_3$, CH_3OH and C_6H_{14} (hexane) [82–84]. Figure 6.8a shows a chemiresistive sensor employing rGO-SO_3H decorated with Ag nanoparticles [84]. The sensors were fabricated by gravure printing the functional ink over interdigitated metal electrodes. As presented in Fig. 6.8b, c, the sensors exhibited a fast response over NO_2 and selectivity for NO_2 and NH_3 as well as linear response to humidity [84].

Figure 6.8d shows a micrograph of a CMOS integrated graphene humidity sensor [13]. The sensor was fabricated by inkjet printing a graphene-PVP sensing layer over the interdigitated electrode of the CMOS chip. The sensor exhibited resistance changes for relative humidity changes between 10–80% [13]. The authors proposed that the response (i.e. resistance change) arose from the change in the graphene inter-flake percolative network in the swelling hygroscopic PVP upon exposure to humidity [13]. This approach was the first to combine inkjet printing of 2D materials and the scalability of CMOS, foreseeing a prospect of seamless integration of printed 2D materials with existing silicon technologies for flexible and miniaturised 2D material based devices.

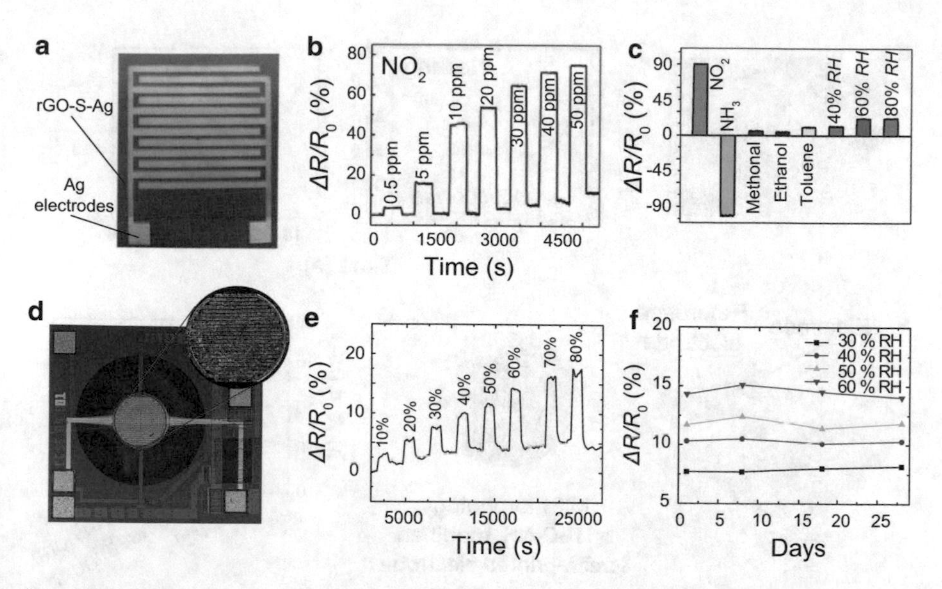

Fig. 6.8 Printed chemical sensors: (**a**) Photograph of gravure-printed rGO/Ag gas sensors, and (**b**) the corresponding response with respect to NO_2 concentration and (**c**) selective response to various gases and humidity levels. Reproduced with permission from Ref. [84]. Copyright (2014), American Chemical Society. (**d**) CMOS integration of inkjet-printed graphene/PVP polymer composites for humidity sensing, Inset: the micro hotplate area with printed graphene/PVP polymer and (**e**) sensor response and (**f**) sensing stability at different humidity (*RH*) levels. Reproduced with permission from Ref. [13]. Copyright (2015), Nature Publishing Group

6.3.2 Temperature Sensors

Temperature sensors are widely utilised in research and industrial process monitoring and diagnostics, as well as within smart buildings, electrical or electronic products and thermal management [85, 86]. There is a growing interest in wearable temperature sensors that are capable of monitoring on-skin temperature in order to provide insights into general wellbeing and physical activities [86, 87]. Composites of graphene and thermoelectrics [88], for instance, thermoelectric polymers (e.g. poly(3,4-ethylenedioxythiophene) polystyrene sulfonate (PEDOT:PSS)) [89, 90] and 2D materials (e.g. Bi_2Te_3) [91], are a promising material platform for printable temperature sensors. However, to date only limited developments using 2D materials have been seen in this field.

Figure 6.9a is photograph of an inkjet-printed epidermal temperature sensor [92]. The sensors were fabricated by inkjet printing graphene/PEDOT:PSS onto a stretchable polyurethane substrate. As shown in Fig. 6.9b, when in operation, the temperature sensors exhibited changes in resistance in the heating and cooling processes between 34 and 40 °C; the 90th percentile response time was 18 s, and the 90th percentile recovery time was 20 s. This particular example demonstrated by Vuorinen et al. coupled with the small form factors achievable with 2D material-based sensors shows promise for wearable technology applications.

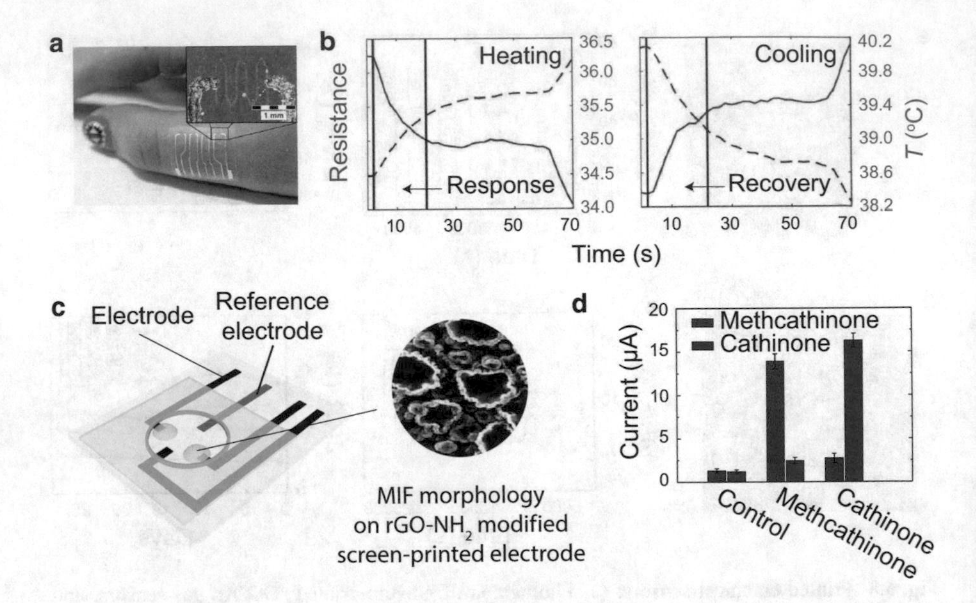

Fig. 6.9 Printed temperature and biosensors: (**a**) Photograph of graphene/PEDOT:PSS temperature sensors attached on to the skin, and (**b**) the corresponding response to temperature change. Reproduced with permission from Ref. [92]. Copyright (2016), Nature Publishing Group. (**c**) Schematic of biosensors with SEM micrograph of the molecularly imprinted film (MIF) morphology on rGO-NH_2 modified electrode, and (**d**) the corresponding device response. Reproduced with permission from Ref. [93]. Copyright (2013) The Royal Society of Chemistry

6.3.3 *Strain, Pressure and Touch Sensors*

Strain sensors, or strain gauges, are used across a variety of applications [94]. A strain sensor typically consists of a conductive pattern that is capable of producing a change in the electrical read-out upon geometric deformation [94–96]. This change can be either resistive or capacitive depending on the sensor design. Graphene has been widely reported in compact strain sensor fabrication as a standalone material [97–99] or in the form of composites (e.g. polyvinylidene fluoride (PVDF) [100] and nanocellulose [101]). For example, wearable strain sensors made possible by 2D materials can be used for human motion detection and as control units for robotics or mixed-reality interfaces [98]. However, reports on printed 2D material strain sensors remain scarce.

Casiraghi et al. recently reported an inkjet printed graphene strain sensor on paper [102]. The schematic sensor configuration is in the inset of Fig. 6.10a, showing that the sensors were composed of inkjet-printed conductive electrodes and an active strain sensing line in between. As presented, the printed graphene strain sensors exhibited fast, prominent changes in electrical resistance under bending tests. The authors further employed the printed sensor in an electrical circuit as a variable resistance to tune the luminosity of an LED under tensile or compressive strains; Fig. 6.10c–e.

Pressure and touch sensors can either detect physical pressure, contact or even proximity. They represent another interactive technology widely applied in consumer portable devices such as smart phones or touch pads. Pressure sensors can be created via structures where an active sensing layer is 'sandwiched' between contacts, or where the active sensing spots are connected to contacts, or as a combination of both. Published reports on printed 2D material pressure and touch sensors are scarce although there have been a number of demonstrations of such devices by commercial entities. Figure 6.10f shows a printed graphene thin film pressure sensing panel demonstrated by *Haydale*, a company dealing in graphene-based applications. In this panel design, a printed graphene/binder composite was 'sandwiched' between a printed electrode matrix, such that the matrix could detect resistance changes of the graphene/binder composite upon pressure and indicate the location of touch.

The flexographic-printed graphene based circuits as presented in Fig. 6.10g were designed for the demonstration of large-scale production of capacitive touch surfaces for interactive devices. In the example shown in Fig. 6.10g, the touch surface was attached to a PCB programmed to process the received capacitive changes upon touch, and hence to trigger a sound output through a speaker. Capacitive sensors work by detecting capacitive changes, and therefore the active surface area (printed graphene) may also be laminated behind thin paper or plastic substrates, enabling the development of a wide variety of interactive surface applications ranging from wireless keyboards, pianos and educational posters.

Although present demonstrations are based mostly on books and toys, the future potential for strain, pressure and touch sensors is huge. In particular, the use of stretchable and flexible sensors attached onto the surface of the human body can perceive external stimuli [103] may have a wide variety of applications in robotics, wearable electronics as well as prosthetics and other medical devices [94].

6.3.4 Biosensors

Biosensors can be described as integrated miniaturised devices employing biological recognition elements (e.g. enzymes, antibodies, oligonucleotides, receptor proteins, microorganisms or cells) as active sensing components [104]. With an ever-increasing demand for miniaturised, smart healthcare systems, the field of biosensor technology demands advances in the underlying materials technology and engineering for high-level, novel biosensing performance [54, 105]. 2D materials offer a promising carrier platform for the biological recognition elements, as pioneered by Ref. [75] where CVD graphene incorporating antimicrobial peptides was used for in-vitro bioselective detection of bacteria. The reports on printed 2D material biosensors, however, are very limited, with the majority of them are being

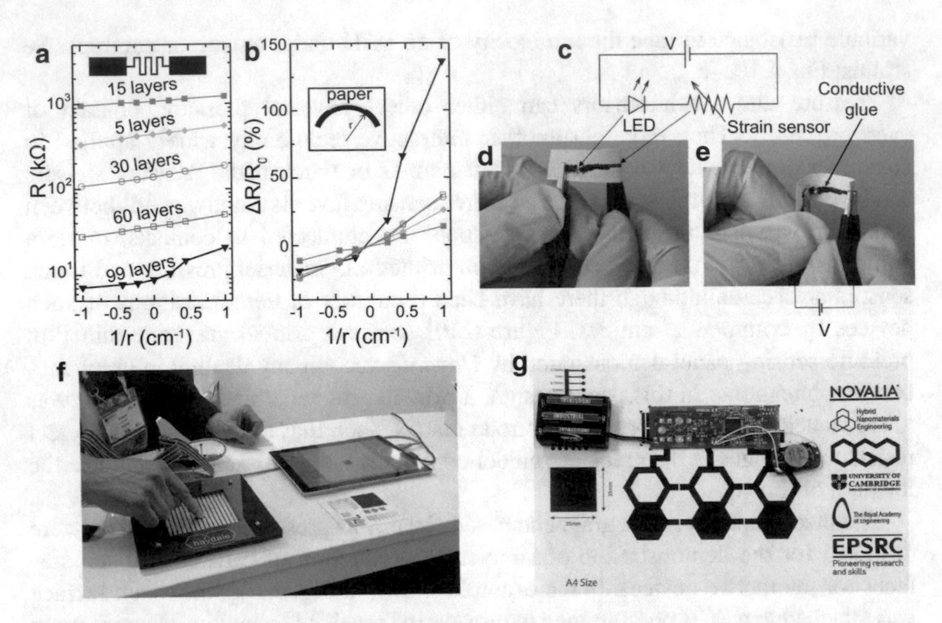

Fig. 6.10 Printed strain, pressure and touch sensors: (**a**) Resistance and (**b**) sensitivity of inkjet-printed graphene strain sensor (inset) as a function of the inverse of the bending radius and for different layer thicknesses. (**c**) Schematic circuit employing the graphene strain sensor as a variable resistor under (**d**) tensile and (**e**) compressive strains [102]. Reproduced with permission from Ref [102]. Copyright (2017) Elsevier. (**f**) *Haydale* graphene pressure sensors demonstration, source – ieee.org. (**g**) Flexographic-printed graphene-based capacitive touchpads from the University of Cambridge

based on rGO. This is due to the fact that rGO can be readily decorated with a wide variety of biological recognition elements.

Figure 6.9c schematically shows a molecularly imprinted film (MIF) based biosensor for psychotropic compound detection, where MIF was synthesised by electro-polymerisation on rGO decorated electrodes [93]. MIF is a widely applied process in biosensor fabrication where functional and cross-linking monomers are polymerised in the presence of a target imprint molecule that acts as a molecular template [106]. This template can then be used to detect similar biological compounds [106]. For sensor fabrication, rGO was functionalised with $-NH_2$, and was subsequently dropcast onto screen printed graphite electrodes. It was shown that the $rGO\text{-}NH_2$ flakes were flat with the $-NH_2$ exposed, such that MIF could be synthesised on top. The inset shows the morphology of the synthesised MIF. As shown in Fig. 6.9d, the fabricated sensors based on methcathinone and cathinone MIFs exhibited a highly selective sensitivity for methcathinone and cathinone, respectively [93]. Kong et al. later followed a similar strategy to demonstrate a blood glucose biosensor [107].

Instead of making use of biological recognition elements, hybrids of 2D materials and other functional materials (e.g. metals, metal oxides and quantum dots) are also emerging as a promising material platform [104, 108, 109]. For instance, Zhang et al. demonstrated ultrasensitive non-enzymatic glucose sensors based on screen-printed rGO/copper oxide (CuO) [108]. The authors showed that the addition of rGO to CuO significantly improved the sensitivity to glucose, which reached similar sensitivities to those of commercial enzymatic glucose sensors. Such non-enzymatic glucose sensors show interesting alternatives for the current incumbent generation of glucose sensors which use enzymes as a sensing material. However, they do not presently have any commercial applications. The use of chemically pristine or functionalised 2D materials as a sensing layer acts largely as a replacement for biological compounds in traditional biosensors. This potentially expands the application of biosensors into environments where conventional biological sensing elements are not suitable for use.

6.4 Printed Energy Storage

The pressure to limit or replace the use of fossil fuels and advent of portable devices means that battery and supercapacitor technology as energy storage solutions are receiving a significantly increased focus. A battery relies on electrochemical reactions at the electrodes to deliver energy [110–112]. A supercapacitor (i.e. electric double-layer capacitor), utilises a double-layer effect on the electrodes to store electrical energy [113, 114]. Achieving highly efficient and dense energy storage requires optimal underlying electrode materials technology and engineering [110, 113–116]. 2D materials like graphene and MXenes are ideal as next generation electrodes for these devices. The proliferation of 2D materials in functional printing adds to the versatility of energy storage solutions by allowing batteries and supercapacitors to be fabricated in a low-cost, high-throughput manner.

6.4.1 Batteries

Batteries are energy devices that are made up of electrochemical cells. Each electrochemical cell consists of a negative electrode (anode), a positive electrode (cathode) and an electrolyte [117]. Batteries have high energy densities (~120–170 Wh kg^{-1}) and are capable of storing energy for long periods of time. Printable batteries will enable the design and manufacture of low-cost and flexible electronics such as wearables, compact and flexible displays and small devices.

At present, the most common rechargeable lithium-ion batteries are based on a $LiCoO_2$ cathode and a graphite anode. Such a combination has a theoretical energy density of 387 Wh kg^{-1} and a measured density of ~120–150 Wh kg^{-1} [118]. 2D materials have the potential to improve the energy density of batteries due to their electrical conductivity, high specific surface area, the large number of active sites for Li^+ storage and short Li^+ diffusion distances by being used as alternatives to existing anodic and cathodic materials [53, 63, 114, 119–121].

Indeed, there are already many successful demonstrations with solution-processed graphene [122–124] and other 2D materials such as MoS_2 [125, 126] and in particular, MXenes [127–131].

A parameter to take note of when characterising the performance of a battery is its specific capacity. The theoretical specific capacity is the total ampere-hours (Ah) available when the battery is discharged at a certain discharge current, per unit weight. For instance, pristine single-layer graphene as an electrode has a theoretical specific capacity of 744 mAh g^{-1} assuming Li^+ is adsorbed on both sides of graphene to form Li_2C_6 [132], while the specific capacity of conventional graphite anodes is 372 mAh g^{-1} [133]. Indeed it has been demonstrated that UALPE graphene, deposited as an ink, has achieved a specific capacity of ~1,500 mAh g^{-1} by Hassoun et al. [119]. The authors proposed that the higher specific capacity was largely due to the edges of the graphene flakes within the inks which were <100 nm, hence maximising specific surface area [119]. Similarly, MXenes have also been shown to be particularly suited as an electrode material for both Lithium-ion batteries (LiBs) and non-lithium-ion-batteries (NLiBs). This is because MXenes have been shown to accommodate various ions of different sizes, significantly expanding the current selection of electrode materials. Xie et al. theoretically demonstrated this with the results presented in Fig. 6.11 [134]. Indeed MXenes like Ti_2CT_x and $Ti_3C_2T_x$ have been used in Li-S batteries as a viable electrode material [135, 136]. The encapsulation of Sn nanoparticles within layers of $Ti_3C_2T_x$ resulted in a specific capacity approaching 2,000 mAh g^{-1} [137]. MoS_2 is also a promising potential electrode material for battery applications. Hwang et al. demonstrated MoS_2 anodes, achieving a specific capacity of 700 mAh g^{-1} [126].

In the demonstrations for graphene, MoS_2 and MXenes mentioned, the 2D materials were prepared either through chemical synthesis or LPE. This is of particular relevance for printing as such material in liquid phase can easily be incorporated into binder materials to produce 2D material inks.

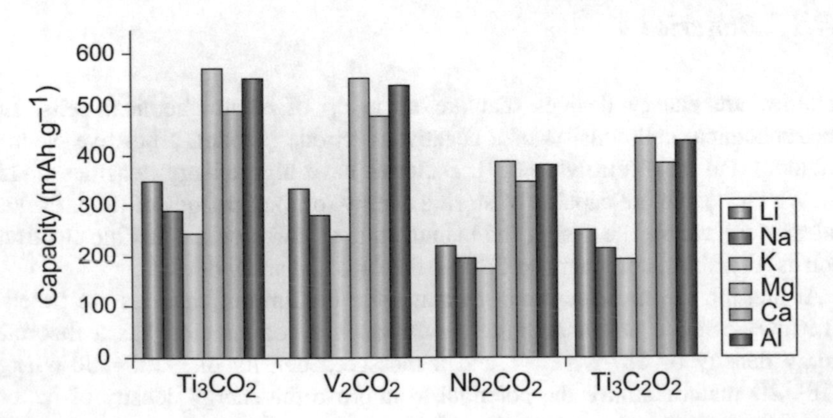

Fig. 6.11 Lithium and non-lithium ion theoretical capacities on O-terminated MXene nanosheets. Reproduced with permission from Ref. [120]. Copyright (2017) Nature Publishing Group

6.4.2 *Supercapacitors*

Supercapacitors are energy devices with ultrahigh power densities. They are essentially maintenance-free, possess longer life-cycles than batteries, require a very simple charging circuit, experience no memory effect, and are generally safer without risk of explosion [138]. However supercapacitors have lower energy densities, in the order of ($\sim$5–10 Wh kg^{-1}), when compared with traditional LiBs ($\sim$120–170 Wh kg^{-1}).

A supercapacitor requires mere seconds to charge when compared with batteries that usually take hours [138]. They are suited to applications that require many rapid charge/discharge cycles rather than long-term compact energy storage. Applications include automotive, trains and elevators, where they can be used for regenerative braking, short-term energy storage or burst-mode power delivery, typically applications that have short load cycles and high reliability requirements [139]. A supercapacitor structure consists of a current collector, electrodes, electrolyte, and separator. It is the electrode that is key to improving device performance. Graphene, with a theoretical specific surface area of 2,630 m^2 g^{-1}, is a promising material for supercapacitors [138, 139].

Printing technologies have been explored as a high-precision deposition method for the fabrication of supercapacitor electrodes. For example, Le et al. reported graphene supercapacitors using inkjet-printed and subsequently thermally reduced GO electrodes [140]. More recently, Hyun et al. demonstrated all-printed micro-supercapacitors that employed inkjet-printed graphene electrodes [141]. As schematically shown in Fig. 6.12a, the micro-supercapacitors were fabricated in three steps: an imprinted ink receiver was prepared at first. Next, a graphene/ethyl cellulose ink was inkjet-printed to pattern the electrodes, followed by binder decomposition through photonic annealing. Electrolyte was then deposited by inkjet. As presented in Fig. 6.12b, the average specific capacitance of this micro-supercapacitor array was $\sim$221 $\pm$ 16 μF cm^{-2} at a cyclic voltammetry scan rate of 100 mV s^{-1}, with a device performance variation of $<$10%. This micro-supercapacitor array exhibited good mechanical durability, withstanding over 1,000 bending cycles; Fig. 6.12c.

Besides inkjet printing, there have also been demonstrations of supercapacitors using other printing processes. Hyun et al. reported supercapacitors based on screen-printed graphene/PANI electrodes [8] and Yeates et al. demonstrated screen-printed graphene supercapacitors on textiles for wearable electronics applications [142].

On-chip micro-supercapacitors are also attractive for applications in wearables and remote sensors. In 2012, El-Kady et al. first fabricated supercapacitors via a laser scribing process [143]. The films produced were mechanically robust with high electrical conductivity (1,738 S m^{-1}) and specific surface area (1,520 m^2 g^{-1}), which could be used directly as electrodes without binders or current collectors. This work took advantage of the laser scribing technique to fabricate rGO interdigital electrodes, which enabled them to 'write' or 'print' on-chip microsupercapacitor arrays [143].

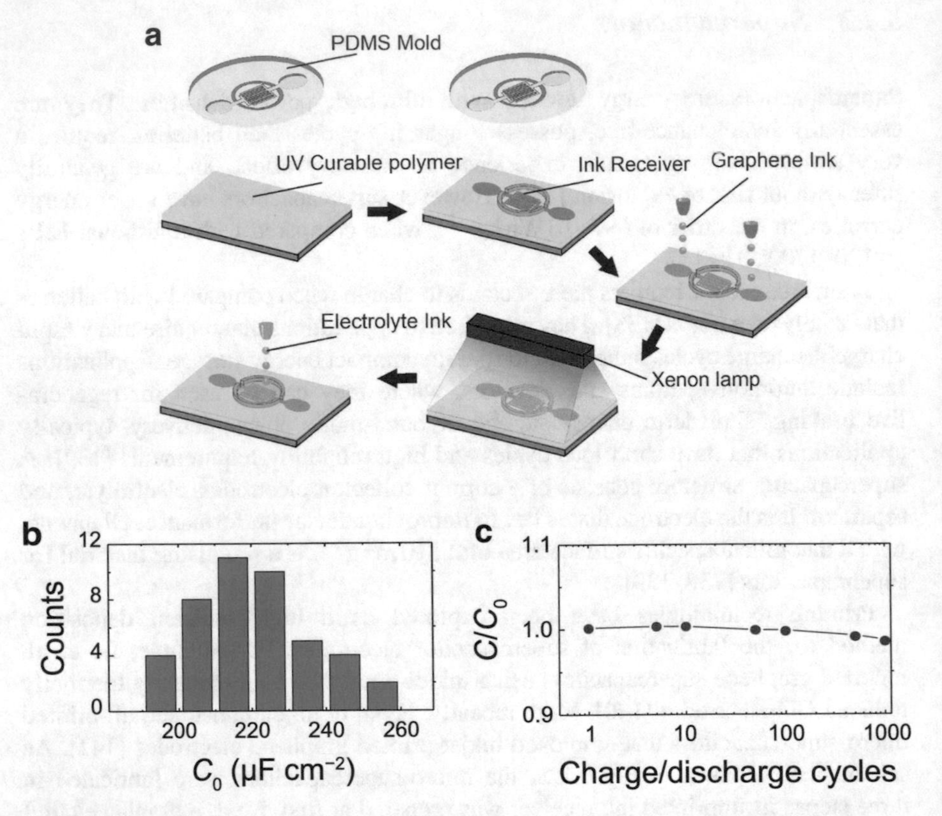

Fig. 6.12 Printed energy storage using 2D materials: (**a**) Schematic of the fabrication process for inkjet-printed graphene supercapacitors. (**b**) histogram of the specific capacitance and (**c**) typical relative specific capacitance over 1,000 charge/discharge cycles at a scan rate of 1,000 mV s^{-1}. Adapted from Ref. [141]

As previously stated, applications that require short load cycles and high reliability are ideal for supercapacitors. This is especially so for applications such as wearables and miniaturised energy storage. Printing offers an ideal method of fabrication for such small-factor supercapacitor devices. Moving forward, such printed supercapacitors capable of storing as much energy as a battery and yet that can be fully recharged in 1 or 2 minutes would be considered a revolutionary advancement in energy technology and a goal to be desired [138].

6.4.3 3D Printing of 2D Materials for Energy Storage

3D printing of 2D materials shows particular promise for energy storage. As discussed in Sect. 5.6.1, 3D printing is a process whereby plastic filaments, such as polycarbonate (PC), acrylonitrile butadiene styrene (ABS) and polyactic acid (PLA) are extruded and deposited layer-by-layer onto the build platform.

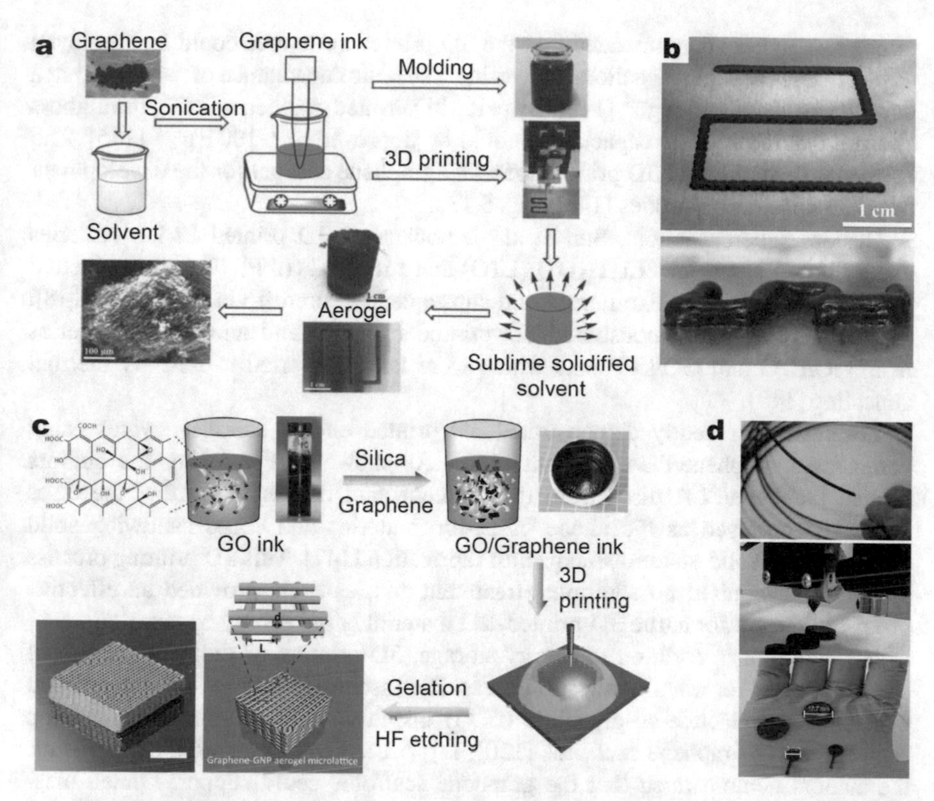

Fig. 6.13 3D printing of 2D materials: (**a**) Schematic of the freeze gelation of graphene process. (**b**) Top-view (one layer) and side-view (three layer) photographs of 3D-printed graphene patterns. Reproduced with permission from Ref. [145]. Copyright (2016), WILEY-VCH Verlag GmbH & Co. KGaA, Weinheim. (**c**) Schematic of the fabrication process of 3D-printed GO/graphene supercapacitors. Reproduced with permission from Ref. [146]. Copyright (2016), American Chemical Society. (**d**) Photographs of 3D-printed graphene based batteries and supercapacitors from printed filaments. Reproduced with permission from Ref. [147]. Copyright (2017), Nature Publishing Group

For the 3D printing of 2D materials this 'direct-ink' writing typically requires in-situ formulation, gelation and curing of the inks, negating some of the relative throughput advantages that 3D printing offers. For example, García-Tuñon et al. developed an aqueous GO ink using a branched copolymer surfactant for 3D printing [144]. This approach allowed the ink to form self-supporting 3D structures after deposition. The printed 3D structures were then frozen in liquid nitrogen and subsequently freeze-dried for thermal reduction of GO.

Instead of using surfactants, Lin et al. developed a graphene aerogel through room temperature freeze gelation of graphene dispersions in phenol (C_6H_6O) or camphene ($C_{10}H_{16}$); Fig. 6.13a [145]. This approach allowed self-solidified structures at room temperature after jetting, as shown in Fig. 6.13b. The 3D printed structures could be removed from the solvents through sublimation at room temperature. The authors demonstrated that the 3D printed structures could be employed as electrodes for supercapacitors, delivering a specific capacitance of ~75 F g^{-1} at a current density of 100 A g^{-1} [145]. By using 3D printed graphene/CNTs, the authors showed that the specific capacitance could be improved to ~100 F g^{-1} [145]. Zhu et al. also demonstrated 3D printing of a GO/graphene aerogel for the development of supercapacitor electrodes [146]; Fig. 6.13c.

Besides supercapacitors, Sun et al. demonstrated 3D printed Li-ion batteries using lithium-based inks $Li_4Ti_5O_{12}$ (LTO) and $LiFePO_4$ (LFP). The authors jetted the ink from a syringe-like nozzle and subsequently cured it via annealing [148]. Similarly, Fu et al. demonstrated 3D printed cathodes and anodes for batteries from GO/LTO and GO/LFP inks which were later converted to rGO by thermal annealing [149].

Foster et al. recently demonstrated 3D printed energy storage devices using commercial graphene/PLA filament [147]. As shown in Fig. 6.13d, the authors printed graphene/PLA discs from the filament, and demonstrated that the discs could be employed as the anode of Li-ion batteries and could sandwich solid electrolyte for solid-state supercapacitor fabrication [147]. This 3D printing process of filaments required no additional treatment and, as such, provided an effective proof-of-concept for a true 3D-printed, 2D material based energy application.

We note that in addition to energy storage, 3D printing of graphene is also of special interest for other application areas. For instance, Jakus et al. demonstrated a graphene/polylactide-co-glycolide (PLG) ink that could be printed to develop robust, flexible graphene scaffolds [150]; Fig. 6.14. Through in vitro experiments, the authors demonstrated that the graphene scaffolds could support human mesenchymal stem cell adhesion, viability, proliferation, and neurogenic differentiation with significant upregulation of glial and neuronal genes [150]. 3D printing of graphene and other 2D materials therefore promises an attractive pathway towards miniaturised and customised device prototyping for energy, sensing and biomedical applications [150].

6.5 Applications in Membranes and Barriers

The use of 2D materials for physical properties in barrier, shielding and membrane applications also holds great interest for 2D material inks. Due to the high mechanical strength, thermal and chemical stabilities, and tightly packed crystal structure in two dimensions, 2D materials have been considered in a variety of membrane and barrier applications to date [151–154]. Indeed, dispersions and inks of 2D materials have been widely demonstrated in barrier coatings [155, 156], filtration [157], anti-biofouling [156, 158] and EMI shielding [159].

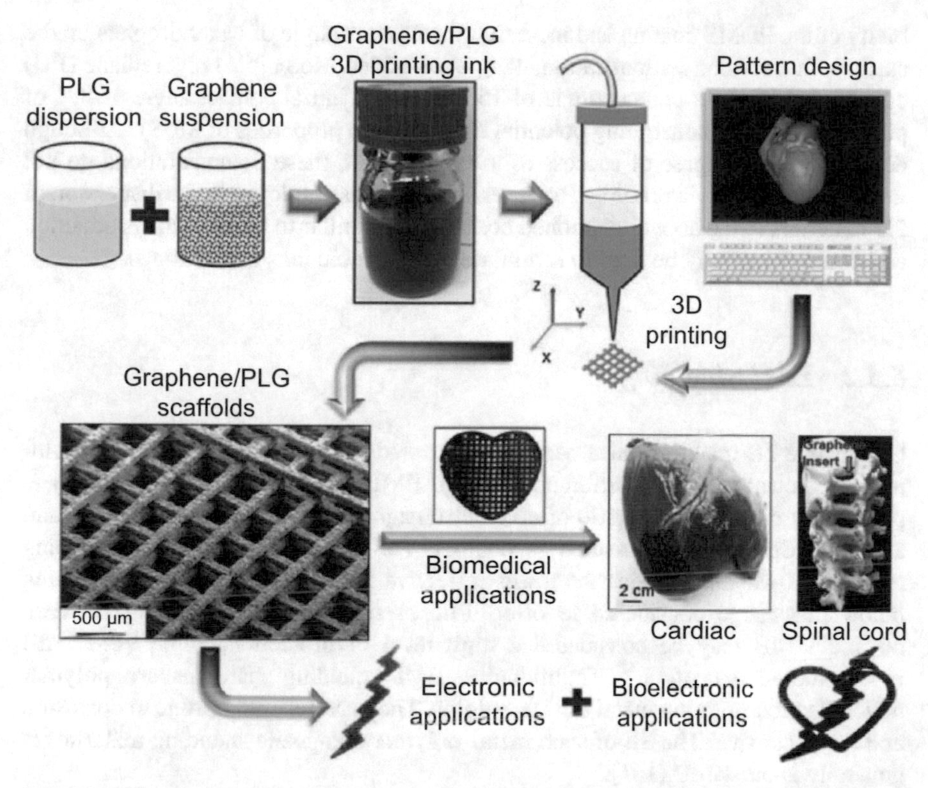

Fig. 6.14 3D-printed graphene scaffolds for electronic and biomedical applications. Reproduced with permission from Ref. [150]. Copyright (2015), American Chemical Society

Fig. 6.15 (**a**) Spray-coated rGO-TiO_2-DE superhydrophobic coating (**b**) exhibiting high degrees of superhydrophobicity before abrasion and (**c**) after abrasion. Reprinted with permission from Ref. [156]. Copyright (2015) American Chemical Society

6.5.1 Barrier

For instance, Nine et al. demonstrated a superhydrophobic graphene-based composite coating with self-cleaning and corrosion barrier properties made of mixture of rGO, diatomaceous earth (DE) and polydimethylsiloxane (PDMS); [156]; Fig. 6.15a–c. It was found that the addition of rGO and DE enhanced the hydrophobicity of the PDMS coating and increased the contact angle of water droplets on the surface. In a similar demonstration, Tang et al. added MoS_2 to a polyurethane (PU) coating, achieving a contact angle of 155° from an initial contact angle of 87° of pure PU [160], demonstrating potential hydrophobic properties of MoS_2. Although achieving some degree of success as a barrier layer, these demonstrations do not involve deposition via printing. However, the methods employed in the dispersion of 2D materials in the above-mentioned coatings are similar to those in ink production and could potentially be used to formulate hydrophobic inks and coatings.

6.5.2 EMI Shielding

Conductive 2D materials have been demonstrated as potential materials for shielding in electromagnetic interference (EMI). EMI shielding effectiveness (SE) is principally measured as a ratio of an electromagnetic signal's intensity before and after shielding and is measured in decibels (dB) [161]. In general, a shielding range of 10–30 dB provides the lowest effective level of shielding, with anything below that range considered to offer little or no shielding. Shielding between 60 and 90 dB may be considered a high level of protection, while 90–120 dB is considered exceptional. Traditionally, EMI shielding materials are polymer materials incorporating metal foils as shields. They are generally prone to corrosion and are expensive. The SE of such metal-polymer composite shielding materials is generally about 40 dB [161].

Graphene, with its low density and conductive properties has attracted interest as a lightweight, low-cost and corrosion-resistant material for EMI shielding. Song et al. demonstrated this by incorporating graphene sandwiched in a polymer structure that achieved a shielding effectiveness of 27 dB; Fig. 6.16a [162]. Similarly, Liang et al. incorporated graphene directly into a graphene-epoxy composite that achieved up to 21 dB of EMI protection at 8.2 GHz with a 15 wt% loading; Fig 6.16b.

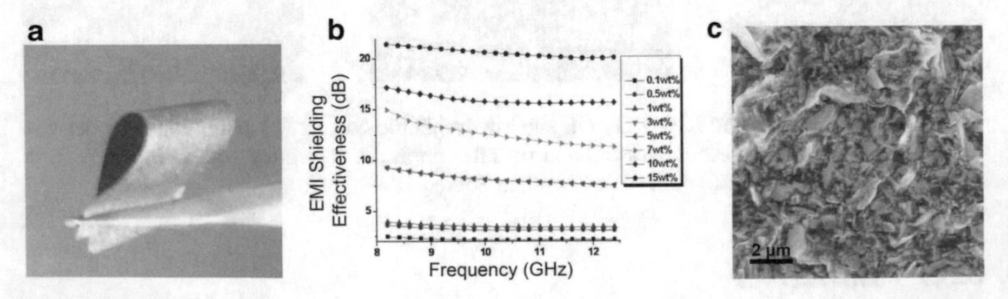

Fig. 6.16 (**a**) Photographs of graphene/polymer composite films for EMI shielding. Reprinted with permission from Ref. [162]. Copyright (2014) Elsevier. (**b**) EMI shielding effectiveness of graphene/epoxy composites with respect to different loadings of graphene and an (**c**) SEM image of the graphene/epoxy composite. Reprinted with permission from Ref. [163]. Copyright (2009) Elsevier

MXenes, with their combination of metallic conductivity, and easy coating capabilities have attracted attention as a possible EMI shielding material. As with graphene, Shahzad et al. achieved a 92 dB SE with a 45-μm-thick Ti_3C_2Tx-sodium alginate composite, which is the highest among synthetic materials of comparable thickness produced yet.

As with barrier applications, despite no fully printed EMI coating applications demonstrated to date, the techniques employed in these demonstrations are similar to ink production and therefore hold potential to be printed.

6.5.3 Membrane

Graphene-based membranes have well-defined nanometer pores and can exhibit low frictional water flow, making them of interest for filtration and separation applications [164]. This is the case for GO with its inter-layer spacing of 13.5 Å. Whilst this is capable of separating larger molecules, the inter-layer spacing is still larger than the diameters of many common salts, making intrinsic, unmodified GO membranes ineffective for desalination. However, Abraham et al. showed that GO membranes may be 'tunable' by the fabrication of selectively permeable membranes. Vacuum-filtered GO strips left in different relative humidities over 2 weeks were able to achieve varied interlayer-spacing from 6.4 to 9.8 Å, which is capable of filtering out most salts [165]. The GO laminates were physically cut out and sandwiched in between epoxy strips which were then used as a 'physically-confined' GO (PCGO) filtration membrane; Fig. 6.17a. A schematic and a photograph is given in Fig. 6.17b, c. The resulting filter exhibited a 97% rejection for NaCl, proving potential for water desalination. In a similar vein, Chen et al. reported the ability to control the pore size of GO membranes all the way down to 1 Å using K^+, Na^+, Ca^+, Li^+ or Mg^+ cations [166]. This refinement of GO membrane tuning could potentially allow for the sieving of gases, solvent dehydration and molecular sieving. For printed 2D material applications, formulated GO inks deposited via printing might be tuned in a similar manner in order to achieve sieving capabilities in future.

6.6 Future Development Possibilities for 2D Material Printed Applications

This book summarises the current state of development of 2D material inks and printing. Since the first demonstration of printed graphene electronics in 2012, this field has witnessed four major advances: the functional material set has evolved from graphene alone to a wider range of available 2D materials; 2D material inks have advanced from simple solution-processed dispersions into formulated inks

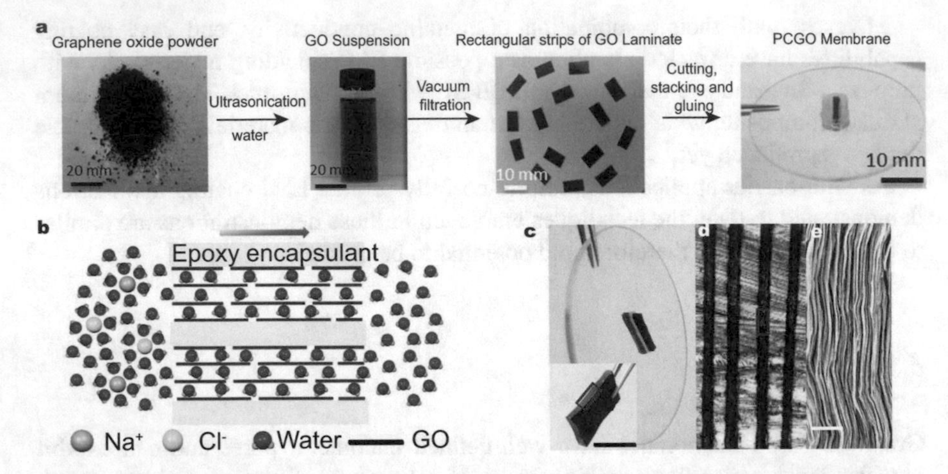

Fig. 6.17 Physically confined GO membranes with tunable interlayer spacing. (**a**) Schematic illustrating the direction of ion/water permeation along graphene planes. (**b**) Schematic of the desalination GO filter. (**c**) Photograph of a PCGO membrane glued into a rectangular slot within a plastic disk of 5 cm in diameter. Inset: photo of the GO stack before it was placed inside the slot. Scale bar, 5 mm. (**d**) Optical micrograph of the cross-sectional area marked by a red rectangle in (**c**), which shows 100 μm-thick GO laminates (black) embedded in epoxy. Epoxy is seen in light yellow with dark streaks because of surface scratches. (**e**), Scanning electron microscopy image from the marked region in (**c**). Scale bar, 1 μm. Reproduced with permission from Ref. [165]. Copyright (2017) Springer Nature

adapted for specific print processes; the exploitable printing technologies have expanded from inkjet printing to a broader printing platform (including screen, gravure and flexographic printing, as well as 3D printing) offering faster production speeds and more complex device structures; the investigated applications have extended to several major technology areas, including optoelectronics, photonics, sensors and energy storage. Taken together, these represent significant advances for the scope of 2D material inks, and associated printing, as a promising platform for a next generation of disruptive technologies [1].

Despite these advances, there are huge opportunities to improve the reliability and performance of these applications [1]. It is necessary to scale up from laboratory demonstrations to device manufacturing ready for real-world applications. Realising this requires significant improvements in materials processing and engineering, ink formulations and device designs. Though notable progress has been made over the last decade, there remains much to be done on solution processing to improve the exfoliation efficiency, to raise the production yield, reduce the production cost and to better sort the exfoliated flakes [53]. Improved ink formulation to optimise fluidic properties, drying dynamics and ink-substrate interactions, and fundamental research into binders, solvent and additive compatibility are also of critical importance. To harness the full potential of 2D materials, optimal device designs are essential for high-level device performance, and expansion of the currently demonstrated applications. It will also be interesting to combine 2D

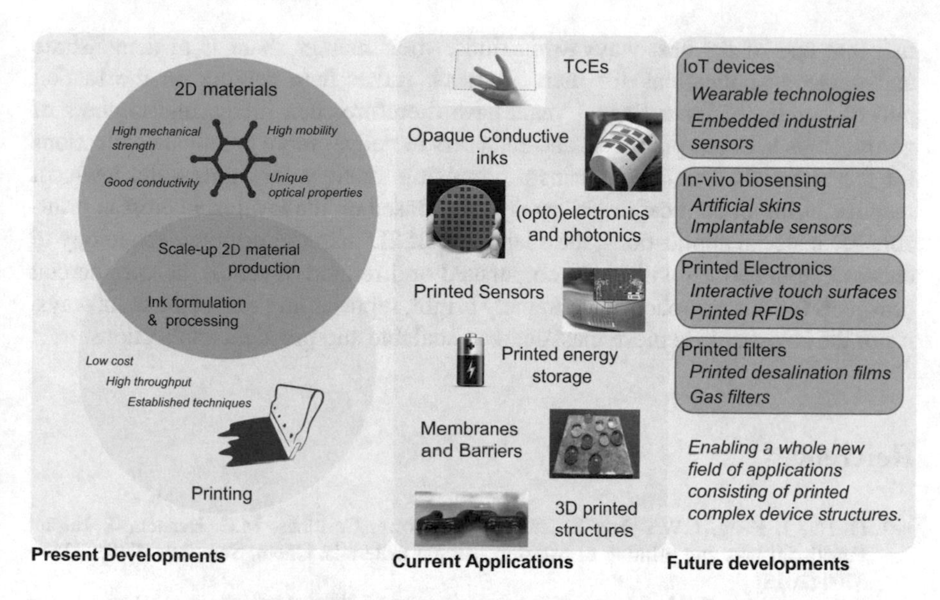

Fig. 6.18 Potential future development directions for 2D material inks

material ink formulations with other functional materials, such as biomolecules, 0D and 2D perovskites and 1D nanowires/nanotubes such as CNTs, for more flexible material selection and device design.

In the short and near term, the advantages offered by printing technologies, such as controlled additive patterning, thin form-factors and mechanical flexibility, may enable improved device performance and even new technologies and applications [1]. Seamless integration of printed 2D materials with the existing bulk semiconductor technologies, optoelectronics and sensing technologies and even integrated circuits may also be possible. Early, rudimentary examples have already been demonstrated. Such technologies may also involve incorporation of other functional materials such that multiple concurrent functions are realised, leading to the development of 'multifunctional' inks [1]; Fig. 6.18.

In the longer term, this fusion of technologies, materials platform and functionality would offer manufacturing flexibility, high-level integration and miniaturisation for 2D material based printable electronics, sensing and interactive devices. This will require precisely controlled, additive patterning of multiple 2D materials and other functional materials over large areas and/or for thousands of devices with tight manufacturing tolerance. This could potentially lead to the development of all-printed, complex device structures (e.g. heterostructures) for unprecedented new applications in light detection and imaging, energy storage, interactive touch surfaces, wearable sensors, RFID tags and even artificial skins [1].

Current progress is now approaching a point where the production cost is reasonably low and consistent. Commercialisation efforts have recently begun a shift towards scaled-up production. In particular, producers are waking up to the fact that one of the best ways to maximise their market share is to demonstrate and develop applications for their products, rather than relying on the market pull of graphene alone. Recent years have therefore seen increasing numbers of partnerships between producers and end users to deliver more practical applications for the materials. Printing remains a promising strategy for large-scale, low-cost manufacturing of application driven devices. Based on the level of interest in printable 2D material applications, the adaptation of 2D material printing technology to end-user applications will likely be pursued and refined. Overall, the commercial prospects for this approach are extremely bright, representing one of the main ways, if not the key way to achieve mass-market, scalable and practical applications.

References

1. G. Hu, J. Kang, L.W.T. Ng, X. Zhu, R.C.T. Howe, C. Jones, M.C. Hersam, T. Hasan, Functional inks and printing of two-dimensional materials. Chem. Soc. Rev. **47**(9), 3265–3300 (2018)
2. E.B. Secor, P.L. Prabhumirashi, K. Puntambekar, M.L. Geier, M.C. Hersam, Inkjet printing of high conductivity, flexible graphene patterns. J. Phys. Chem. Lett. **4**(8), 1347–1351 (2013)
3. E.B. Secor, S. Lim, H. Zhang, C.D. Frisbie, L.F. Francis, M.C. Hersam, Gravure printing of graphene for large-area flexible electronics. Adv. Mater. **26**(26), 4533–4538 (2014)
4. E.B. Secor, B.Y. Ahn, T.Z. Gao, J.A. Lewis, M.C. Hersam, Rapid and versatile photonic annealing of graphene inks for flexible printed electronics. Adv. Mater. **27**(42), 6683–6688 (2015)
5. K. Arapov, G. Bex, R. Hendriks, E. Rubingh, R. Abbel, G. de With, H. Friedrich, Conductivity enhancement of binder-based graphene inks by photonic annealing and subsequent compression rolling. Adv. Eng. Mater. **18**(7), 1234–1239 (2016)
6. X. Huang, T. Leng, X. Zhang, J.C. Chen, K.H. Chang, A.K. Geim, K.S. Novoselov, Z. Hu, Binder-free highly conductive graphene laminate for low cost printed radio frequency applications. Appl. Phys. Lett. **106**(20), 203105 (2015)
7. P.G. Karagiannidis, S.A. Hodge, L. Lombardi, F. Tomarchio, N. Decorde, S. Milana, I. Goykhman, Y. Su, S.V. Mesite, D.N. Johnstone, R.K. Leary, P.A. Midgley, N.M. Pugno, F. Torrisi, A.C. Ferrari, Microfluidization of graphite and formulation of graphene-based conductive inks. ACS Nano **11**, 2742–2755 (2017)
8. W.J. Hyun, E.B. Secor, M.C. Hersam, C.D. Frisbie, L.F. Francis, High-resolution patterning of graphene by screen printing with a silicon stencil for highly flexible printed electronics. Adv. Mater. **27**(1), 109–115 (2015)
9. R.C.T. Howe, G. Hu, Z. Yang, T. Hasan, Functional inks of graphene, metal dichalcogenides and black phosphorus for photonics and (opto)electronics. Proc. SPIE **9553**, 95530R (2015)
10. D. Stauffer, A. Aharony, *Introduction to Percolation Theory* (Taylor and Francis, London, 1992)
11. J. Li, J.K. Kim, Percolation threshold of conducting polymer composites containing 3D randomly distributed graphite nanoplatelets. Compos. Sci. Technol. **67**, 2114–2120 (2007)
12. P.J. Brigandi, J.M. Cogen, R.A. Pearson, Electrically conductive multiphase polymer blend carbon-based composites. Polym. Eng. Sci. **54**(1), 1–16 (2014)
13. S. Santra, G. Hu, R.C.T. Howe, A. De Luca, S.Z. Ali, F. Udrea, J.W. Gardner, S.K. Ray, P.K. Guha, T. Hasan, CMOS integration of inkjet-printed graphene for humidity sensing. Sci. Rep. **5**, 17374 (2015)

14. A. Capasso, A.E. Del Rio Castillo, H. Sun, A. Ansaldo, V. Pellegrini, F. Bonaccorso, Ink-jet printing of graphene for flexible electronics: an environmentally-friendly approach. Solid State Commun. **224**, 53–63 (2015)
15. G.R. Ruschau, S. Yoshikawa, R.E. Newnham, Resistivities of conductive composites. J. Appl. Phys. **72**(3), 953–959 (1992)
16. A. Dani, A.A. Ogale, Electrical percolation behavior of short-fiber composites: experimental characterization and modeling. Compos. Sci. Technol. **56**(8), 911–920 (1996)
17. A.R. Madaria, A. Kumar, F.N. Ishikawa, C. Zhou, Uniform, highly conductive, and patterned transparent films of a percolating silver nanowire network on rigid and flexible substrates using a dry transfer technique. Nano Res. **3**(8), 564–573 (2010)
18. G.E. Pike, C.H. Seager, Percolation and conductivity: a computer study. I. Phys. Rev. B **10**(4), 1421–1434 (1974)
19. F. Bonaccorso, Z. Sun, T. Hasan, A.C. Ferrari, Graphene photonics and optoelectronics. Nat. Photon. **4**(9), 611–622 (2010)
20. F. Bonaccorso, A. Lombardo, T. Hasan, Z. Sun, L. Colombo, A.C. Ferrari, Production and processing of graphene and 2D crystals. Mater. Today **15**(12), 564–589 (2012)
21. S. De, J.N. Coleman, Are there fundamental limitations on the sheet resistance and transmittance of thin graphene films? ACS Nano **4**(5), 2713–2720 (2010)
22. S. Bae, H. Kim, Y. Lee, X. Xu, J.-S. Park, Y. Zheng, J. Balakrishnan, T. Lei, H.R. Kim, Y.I. Song, Y.-J. Kim, K.S. Kim, B. Ozyilmaz, J.-H. Ahn, B.H. Hong, S. Iijima, Roll-to-roll production of 30-inch graphene films for transparent electrodes. Nat. Nanotechnol. **5**(8), 574–578 (2010)
23. S. De, P.J. King, M. Lotya, A. O'Neill, E.M. Doherty, Y. Hernandez, G.S. Duesberg, J.N Coleman, Flexible, transparent, conducting films of randomly stacked graphene from surfactant stabilized, oxide-free graphene dispersions. Small **6**(3), 458–464 (2010)
24. K.S. Kim, Y. Zhao, H. Jang, S.Y. Lee, J.M. Kim, K.S. Kim, J.-H. Ahn, P. Kim, J.-Y. Choi, B.H. Hong, Large-scale pattern growth of graphene films for stretchable transparent electrodes. Nature **457**(7230), 706–710 (2009)
25. C. Mattevi, G. Eda, S. Agnoli, S. Miller, K.A. Mkhoyan, O. Celik, D. Mastrogiovanni, G. Granozzi, E. Garfunkel, M. Chhowalla, Evolution of electrical, chemical, and structural properties of transparent and conducting chemically derived graphene thin films. Adv. Funct. Mater. **19**(16), 2577–2583 (2009)
26. M.S. Kang, K.T. Kim, J.U. Lee, W.H. Jo, Direct exfoliation of graphite using a non-ionic polymer surfactant for fabrication of transparent and conductive graphene films. J. Mater. Chem. C **1**(9), 1870 (2013)
27. G. Eda, G. Fanchini, M. Chhowalla, Large-area ultrathin films of reduced graphene oxide as a transparent and flexible electronic material. Nat. Nanotechnol. **3**(5), 270–274 (2008)
28. X. Ho, J. Wei, Films of carbon nanomaterials for transparent conductors. Materials **6**(6), 2155–2181 (2013)
29. D.S. Hecht, L. Hu, G. Irvin, Emerging transparent electrodes based on thin films of carbon nanotubes, graphene, and metallic nanostructures. Adv. Mater. **23**(13), 1482–1513 (2011)
30. X. Huang, Z. Zeng, Z. Fan, J. Liu, H. Zhang, Graphene-based electrodes. Adv. Mater. **24**(45), 5979–6004 (2012)
31. C.G. Granqvist, Transparent conductors as solar energy materials: a panoramic review. Sol. Energy Mater. Sol. Cells **91**(17), 1529–1598 (2007)
32. I. Hamberg, C.G. Granqvist, Evaporated Sn-doped In_2O_3 films: basic optical properties and applications to energy-efficient smart windows. J. Appl. Phys. **60**, 123–159 (1986)
33. T. Minami, Transparent conducting oxide semiconductors for transparent electrodes. Semicond. Sci. Technol. **20**(4), S35–S44 (2005)
34. L. Holland, G. Siddall, The properties of some reactively sputtered metal oxide films. Vacuum **3**(4), 375–391 (1953)
35. D. Ginley, H. Hosono, D.C. Paine (eds.), *Handbook of Transparent Conductors* (Springer, Berlin, 2011)

36. X. Wang, L. Zhi, K. Müllen, Transparent, conductive graphene electrodes for dye-sensitized solar cells. Nano Lett. **8**(1), 323–327 (2008)
37. P. Blake, P.D. Brimicombe, R.R. Nair, T.J. Booth, D. Jiang, F. Schedin, L.A. Ponomarenko, S.V. Morozov, H.F. Gleeson, E.W. Hill, A.K. Geim, K.S. Novoselov, Graphene-based liquid crystal device. Nano Lett. **8**(6), 1704–1708 (2008)
38. Y. Hernandez, V. Nicolosi, M. Lotya, F.M. Blighe, Z. Sun, S. De, I.T. McGovern, B. Holland, M. Byrne, Y.K. Gun'Ko, J.J. Boland, P. Niraj, G. Duesberg, S. Krishnamurthy, R. Goodhue, J. Hutchison, V. Scardaci, A.C. Ferrari, J.N. Coleman, High-yield production of graphene by liquid-phase exfoliation of graphite. Nat. Nanotechnol. **3**(9), 563–568 (2008)
39. L.J. Cote, F. Kim, J. Huang, Langmuir–Blodgett assembly of graphite oxide single layers. J. Am. Chem. Soc. **131**(3), 1043–1049 (2009)
40. F. Torrisi, T. Hasan, W. Wu, Z. Sun, A. Lombardo, T.S. Kulmala, G.-W. Hsieh, S. Jung, F. Bonaccorso, P.J. Paul, D. Chu, A.C. Ferrari, Inkjet-printed graphene electronics. ACS Nano **6**(4), 2992–3006 (2012)
41. A.A. Green, M.C. Hersam, Solution phase production of graphene with controlled thickness via density differentiation. Nano Lett. **9**(12), 4031–4036 (2009)
42. N.D. Matyba, P. Yamaguchi, H. Eda, G. Chhowalla, M. Edman, L.Y. Robinson, Graphene and mobile ions: the key to all plastic, solution processed light emitting devices. ACS Nano **4**(2), 637–642 (2010)
43. S. Forget, S. Chenais, A. Siove, Organic light-emitting diodes. Photochem. Photophys. Polym. Mater. **4**(1), 309–350 (2010)
44. K. Hantanasirisakul, M.-ŘQ. Zhao, P. Urbankowski, J. Halim, B. Anasori, S. Kota, C.E. Ren, M.W. Barsoum, Y. Gogotsi, Fabrication of $Ti_3C_2T_x$ MXene transparent thin films with tunable optoelectronic properties. Adv. Electron. Mater. **2**, 1600050 (2016)
45. R.A. Matula, Electrical resistivity of copper, gold, palladium, and silver. J. Phys. Chem. Ref. Data **8**(4), 1147–1298 (1979)
46. E.B. Secor, T.Z. Gao, A.E. Islam, R. Rao, S.G. Wallace, J. Zhu, K.W. Putz, B. Maruyama, M.C. Hersam, Enhanced conductivity, adhesion, and environmental stability of printed graphene inks with nitrocellulose. Chem. Mater. **29**(5), 2332–2340 (2017)
47. R.H. Leach, R.J. Pierce, E.P. Hickman, M.J. Mackenzie, H.G. Smith (eds.), *The Printing Ink Manual*, 5th edn. (Springer, Dordrecht, 1993)
48. H. Kipphan (ed.), *Handbook of Print Media: Technologies and Production Methods* (Springer, Berlin, 2001)
49. F.C. Krebs, Fabrication and processing of polymer solar cells: a review of printing and coating techniques. Sol. Energy Mater. Sol. Cells **93**(4), 394–412 (2009)
50. M.A.M. Leenen, V. Arning, H. Thiem, J. Steiger, R. Anselmann, Printable electronics: flexibility for the future. Phys. Status Solidi **206**(4), 588–597 (2009)
51. F. Xia, H. Wang, D. Xiao, M. Dubey, A. Ramasubramaniam, Two-dimensional material nanophotonics. Nat. Photon. **8**(12), 899–907 (2014)
52. A.K. Geim, K.S. Novoselov, The rise of graphene. Nat. Mater. **6**, 183–191 (2007)
53. V. Nicolosi, M. Chhowalla, M.G. Kanatzidis, M.S. Strano, J.N. Coleman. Liquid exfoliation of layered materials. Science **340**(6139), 1226419–1226437 (2013)
54. A.C. Ferrari, F. Bonaccorso, V. Fal'ko, K.S. Novoselov, S. Roche, P. Bøggild, S. Borini, F.H.L. Koppens, V. Palermo, N. Pugno, J.A. Garrido, R. Sordan, A. Bianco, L. Ballerini, M. Prato, E. Lidorikis, J. Kivioja, C. Marinelli, T. Ryhänen, A. Morpurgo, J.N. Coleman, V. Nicolosi, L. Colombo, A. Fert, M. Garcia-Hernandez, A. Bachtold, G.F. Schneider, F. Guinea, C. Dekker, M. Barbone, Z. Sun, C. Galiotis, A.N. Grigorenko, G. Konstantatos, A. Kis, M. Katsnelson, L. Vandersypen, A. Loiseau, V. Morandi, D. Neumaier, E. Treossi, V. Pellegrini, M. Polini, A. Tredicucci, G.M. Williams, B. Hee Hong, J.-H. Ahn, J. Min Kim, H. Zirath, B.J. van Wees, H. van der Zant, L. Occhipinti, A. Di Matteo, I.A. Kinloch, T. Seyller, E. Quesnel, X. Feng, K. Teo, N. Rupesinghe, P. Hakonen, S.R.T. Neil, Q. Tannock, T. Löfwander, J. Kinaret, Science and technology roadmap for graphene, related two-dimensional crystals, and hybrid systems. Nanoscale **7**(11), 4598–4810 (2015)

55. A.G. Kelly, T. Hallam, C. Backes, A. Harvey, A.S. Esmaeily, I. Godwin, J. Coelho, V. Nicolosi, J. Lauth, A. Kulkarni, S. Kinge, L.D.A. Siebbeles, G.S. Duesberg, J.N. Coleman, All-printed thin-film transistors from networks of liquid-exfoliated nanosheets. Science **356**(6333), 69–73 (2017)
56. D. McManus, S. Vranic, F. Withers, V. Sanchez-Romaguera, M. Macucci, H. Yang, R. Sorrentino, K. Parvez, S.-K. Son, G. Iannaccone, K. Kostarelos, G. Fiori, C. Casiraghi, Water-based and biocompatible 2D crystal inks for all-inkjet-printed heterostructures. Nat. Nanotechnol. **12**(4), 343–350 (2017)
57. G. Hu, T. Albrow-Owen, X. Jin, A. Ali, Y. Hu, R.C.T. Howe, K. Shehzad, Z. Yang, X. Zhu, R.I. Woodward, T.-C. Wu, H. Jussila, J.-B. Wu, P. Peng, P.-H. Tan, Z. Sun, E.J.R. Kelleher, M. Zhang, Y. Xu, T. Hasan, Black phosphorus ink formulation for inkjet printing of optoelectronics and photonics. Nat. Commun. **8**(1), 278 (2017)
58. T. Carey, S. Cacovich, G. Divitini, J. Ren, A. Mansouri, J.M. Kim, C. Wang, C. Ducati, R. Sordan, F. Torrisi, Fully inkjet-printed two-dimensional material field-effect heterojunctions for wearable and textile electronics. Nat. Commun. **8**(1), 1202 (2017)
59. H. Sirringhaus, 25th anniversary article: organic field-effect transistors: the path beyond amorphous silicon. Adv. Mater. **26**(9), 1319–1335 (2014)
60. K. Fukuda, Y. Takeda, Y. Yoshimura, R. Shiwaku, L.T. Tran, T. Sekine, M. Mizukami, D. Kumaki, S. Tokito, Fully-printed high-performance organic thin-film transistors and circuitry on one-micron-thick polymer films. Nat. Commun. **5**, 4147 (2014)
61. S. Oktyabrsky, P. Ye (eds.), *Fundamentals of III-V Semiconductor MOSFETs* (Springer US, Boston, 2010)
62. B. Radisavljevic, A. Radenovic, J. Brivio, V. Giacometti, A. Kis, Single-layer MoS_2 transistors. Nat. Nanotechnol. **6**(3), 147–150 (2011)
63. M. Xu, T. Liang, M. Shi, H. Chen, Graphene-like two-dimensional materials. Chem. Rev. **113**(5), 3766–3798 (2013)
64. D.J. Finn, M. Lotya, G. Cunningham, R.J. Smith, D. McCloskey, J.F. Donegan, J.N. Coleman, Inkjet deposition of liquid-exfoliated graphene and MoS_2 nanosheets for printed device applications. J. Mater. Chem. C **2**(5), 925–932 (2014)
65. F. Withers, H. Yang, L. Britnell, A.P. Rooney, E. Lewis, A. Felten, C.R. Woods, V. Sanchez Romaguera, T. Georgiou, A. Eckmann, Y.J. Kim, S.G. Yeates, S.J. Haigh, A.K. Geim, K.S. Novoselov, C. Casiraghi, Heterostructures produced from nanosheet-based inks. Nano Lett. **14**(7), 3987–3992 (2014)
66. Z. Sun, D. Popa, T. Hasan, F. Torrisi, F. Wang, E.J.R. Kelleher, J.C. Travers, V. Nicolosi, A.C. Ferrari, A stable, wideband tunable, near transform-limited, graphene-mode-locked, ultrafast laser. Nano Res. **3**(9), 653–660 (2010)
67. M. Trushin, E.J.R. Kelleher, T. Hasan, Theory of edge-state optical absorption in two-dimensional transition metal dichalcogenide flakes. Phys. Rev. B **94**, 155301 (2016)
68. U. Keller, Recent developments in compact ultrafast lasers. Nature **424**(6950), 831–838 (2003)
69. R.I. Woodward, E.J. Kelleher, T.H. Runcorn, S.V. Popov, F. Torrisi, R.C.T. Howe, T. Hasan, Q-switched fiber laser with MoS_2 saturable absorber, in *CLEO: 2014*, paper SM3H.6. OSA (2014)
70. M. Zhang, L. Huang, J. Chen, C. Li, G. Shi, Ultratough, ultrastrong, and highly conductive graphene films with arbitrary sizes. Adv. Mater. **26**(45), 7588–7592 (2014)
71. R.I. Woodward, R.C.T. Howe, T.H. Runcorn, G. Hu, F. Torrisi, E.J.R. Kelleher, T. Hasan, Wideband saturable absorption in few-layer molybdenum diselenide ($MoSe_2$) for Q-switching Yb-, Er- and Tm-doped fiber lasers. Opt. Express **23**(15), 20051 (2015)
72. M. Zhang, G. Hu, G. Hu, R.C.T. Howe, L. Chen, Z. Zheng, T. Hasan, Yb- and Er-doped fiber laser Q-switched with an optically uniform, broadband WS_2 saturable absorber. Sci. Rep. **5**, 17482 (2015)
73. F. Yavari, N. Koratkar, Graphene-based chemical sensors. J. Phys. Chem. Lett. **3**(13), 1746–1753 (2012)

74. F. Schedin, A.K. Geim, S.V. Morozov, E.W. Hill, P. Blake, M.I. Katsnelson, K.S. Novoselov, Detection of individual gas molecules adsorbed on graphene. Nat. Mater. **6**(9), 652–655 (2007)
75. M.S. Mannoor, H. Tao, J.D. Clayton, A. Sengupta, D.L. Kaplan, R.R. Naik, N. Verma, F.G. Omenetto, M.C. McAlpine, Graphene-based wireless bacteria detection on tooth enamel. Nat. Commun. **3**, 763 (2012)
76. K. Shehzad, T. Shi, A. Qadir, X. Wan, H. Guo, A. Ali, W. Xuan, H. Xu, Z. Gu, X. Peng, J. Xie, L. Sun, Q. He, Z. Xu, C. Gao, Y.-S. Rim, Y. Dan, T. Hasan, P. Tan, E. Li, W. Yin, Z. Cheng, B. Yu, Y. Xu, J. Luo, X. Duan, Designing an efficient multimode environmental sensor based on graphene-silicon heterojunction. Adv. Mater. Technol. **2**(4), 1600262 (2017)
77. Y. Yao, L. Tolentino, Z. Yang, X. Song, W. Zhang, Y. Chen, C.-P. Wong, High-concentration aqueous dispersions of MoS_2. Adv. Funct. Mater. **23**(28), 3577–3583 (2013)
78. S.-Y. Cho, Y. Lee, H.-J. Koh, H. Jung, J.-S. Kim, H.-W. Yoo, J. Kim, H.-T. Jung, Superior chemical sensing performance of black phosphorus: comparison with MoS_2 and graphene. Adv. Mater. **28**(32), 7020–7028 (2016)
79. P. He, J.R. Brent, H. Ding, J. Yang, D.J. Lewis, P. O'Brien, B. Derby, Fully printed high performance humidity sensors based on two-dimensional materials. Nanoscale **10**, 5599 (2018)
80. X.-F. Yu, Y.-C. Li, J.-B. Cheng, Z.-B. Liu, Q.-Z. Li, W.-Z. Li, X. Yang, B. Xiao, Monolayer Ti_2CO_2: a promising candidate for NH_3 sensor or capturer with high sensitivity and selectivity. ACS Appl. Mater. Interfaces **7**(24), 13707–13713 (2015)
81. S.J. Kim, H.-J. Koh, C.E. Ren, O. Kwon, K. Maleski, S.-Y. Cho, B. Anasori, C.-K. Kim, Y.-K. Choi, J. Kim, Y. Gogotsi, H.-T. Jung, Metallic $Ti_3C_2T_x$ MXene gas sensors with ultrahigh signal-to-noise ratio. ACS Nano **12**, 986–993 (2018)
82. V. Dua, S.P. Surwade, S. Ammu, S.R. Agnihotra, S. Jain, K.E. Roberts, S. Park, R.S. Ruoff, S.K. Manohar, All-organic vapor sensor using inkjet-printed reduced graphene oxide. Angew. Chem. Int. Ed. **49**(12), 2154–2157 (2010)
83. M.G. Chung, D.H. Kim, H.M. Lee, T. Kim, J.H. Choi, D.K. Seo, J.-B. Yoo, S.-H. Hong, T.J. Kang, Y.H. Kim, Highly sensitive NO_2 gas sensor based on ozone treated graphene. Sens. Actuators B Chem. **166–167**, 172–176 (2012)
84. L. Huang, Z. Wang, J. Zhang, J. Pu, Y. Lin, S. Xu, L. Shen, Q. Chen, W. Shi, Fully printed, rapid-response sensors based on chemically modified graphene for detecting NO_2 at room temperature. ACS Appl. Mater. Interfaces **6**(10), 7426–7433 (2014)
85. B.W. Kennedy, Thin film temperature sensor. Rev. Sci. Instrum. **40**(9), 1169–1172 (1969)
86. W.-H. Yeo, Y.-S. Kim, J. Lee, A. Ameen, L. Shi, M. Li, S. Wang, R. Ma, S.H. Jin, Z. Kang, Y. Huang, J.A. Rogers, Multifunctional epidermal electronics printed directly onto the skin. Adv. Mater. **25**(20), 2773–2778 (2013)
87. D. Son, J. Lee, S. Qiao, R. Ghaffari, J. Kim, J.E. Lee, C. Song, S.J. Kim, D.J. Lee, S.W. Jun, S. Yang, M. Park, J. Shin, K. Do, M. Lee, K. Kang, C.S. Hwang, N. Lu, T. Hyeon, D.-H. Kim, Multifunctional wearable devices for diagnosis and therapy of movement disorders. Nat. Nanotechnol. **9**(5), 397–404 (2014)
88. T. Juntunen, H. Jussila, M. Ruoho, S. Liu, G. Hu, T. Albrow-Owen, L.W.T. Ng, R.C.T. Howe, T. Hasan, Z. Sun, I. Tittonen, Inkjet printed large-area flexible few-layer graphene thermoelectrics. *Adv. Funct. Mat.* **28**(22), 1800480 (2018)
89. C. Bali, A. Brandlmaier, A. Ganster, O. Raab, J. Zapf, A. Hübler, Fully inkjet-printed flexible temperature sensors based on carbon and PEDOT:PSS Mater. Today Proc. **3**(3), 739–745 (2016)
90. J.Y. Hong, W. Kim, D. Choi, J. Kong, H.S. Park, Omnidirectionally stretchable and transparent graphene electrodes. ACS Nano **10**, 9446–9455 (2016)
91. K. Agarwal, V. Kaushik, D. Varandani, A. Dhar, B.R. Mehta, Nanoscale thermoelectric properties of Bi_2Te_3-graphene nanocomposites: conducting atomic force, scanning thermal and kelvin probe microscopy studies. J. Alloys Compd. **681**, 394–401 (2016)
92. T. Vuorinen, J. Niittynen, T. Kankkunen, T.M. Kraft, M. Mäntysalo, Inkjet-printed graphene/PEDOT:PSS temperature sensors on a skin-conformable polyurethane substrate. Sci. Rep. **6**(1), 35289 (2016)

93. D. Zang, M. Yan, S. Ge, L. Ge, J. Yu, A disposable simultaneous electrochemical sensor array based on a molecularly imprinted film at a NH_2-graphene modified screen-printed electrode for determination of psychotropic drugs. Analyst **138**, 2704–2711 (2013)
94. M. Amjadi, K.U. Kyung, I. Park, M. Sitti, Stretchable, skin-mountable, and wearable strain sensors and their potential applications: a review. Adv. Funct. Mater. **26**(11), 1678–1698 (2016)
95. E.E. Simmons Jr., Patent, US2393714A, Strain Gauge, 1941-07-23
96. B. Stephen, E. Graham, K. Michael, W. Neil, *MEMS Mechanical Sensors* (Artech House, London, 2004)
97. S.-H. Bae, Y. Lee, B.K. Sharma, H.-J. Lee, J.-H. Kim, J.-H. Ahn, Graphene-based transparent strain sensor. Carbon **51**, 236–242 (2013)
98. Y. Wang, L. Wang, T. Yang, X. Li, X. Zang, M. Zhu, K. Wang, D. Wu, H. Zhu, Wearable and highly sensitive graphene strain sensors for human motion monitoring. Adv. Funct. Mater. **24**(29), 4666–4670 (2014)
99. M. Hempel, D. Nezich, J. Kong, M. Hofmann, A novel class of strain gauges based on layered percolative films of 2D materials. Nano Lett. **12**(11), 5714–5718 (2012)
100. V. Eswaraiah, K. Balasubramaniam, S. Ramaprabhu, Functionalized graphene reinforced thermoplastic nanocomposites as strain sensors in structural health monitoring. J. Mater. Chem. **21**(34), 12626 (2011)
101. C. Yan, J. Wang, W. Kang, M. Cui, X. Wang, C.Y. Foo, K.J. Chee, P.S. Lee, Highly stretchable piezoresistive graphene-nanocellulose nanopaper for strain sensors. Adv. Mater. **26**(13), 2022–2027 (2014)
102. C. Casiraghi, M. Macucci, K. Parvez, R. Worsley, Y. Shin, F. Bronte, C. Borri, M. Paggi, G. Fiori, Inkjet printed 2D-crystal based strain gauges on paper. Carbon **129**, 462–467 (2018)
103. S. Zhao, J. Li, D. Cao, G. Zhang, J. Li, K. Li, Y. Yang, W. Wang, Y. Jin, R. Sun, C.P. Wong, Recent advancements in flexible and stretchable electrodes for electromechanical sensors: strategies, materials, and features. ACS Appl. Mater. Interfaces **9**(14), 12147–12164 (2017)
104. K. Shavanova, Y. Bakakina, I. Burkova, I. Shtepliuk, R. Viter, A. Ubelis, V. Beni, N. Starodub, R. Yakimova, V. Khranovskyy, Application of 2D non-graphene materials and 2D oxide nanostructures for biosensing technology. Sensors **16**(2), 223 (2016)
105. J. Li, F. Rossignol, J. Macdonald, Inkjet printing for biosensor fabrication: combining chemistry and technology for advanced manufacturing. Lab Chip **15**(12), 2538–2558 (2015)
106. K. Haupt Mosbach, Molecularly imprinted polymers and their use in biomimetic sensors. Chem. Rev. **100**, 2495–2504 (2000)
107. F.Y. Kong, S.X. Gu, W.W. Li, T.T. Chen, Q. Xu, W. Wang, A paper disk equipped with graphene/polyaniline/Au nanoparticles/glucose oxidase biocomposite modified screen-printed electrode: toward whole blood glucose determination. Biosens. Bioelectron. **56**, 77–82 (2014)
108. Z. Zhang, P. Pan, X. Liu, Z. Yang, J. Wei, Z. Wei, 3D-copper oxide and copper oxide/few-layer graphene with screen printed nanosheet assembly for ultrasensitive non-enzymatic glucose sensing. Mater. Chem. Phys. **187**, 28–38 (2017)
109. A. Ambrosi, C.K. Chua, A. Bonanni, M. Pumera, Electrochemistry of graphene and related materials. Chem. Rev. **114**(14), 7150–7188 (2014)
110. R. Marom, S.F. Amalraj, N. Leifer, D. Jacob, D. Aurbach, A review of advanced and practical lithium battery materials. J. Mater. Chem. **21**(27), 9938 (2011)
111. M. Yilmaz, P.T. Krein, Review of battery charger topologies, charging power levels, and infrastructure for plug-in electric and hybrid vehicles. IEEE Trans. Power Electron. **28**(5), 2151–2169 (2013)
112. L. Lu, X. Han, J. Li, J. Hua, M. Ouyang, A review on the key issues for lithium-ion battery management in electric vehicles. J. Power Sources **226**, 272–288 (2013)
113. P. Simon, Y. Gogotsi, Materials for electrochemical capacitors. Nat. Mater. **7**(11), 845–854 (2008)
114. Y. Shao, M.F. El-Kady, L.J. Wang, Q. Zhang, Y. Li, H. Wang, M.F. Mousavi, R.B. Kaner, Graphene-based materials for flexible supercapacitors, Chem. Soc. Rev. **44**(11), 3639–3665 (2015)

115. A. Manthiram, A. Vadivel Murugan, A. Sarkar, T. Muraliganth, Nanostructured electrode materials for electrochemical energy storage and conversion. Energy Environ. Sci. **1**(6), 621 (2008)
116. A.S. Aricò, P. Bruce, B. Scrosati, J.-M. Tarascon, W. van Schalkwijk, Nanostructured materials for advanced energy conversion and storage devices. Nat. Mater. **4**(5), 366–377 (2005)
117. C. Daniel, J.O. Besenhard (eds.), *Handbook of Battery Materials* (Wiley-VCH Verlag GmbH & Co. KGaA, Weinheim, 2011)
118. P.G. Bruce, S.A. Freunberger, L.J. Hardwick, J.-M. Tarascon, Li-O_2 and Li-S batteries with high energy storage. Nat. Mater. **11**, 172 (2012)
119. J. Hassoun, F. Bonaccorso, M. Agostini, M. Angelucci, M.G. Betti, R. Cingolani, M. Gemmi, C. Mariani, S. Panero, V. Pellegrini, B. Scrosati, An advanced lithium-ion battery based on a graphene anode and a lithium iron phosphate cathode. Nano Lett. **14**(8), 4901–4906 (2014)
120. B. Anasori, M.R. Lukatskaya, Y. Gogotsi, 2D metal carbides and nitrides (MXenes) for energy storage. Nat. Rev. Mater. **2**(2), 16098 (2017)
121. R. Raccichini, A. Varzi, S. Passerini, B. Scrosati, The role of graphene for electrochemical energy storage. Nat. Mater. **14**(3), 271–279 (2014)
122. X. Yang, C. Cheng, Y. Wang, L. Qiu, D. Li, Liquid-mediated dense integration of graphene materials for compact capacitive energy storage. Science **341**(6145), 534–537 (2013)
123. M.F. El-Kady, R.B. Kaner, Scalable fabrication of high-power graphene micro-supercapacitors for flexible and on-chip energy storage. Nat. Commun. **4**, 1475 (2013)
124. T. Liu, M. Leskes, W. Yu, A.J. Moore, L. Zhou, P.M. Bayley, G. Kim, C.P. Grey, Cycling LiO_2 batteries via LiOH formation and decomposition. Science **350**(6260), 530–533 (2015)
125. J. Xiao, D. Choi, L. Cosimbescu, P. Koech, J. Liu, J.P. Lemmon, Exfoliated MoS_2 nanocomposite as an anode material for lithium ion batteries. Chem. Mater. **22**(16), 4522–4524 (2010)
126. H. Hwang, H. Kim, J. Cho, MoS_2 nanoplates consisting of disordered graphene-like layers for high rate lithium battery anode materials. Nano Lett. **11**(11), 4826–4830 (2011)
127. Y. Dall'Agnese, P.-L. Taberna, Y. Gogotsi, P. Simon, Two-dimensional vanadium carbide (MXene) as positive electrode for sodium-ion capacitors. J. Phys. Chem. Lett. **6**(12), 2305–2309 (2015)
128. G. Zou, Z. Zhang, J. Guo, B. Liu, Q. Zhang, C. Fernandez, Q. Peng, Synthesis of MXene/Ag composites for extraordinary long cycle lifetime lithium storage at high rates. ACS Appl. Mater. Interfaces **8**(34), 22280–22286 (2016)
129. Z. Ling, C.E. Ren, M.-Q. Zhao, J. Yang, J.M. Giammarco, J. Qiu, M.W. Barsoum, Y. Gogotsi, Flexible and conductive MXene films and nanocomposites with high capacitance. Proc. Natl. Acad. Sci. U.S.A. **111**(47), 16676–16681 (2014)
130. N. Kurra, B. Ahmed, Y. Gogotsi, H.N. Alshareef, MXene-on-paper coplanar microsupercapacitors. Adv. Energy Mater. **6**, 1601372 (2016)
131. J. Li, V. Mishukova, M. Östling, All-solid-state micro-supercapacitors based on inkjet printed graphene electrodes. Appl. Phys. Lett. **109**(12), 123901 (2016)
132. G. Wang, X. Shen, J. Yao, J. Park, Graphene nanosheets for enhanced lithium storage in lithium ion batteries. Carbon **47**(8), 2049–2053 (2009)
133. E.J. Yoo, J. Kim, E. Hosono, H.S. Zhou, T. Kudo, I. Honma, Large reversible Li storage of graphene nanosheet families for use in rechargeable lithium ion batteries. Nano Lett. **8**, 2277–2282 (2008)
134. Y. Xie, Y. Dall'Agnese, M. Naguib, Y. Gogotsi, M.W. Barsoum, H.L. Zhuang, P.R.C. Kent, Prediction and characterization of mxene nanosheet anodes for non-lithium-ion batteries. ACS Nano **8**(9), 9606–9615 (2014)
135. J. Luo, X. Tao, J. Zhang, Y. Xia, H. Huang, L. Zhang, Y. Gan, C. Liang, W. Zhang, Sn^{4+}Ion decorated highly conductive Ti_3C_2MXene: promising lithium-ion anodes with enhanced volumetric capacity and cyclic performance. ACS Nano **10**(2), 2491–2499 (2016)
136. X. Liang, A. Garsuch, L.F. Nazar, Sulfur cathodes based on conductive MXene nanosheets for high-performance lithium-sulfur batteries. Angew. Chem. Int. Ed. **54**(13), 3907–3911 (2015)

137. X. Zhao, M. Liu, Y. Chen, B. Hou, N. Zhang, B. Chen, N. Yang, K. Chen, J. Li, L. An, Fabrication of layered Ti_3C_2 with an accordion-like structure as a potential cathode material for high performance lithium-sulfur batteries. J. Mater. Chem. A **3**(15), 7870–7876 (2015)
138. C. Liu, Z. Yu, D. Neff, A. Zhamu, B.Z. Jang, Graphene-based supercapacitor with an ultrahigh energy density. Nano Lett. **10**(12), 4863–4868 (2010)
139. M.D. Stoller, S. Park, Z. Yanwu, J. An, R.S. Ruoff, Graphene-based ultracapacitors. Nano Lett. **8**(10), 3498–3502 (2008)
140. L.T. Le, M.H. Ervin, H. Qiu, B.E. Fuchs, W.Y. Lee, Graphene supercapacitor electrodes fabricated by inkjet printing and thermal reduction of graphene oxide. Electrochem. Commun. **13**(4), 355–358 (2011)
141. W.J. Hyun, E.B. Secor, C.-H. Kim, M.C. Hersam, L.F. Francis, C.D. Frisbie, Scalable, self-aligned printing of flexible graphene micro-supercapacitors. Adv. Energy Mater. **7**(17), 1700285 (2017)
142. A.M.A. Yeates, N. Karim, C. Vallés, S. Afroj, K.S. Novoselov, G. Stephen, Ultraflexible and robust graphene supercapacitors printed on textiles for wearable electronics applications. 2D Mater. **4**(3), 35016 (2017)
143. M.F. El-Kady, V. Strong, S. Dubin, R.B. Kaner, Laser scribing of high-performance and flexible graphene-based electrochemical capacitors. Science **335**(6074), 1326–1330 (2012)
144. G.-T. Esther, B. Suelen, F. Jaime, B. Robert, E. Salvador, D. Eleonora, M.R. Christopher, G. Francisco, S. Eduardo, Printing in three dimensions with graphene. Adv. Mater. **27**(10), 1688–1693 (2015)
145. Y. Lin, F. Liu, G. Casano, R. Bhavsar, I.A. Kinloch, B. Derby, Pristine graphene aerogels by room-temperature freeze gelation. Adv. Mater. **28**(36), 7993–8000 (2016)
146. C. Zhu, T. Liu, F. Qian, T.Y.J. Han, E.B. Duoss, J.D. Kuntz, C.M. Spadaccini, M.A Worsley, Y. Li, Supercapacitors based on three-dimensional hierarchical graphene aerogels with periodic macropores. Nano Lett. **16**(6), 3448–3456 (2016)
147. C.W. Foster, M.P. Down, Y. Zhang, X. Ji, S.J. Rowley-Neale, G.C. Smith, P.J. Kelly, C.E. Banks, 3D printed graphene based energy storage devices. Sci. Rep. **7**, 42233 (2017)
148. S. Ke Wei, T.S. Ahn, B.Y. Seo, J.Y. Dillon, S.J. Lewis, A. Jennifer, 3D printing of interdigitated Li-ion microbattery architectures. Adv. Mater. **25**(33), 4539–4543 (2013)
149. F.K. Wang, Y. Yan, C. Yao, Y. Chen, Y. Dai, J. Lacey, S. Wang, Y. Wan, J. Li, T. Wang, Z. Xu, Y. Hu, Y. Liangbing, Graphene oxide-based electrode inks for 3D-printed lithium-ion batteries. Adv. Mater. **28**(13), 2587–2594 (2016)
150. A.E. Jakus, E.B. Secor, A.L. Rutz, S.W. Jordan, M.C. Hersam, R.N. Shah, Three-dimensional printing of high-content graphene scaffolds for electronic and biomedical applications. ACS Nano **9**(4), 4636–4648 (2015)
151. B.P. Singh, S. Nayak, K.K. Nanda, B.K. Jena, S. Bhattacharjee, L. Besra, The production of a corrosion resistant graphene reinforced composite coating on copper by electrophoretic deposition. Carbon, **61**, 47–56 (2013)
152. R.K. Singh Raman, P. Chakraborty Banerjee, D.E. Lobo, H. Gullapalli, M. Sumandasa, A. Kumar, L. Choudhary, R. Tkacz, P.M. Ajayan, M. Majumder, Protecting copper from electrochemical degradation by graphene coating. Carbon **50**(11), 4040–4045 (2012)
153. N.T. Kirkland, T. Schiller, N. Medhekar, N. Birbilis, Exploring graphene as a corrosion protection barrier. Corros. Sci. **56**, 1–4 (2012)
154. M. Topsakal, H. Aahin, S. Ciraci, Graphene coatings: an efficient protection from oxidation. Phys. Rev. B: Condens. Matter Mater. Phys. **85**(15), 155445 (2012)
155. M. Schriver, W. Regan, W.J. Gannett, A.M. Zaniewski, M.F. Crommie, A. Zettl, Graphene as a long-term metal oxidation barrier: worse than nothing. ACS Nano **7**(7), 5763–5768 (2013)
156. M.J. Nine, M.A. Cole, L. Johnson, D.N.H. Tran, D. Losic, Robust superhydrophobic graphene-based composite coatings with self-cleaning and corrosion barrier properties. ACS Appl. Mater. Interfaces **7**(51), 28482–28493 (2015)
157. S.P. Koenig, L. Wang, J. Pellegrino, J.S. Bunch, Selective molecular sieving through porous graphene. Nat. Nanotechnol. **7**(11), 728–732 (2012)

158. V.R.S.S. Mokkapati, D.Y. Koseoglu-Imer, N. Yilmaz-Deveci, I. Mijakovic, I. Koyuncu, Membrane properties and anti-bacterial/anti-biofouling activity of polysulfone-graphene oxide composite membranes phase inversed in graphene oxide non-solvent RSC Adv. **7**(8), 4378–4386 (2017)
159. B. Wen, M. Cao, M. Lu, W. Cao, H. Shi, J. Liu, X. Wang, H. Jin, X. Fang, W. Wang, J. Yuan, Reduced graphene oxides: light-weight and high-efficiency electromagnetic interference shielding at elevated temperatures. Adv. Mater. **26**(21), 3484–3489 (2014)
160. L.C. Tang, Y.J. Wan, D. Yan, Y.B. Pei, L. Zhao, Y.B. Li, L.B. Wu, J.X. Jiang, G.Q. Lai, The effect of graphene dispersion on the mechanical properties of graphene/epoxy composites. Carbon **60**, 16–27 (2013)
161. S. Geetha, K.K. Satheesh Kumar, C.R.K. Rao, M. Vijayan, D.C. Trivedi, EMI shielding: methods and materials-a review. J. Appl. Polym. Sci. **112**(4), 2073–2086 (2009)
162. W.L. Song, M.S. Cao, M.M. Lu, S. Bi, C.Y. Wang, J. Liu, J. Yuan, L.Z. Fan, Flexible graphene/polymer composite films in sandwich structures for effective electromagnetic interference shielding. Carbon **66**, 67–76 (2014)
163. J. Liang, Y. Wang, Y. Huang, Y. Ma, Z. Liu, J. Cai, C. Zhang, H. Gao, Y. Chen, Electromagnetic interference shielding of graphene/epoxy composites. Carbon **47**(3), 922–925 (2009)
164. R.K. Joshi, P. Carbone, F.C. Wang, V.G. Kravets, Y. Su, I.V. Grigorieva, H.A. Wu, A.K. Geim, R.R. Nair, Precise and ultrafast molecular sieving through graphene oxide membranes. Science **343**(6172), 752–754 (2014)
165. J. Abraham, K.S. Vasu, C.D. Williams, K. Gopinadhan, Y. Su, C.T. Cherian, J. Dix, E. Prestat, S.J. Haigh, I.V. Grigorieva, P. Carbone, A.K. Geim, R.R. Nair, Tunable sieving of ions using graphene oxide membranes. Nat. Nanotechnol. **12**(6), 546–550 (2017)
166. L. Chen, G. Shi, J. Shen, B. Peng, B. Zhang, Y. Wang, F. Bian, J. Wang, D. Li, Z. Qian, G. Xu, G. Liu, J. Zeng, L. Zhang, Y. Yang, G. Zhou, M. Wu, W. Jin, J. Li, H. Fang, Ion sieving in graphene oxide membranes via cationic control of interlayer spacing. Nature **550**(7676), 380–383 (2017)

Index

L. W. T. Ng et al., *Printing of Graphene and Related 2D Materials*,
https://doi.org/10.1007/978-3-319-91572-2

Printed in the United States
By Bookmasters